The Design of Climate Policy

CESifo Seminar Series
edited by Hans-Werner Sinn

Labor Market Institutions and Public Regulation
Jonas Agell, Michael Keen, and Alfons J. Weichenrieder, editors

Venture Capital, Entrepreneurship, and Public Policy
Vesa Kanniainen and Christian Keuschnigg, editors

Exchange Rate Economics: Where Do We Stand?
Paul De Grauwe, editor

Prospects for Monetary Unions after the Euro
Paul De Grauwe and Jacques Mélitz, editors

Structural Unemployment in Western Europe: Reasons and Remedies
Martin Werding, editor

Institutions, Development, and Economic Growth
Theo S. Eicher and Cecilia García-Peñalosa, editors

Competitive Failures in Insurance Markets: Theory and Policy Implications
Pierre-André Chiappori and Christian Gollier, editors

Japan's Great Stagnation: Financial and Monetary Policy Lessons for Advanced Economies
Michael M. Hutchison and Frank Westermann, editors

Tax Policy and Labor Market Performance
Jonas Agell and Peter Birch Sørensen, editors

Privatization Experiences in the European Union
Marko Köthenbürger, Hans-Werner Sinn, and John Whalley, editors

Recent Developments in Antitrust: Theory and Evidence
Jay Pil Choi, editor

Schools and the Equal Opportunity Problem
Ludger Woessmann and Paul E. Peterson, editors

Economics and Psychology: A Promising New Field
Bruno S. Frey and Alois Stutzer, editors

Institutions and Norms in Economic Development
Mark Gradstein and Kai A. Konrad, editors

Pension Strategies in Europe and the United States
Robert Fenge, Georges de Ménil, and Pierre Pestieau, editors

Foreign Direct Investment and the Multinational Enterprise
Steven Brakman and Harry Garretsen, editors

Sustainability of Public Debt
Reinhard Neck and Jan-Egbert Sturm, editors

The Design of Climate Policy
Roger Guesnerie and Henry Tulkens, editors

See http://mitpress.mit.edu for a complete list of titles in this series.

The Design of Climate Policy

edited by Roger Guesnerie and
Henry Tulkens

CESifo Seminar Series

The MIT Press
Cambridge, Massachusetts
London, England

© 2008 Massachusetts Institute of Technology

All rights reserved. No part of this book may be reproduced in any form by any electronic or mechanical means (including photocopying, recording, or information storage and retrieval) without permission in writing from the publisher.

MIT Press books may be purchased at special quantity discounts for business or sales promotional use. For information, please e-mail special_sales@mitpress.mit.edu or write to Special Sales Department, The MIT Press, 55 Hayward Street, Cambridge, MA 02142.

This book was set in Palatino on 3B2 by Asco Typesetters, Hong Kong and was printed and bound in the United States of America.

Library of Congress Cataloging-in-Publication Data

The design of climate policy / edited by Roger Guesnerie and Henry Tulkens.
   p. cm. — (CESifo seminar series)
Includes bibliographical references and index.
ISBN 978-0-262-07302-8 (hardcover : alk. paper) 1. Climatic changes—Government policy. 2. Climatic changes—Economic aspects. I. Guesnerie, R. II. Tulkens, Henry. III. CESifo.
QC981.8.C5D436   2008
363.738′7456—dc22
                                                                      2008017557

10 9 8 7 6 5 4 3 2 1

Contents

Series Foreword ix
Contributors xi

1 Synopsis of the Book 1
Roger Guesnerie and Henry Tulkens

I Design of Climate Institutions

2 Improving on Kyoto: Greenhouse Gas Control as the Purchase of a Global Public Good 13
David F. Bradford

3 The Design of Post-Kyoto Climate Schemes: Selected Questions in Analytical Perspective 37
Roger Guesnerie

4 Design of Climate Change Policies: A Discussion of the GPGP Approach of Bradford and Guesnerie 65
Sushama Murty

5 Untying the Climate-Development Gordian Knot: Economic Options in a Politically Constrained World 75
Jean-Charles Hourcade, P. R. Shukla, and Sandrine Mathy

II Stability of Outcomes

6   Transfer Schemes and Institutional Changes for Sustainable Global Climate Treaties   103
    Johan Eyckmans and Michael Finus

7   Parallel Climate Blocs: Incentives to Cooperation in International Climate Negotiations   137
    Barbara Buchner and Carlo Carraro

8   Cooperation, Stability, and Self-enforcement in International Environmental Agreements: A Conceptual Discussion   165
    Parkash Chander and Henry Tulkens

9   Heterogeneity of Countries in Negotiations of International Environmental Agreements: A Joint Discussion of the Buchner-Carraro, Eyckmans-Finus, and Chander-Tulkens Chapters   187
    Sylvie Thoron

III Policy Design

10   Economics versus Climate Change   201
     William A. Pizer

11   Economics versus Climate Change: A Comment   217
     Richard S. J. Tol

12   Absolute versus Intensity Limits for CO_2 Emission Control: Performance under Uncertainty   221
     Ian Sue Wing, A. Denny Ellerman, and Jaemin Song

13   On Multi-period Allocation of Tradable Emission Permits   253
     Katrin Rehdanz and Richard S. J. Tol

14   Optimal Sequestration Policy with a Ceiling on the Stock of Carbon in the Atmosphere   273
     Gilles Lafforgue, Bertrand Magné, and Michel Moreaux

IV Models and Policies

15 Mind the Rate! Why the Rate of Global Climate Change Matters, and How Much   307
    Philippe Ambrosi

16 Leakage from Climate Policies and Border-Tax Adjustment: Lessons from a Geographic Model of the Cement Industry   333
    Damien Demailly and Philippe Quirion

17 The Global Warming Potential Paradox: Implications for the Design of Climate Policy   359
    Stéphane De Cara, Elodie Galko, and Pierre-Alain Jayet

Index   385

Series Foreword

This book is part of the CESifo Seminar Series. The series aims to cover topical policy issues in economics from a largely European perspective. The books in this series are the products of the papers and intensive debates that took place during the seminars hosted by CESifo, an international research network of renowned economists organised jointly by the Center for Economic Studies at Ludwig-Maximilians-Universität, Munich, and the Ifo Institute for Economic Research. All publications in this series have been carefully selected and refereed by members of the CESifo research network.

Contributors

Philippe Ambrosi
The World Bank

David F. Bradford
Princeton University

Barbara Buchner
International Energy Agency, Paris

Carlo Carraro
Fondazione Eni Enrico Mattei, Venice

Parkash Chander
National University of Singapore

Stéphane De Cara
Institut National de la Recherche Agronomique, Grignon

Damien Demailly
Centre International de Recherche sur l'Environnement et le Développement, Nogent-sur-Marne

A. Denny Ellerman
Massachusetts Institute of Technology

Johan Eyckmans
European University College Brussels

Michael Finus
Fernuniversität in Hagen

Elodie Galko
Ministère de l'économie, des finances et de l'emploi, Paris

Roger Guesnerie
Collège de France, Paris

Jean-Charles Hourcade
Centre International de Recherche sur l'Environnement et le Développement, Nogent-sur-Marne

Pierre-Alain Jayet
Institut National de la Recherche Agronomique, Grignon

Gilles Lafforgue
Université de Toulouse 1

Bertrand Magné
Paul Scherrer Institut, Villingen

Sandrine Mathy
Centre International de Recherche sur l'Environnement et le Développement, Nogent-sur-Marne

Michel Moreaux
University de Toulouse 1

Sushama Murty
University of Warwick

William A. Pizer
Resources for the Future,
Washington DC

Philippe Quirion
Centre International de Recherche
sur l'Environnement et le
Développement, Nogent-sur-
Marne

Katrin Rehdanz
Christian-Albrechts-Universität
zu Kiel

P. R. Shukla
Indian Institute of Management
of Ahmedabad

Jaemin Song
Massachusetts Institute of
Technology

Ian Sue Wing
Boston University

Sylvie Thoron
Groupement de Recherche en
Economie Quantitative d'Aix
Marseille

Richard S. J. Tol
Economic and Social Research
Institute, Dublin

Henry Tulkens
Université Catholique de
Louvain, Louvain-la-Neuve

The Design of Climate Policy

1

Synopsis of the Book

Roger Guesnerie and Henry Tulkens

This book comes as the proceedings of a conference on "The Design of Climate Policies," held at the Venice Summer Institute of CESifo on July 22 and 23, 2005. David Bradford had initially accepted our invitation to give a keynote lecture to the conference. When we learned of his tragic death on February 22, we decided to have his paper read by one of us at the conference as a tribute to him.

The papers are gathered here in four parts, with the unity of each part characterized by the content and/or style. The first two parts are concerned with the conditions for a fruitful international cooperation on climate policies. The papers in part I focus on appropriate institutions for an efficient collective action. The papers in part II deal with the theoretical and practical questions of cohesion of the international community. The papers in parts III and IV bring to the discussion theoretical dimensions (part III) and quantitative tools (part IV) of climate policy design.

Overview of The Chapters

Part I presents the framework within which climate policies are implemented. The opening chapter 2 is the unpublished piece that we had asked David Bradford to contribute for the conference.[1] David presents a proposal for international cooperation that substantially differs from that of the Kyoto agreement. Chapter 3 by Roger Guesnerie has a closely connected theme. Roger begins with comments on David Bradford's proposal, and then revisits the main issues underlying the design of efficient international institutions on climate policies. In chapter 4 Sushama Murty offers her own theoretical insights on the two preceding chapters. Chapter 5 by Jean-Charles Hourcade, P. R. Shukla, and Sandrine Mathy examines the involvement of less developed countries (LDCs) in climate policies and puts the economic viewpoint in a broader political perspective.

Part II puts emphasis on the logic of coalition formation and discusses the stability of international environmental agreements. This is a subject that calls for a game-theoretic analysis. Chapter 6 by Johan Eyckmans and Michael Finus proposes two types of measures to enhance the success of international environmental treaty-making. Chapter 7 by Barbara Buchner and Carlo Carraro assesses the empirical plausibility of emergence of a single or of multiple climate coalitions. Chapter 8 by Parkash Chander and Henry Tulkens take these issues to a more general level and critically review the various notions of stability and related concepts used in the game-theoretic literature on international environmental agreements. Chapter 9 by Sylvie Thoron wraps up this debate.

Part III is devoted to issues of policy design at a general level. In chapter 10 William Pizer pleads for a less ambitious and more decentralized approach to international cooperation in climate affairs than the one currently pursued by the United Nations in the Framework Convention and the Kyoto Protocol. He warns of the multiple dimensions of the design of internal policies. Richard Tol offers his views on these issues in chapter 11. Next specific policy issues are put under theoretical scrutiny, namely in order of appearance: the choice of policy targets, the intertemporal aspects of carbon trade, and the optimal implementation of a sequestration policy. In chapter 12 Ian Sue Wing, A. Denny Ellerman, and Jaemin Song compare absolute and intensity limits for carbon dioxide emission control. They show how these two instruments, although equivalent in a certain world, differ when their performance is under conditions of uncertainty. In chapter 13 Katrin Redhanz and Richard Tol consider how successive permit allocation rules create incentives that accelerate or decelerate emission reduction paths. Chapter 14 by Gilles Lafforgue, Bertrand Magné, and Michel Moreaux, which has been added to the selection of papers presented to the conference, provides an original timing analysis of a sequestration policy that implements a ceiling on the stock of carbon in the atmosphere.

The chapters gathered in part IV take up policy design as well but include elaborate quantitative examinations. In chapter 15 Philippe Ambrosi uses a stochastic optimal control model to evaluate the effect of a constraint on the rate of temperature change for the determination of policies. In chapter 16 Damien Demailly and Philippe Quirion present simulations of a spatial international trade model to discuss leakages from climate policies and border tax adjustments in the cement industry. In chapter 17 Stéphane De Cara, Elodie Galko, and Pierre-Alain Jayet examine how to correct the shortcomings of the standard Global Warming Potential index used for greenhouse gases.

Transversal Debates

We now attempt to put the different contributions in the transversal perspective of some key debates of climate policies. The debates that we single out have their roots in the present situation and refer to the real or imaginary shortcomings of the Kyoto Protocol. We consider two of them: the first refers to *the feasible and desirable extent of international cooperation on climate policies*, and the second to *the optimal effort and the optimal timing of climate policies*.

The existing situation is characterized by a non-unanimous involvement of rich countries and, in the background, the syndrome of US nonratification and a very limited involvement of LDCs. This situation seems particularly detrimental to the effectiveness of the present climate scheme. Indeed, the search for a formula triggering both the participation of all developed countries and the voluntary participation of LDCs has stimulated reflection. This is indeed the preoccupation at the heart of David Bradford's proposal, that we referred to as Global Public Good Purchase (GPGP) as well as in Guesnerie's discussion of post-Kyoto schemes. Under GPGP the public good provision relies on voluntary contributions from developed countries. But an international Bank uses the collected funds to buy emission abatements all over the world. As the emission allowances involve all countries and are set up at a business as usual level, the scheme is expected to trigger participation of all countries, including LDCs. Related mechanisms, nonbinding quotas, might also similarly trigger participation in post Kyoto schemes. However, the present failure to organize what game theorists call a "grand coalition" has led to reflections in different directions.

First, one may see the present situation as reflecting political constraints that have to be taken into account, and therefore adopt a second-best (or third-best) viewpoint. This is a possible reading of the contribution by Hourcade, Shukla and Mathy (chapter 5): in the present context, they look at realistic policies that may reconcile ambitious development objectives with the use of less carbon-intensive technologies in LDCs. A second possible reading of their paper is that without rejecting a "grand plan" à la Kyoto that they may find necessary, they certainly do not find such a "grand plan" sufficient. They plead, for example, for regional cooperation in South Asia, where wider energy trade could both bring more growth and substantial carbon savings.

Pizer (chapter 10) takes a more radical viewpoint and sees intrinsic merits to the present fragmentation. Arguing that an international agreement is not necessary to initiate relevant domestic action on climate change, he is doubtful about the value of international emissions trading at the present stage and favors the heterogeneity of carbon prices across the world.

Does this position reflect some Leibnizian optimism, "tout est pour le mieux dans le meilleur des mondes," as Voltaire had mocked it?[2] No answer can be given without delving into questions of feasibility and desirability: Is a global arrangement between nations feasible, is it desirable?

A starting point to the debate may be found in Buchner and Carraro's argument (chapter 7) that a two-coalition structure—consisting, on one side, of Japan, the European Union, and the former Soviet Union, and on the other side, China and the United States—is politically plausible (although lacking "stability" in both senses to be made clear below), and can improve upon the present fragmented situation for some countries (Japan and the European Union). Eyckmans and Finus (chapter 6) explore more closely the issue of internal and external stability of alternative coalition structures, and whether transfer rules and/or alternative institutional rules as to coalition membership can enhance such stability properties. The answers they provide, based on simulations of a particular numerical model, are often positive, but they do not imply that this kind of stability can be ensured for the coalition of all countries.

In any case, the stability concept under scrutiny is partial. Indeed, for the world considered in these game-theoretical models, the associated game has side payments, and in technical terms is superadditive so that a global arrangement is in a sense always desirable. This means in particular that if the world is split into disjoint coalitions, as it is in Buchner and Carraro (chapter 7), or in any other of the coalition structures considered by Eyckmans and Finus (chapter 6)—other than the coalition of all countries—there does always exist merging plans advantageous for all countries. One such merging plan is the Chander and Tulkens scheme that rests upon the alternative game-theoretical stability concept of the "core" of a cooperative game. Can we therefore conclude that the grand coalition will form?

A first and mostly positive insight is provided by the fact that for the merging plan to be made mutually advantageous, the scheme rests on a formula specifying lump-sum transfers between nations. Now, experience shows that there exists very few examples of explicit transfers between nations. However, is it not the case that in the present context, transfers can be mimicked by changes of quotas? Diminishing the quota of a country while augmenting the quota of another country is like implementing a lump-sum transfer from one country to the other. In chapter 8 (note 12), Chander and Tulkens remind us that, according to their early interpretation of the Kyoto Protocol, this possibility is clearly opened by the "cap and trade" architecture. Yet one may conceivably argue that the extent to which quotas differentiation has been actually used in the Protocol does substantially underscore the amount of transfers required for the sustainability of a core-like stable agreement.

A second argument, apparently negative, is that the core-stable grand coalition may itself be unstable in the internal-external sense. This issue is dominated by the fact that the quality of the climate is a nonexcludable public good: if a nation defects from a global arrangement, either it will still benefit from the effort of the remaining countries if these stick to the agreement, or it will not get any benefit if its defection also entails the defection of these other countries from the global arrangement; in the latter instance, the country realizes that it cannot get any gain from defecting. What the chapter by Chander and Tulkens in this book makes clear is that the alternative just described corresponds to the two stability concepts involved: internal-external stability versus core stability. They may be compatible, but they also may not be, as shown by some simulations of Eyckmans and Finus.

The stability issue thus appears not to be a straightforward one. Also it is more intricate than what is suggested by the simple game-theoretic model under scrutiny in the three chapters just discussed. Therefore let us consider now some directions of complexification in the analysis of stability and fragmentation.

First, the basis of the aggregate models used may be discussed. On the one hand it may the case that, contrary to what is permitted in the simpler game-theoretic setting, large lump-sum transfers, or their counterpart—wide quota differentiation—are not possible in real world negotiations. On the other hand, the analyses of stability in these models are static ones,[3] while the problem is essentially a dynamic one due to the phenomenon of carbon accumulation. While theoretical developments have occurred in that direction,[4] the way in which their results can enrich the interpretations of the Kyoto Protocol remains to be further explored.

Third, the level of aggregation itself may be questioned. It is clear for example that the uniqueness of carbon price assumed in aggregate models will not be reflected at the disaggregate level in real world arrangements. This is clear from the Kyoto scheme itself: even with a world market for quotas, individual countries have a lot of freedom in the determination of internal policies and in the role they give to the world price signal. Hence flexible internal policies, as advocated by Pizer (chapter 10), may not be as incompatible with a strong cooperative agreement as one often believes.

Heterogeneity of carbon prices, however, raises questions that may be the more serious, the less cooperative the choice of internal policies are. The most obvious question is that heterogeneity of carbon prices affects the conditions of international exchange in a way that, in the Kyoto case, is detrimental to the virtuous countries and weakens the return of their effort. This point, made in qualitative terms in Guesnerie (chapter 3), is at the heart of the empirical analysis of Demailly and Quirion (chapter 16). They show, from a detailed model of the cement industry, that imposing in annex-B countries a carbon price of 15 euros per ton of CO_2 leads to a 20 percent internal abatement, but with a significant "leakage" effect on other countries, even though such a leakage might be attenuated or even suppressed with a border tax adjustment.

While the stability discussions raised here bring the trade issue into the picture, it is worth noting that some of the questions previously raised, for example by Pizer (chapter 10), already had a trade flavor or at least counterparts in the discussion of the merits and benefits of international trade: What is a mutually advantageous arrangement that leaves enough freedom and space for the internal policies?

Going one step further, one might wonder why so ambitious an international arrangement on the environment as the Kyoto Protocol should treat the environment separately from trade, and why a deal connecting both fields might not be better. Unless one adopts an arrangement such as Bradford's (chapter 2) that implements a world carbon price and thereby solves simultaneously the problems of participation and "fair competition," the issue of linking environment and trade is inescapable.

The second transversal debate arising from the Kyoto Protocol concerns *the optimal effort and the optimal timing of climate policies*. The contributions in this book do not have much to say on the *level* of effort:

they often derive it from exogenously specified constraints (Lafforgue, Magné, and Moreaux in chapter 14 as well as Ambrosi in chapter 15 take exogenously specified bounds). However, the second issue, *timing*, does creep in many chapters.

In a sense, Lafforgue, Magné, and Moreaux are mainly concerned with optimal timing. They address the question within a model that takes two simplified assumptions: first that the objective of climate policies is summarised in terms of a ceiling of concentration, second that the only means of action is the use of sequestration policy, the timing of which is under scrutiny. In this framework, fossil fuel consumption should be curbed well before the sequestration policy is undertaken.

Ambrosi's contribution puts emphasis on the timing problem under stylized assumptions on the abatement cost: the objective does not only refer to the temperature increase generated by a given profile of emissions but also to the *rate* at which the average temperature changes. The relevance of the rate is a priori obvious: rapid change, for example, would impact the ability of species to migrate. Ambrosi shows effectively that the rate of change has a great influence on the timing of an optimal abatement policy.

The timing issue also appears directly in two other papers. By looking at a two period version of a pollution rights market, Redhanz and Tol (chapter 13) explore an issue of prominent importance for the Kyoto Protocol on which too little seems to have been done: How will the international market for quotas evolve dynamically? Although they consider only a limited set of possibilities of transfer between periods, they identify circumstances under which alternative dynamic allocation rules create incentives to accelerate or decelerate emission reductions.

De Cara, Debove, and Jayet (chapter 17) focus attention on the question of aggregation of greenhouse gases for evaluating their relative impact on climate change. The IPCC has promoted a Global Warming Potential index that answers the aggregation problem. The issue may seem foreign to the timing problem, but as the authors show, following earlier inquiries, this is not the case. The right index depends on the rhythm of emissions abatement, which itself depends on the social objective and the discount rate. These are issues with direct connection to the timing question.

But there is a more indirect and subtler aspect to timing. The price versus quantity issue that has received a lot of attention in the literature does not look, a priori, like a timing issue: the question is whether an environmental policy should aim at controlling the user price of the polluting good or its quantity. Although the two competing tools are equivalent in a certainty world, they are not when costs and benefits are uncertain ex ante. Take a static context. Fixing a quantity may ex post turn out very expensive, if the realized cost is much higher than the expected cost; similarly, fixing a price without having exact knowledge of the realization of costs and/or benefits can lead to quantity choices that are ex post very inappropriate. In a static and uncertain context, a quantity policy and a price policy have different merits. In a dynamical context, the comparison bears on timing. The counterpart of too costly implementation of a quantity objective is too early action (since an excessive move can be corrected tomorrow), whereas the counterpart of a too lenient control of the objective through price corresponds to delayed action.

Whatever the difference in interpretation, the dynamic analysis reproduces the static argument and puts emphasis on the same qualitative factors (slope of the marginal cost curve versus slope of the marginal benefit curve) for which appropriate empirical data are available. Existing studies based on such data suggest that controlling price is better than controlling quantities, as is the aim of the Kyoto Protocol. Although this argument is not unanimously accepted, one may see it as conventional wisdom. It is reiterated with eloquence in Pizer's contribution, and not surprisingly, since Pizer and his co-authors have made this argument academically respectable. While as noted by Guesnerie, Bradford's GPGP is also less subject to the uncertainty bias against quantity, the reader will nevertheless find in the present book a few arguments that challenge the conventional wisdom.

One such argument is made by Sue Wing, Ellerman and Song (chapter 12). They consider controlling under uncertainty the carbon intensity of emissions rather than volume as is the case with the Kyoto Protocol. Controlling the intensity instead of the volume may change the price versus quantity debate, because a lot of the uncertainty about the costs of a quantity policy depends on the rate of growth of GDP, which is a priori difficult to forecast. The cost of an intensity policy would then be less uncertain, and this could open the door to a re-evaluation of the asserted superiority of price policy as compared with an intensity policy. Indeed the authors identify plausible conditions under which an intensity-based limit is to be preferred to an absolute one: positive correlation between emissions and GDP is necessary (but not sufficient). However, they draw cautious conclusions, since in some cases empirical evidence suggests that intensity-based limits may increase the variance of outcomes.

Another and more radical criticism of the conventional wisdom is put forth by Guesnerie (chapter 3). He stresses that the so-called price policy acts through tax but does not directly allow for controlling the price of the polluting goods, namely fossil fuels. Whereas the price change of fossil fuels due to taxation may offset more or less the initial effect, the tax incidence problem remains difficult to ascertain. The simple solution taken by Lafforgue, Magné, and Moreaux (chapter 14) in their intertemporal competitive pricing à la Hotelling gives at least some idea of the extent of the difficulty. Taking into account the uncertainty of prices within a price policy might lead to a drastic reassessment of the conventional wisdom.

In summary, we hope that the variety of insights in this book will motivate future policy debates, particularly with regard to the conditions necessary for global cooperation and appropriate timing of climate policies.

Notes

The editors wish to gratefully acknowledge the kind cooperation of the following persons in the preparation of this book: Larry Karp, Olivier Compte, Claude d'Aspremont, Shlomo Weber, Scott Barrett, Alan Kirman, Michel Le Breton, Jean-Christophe Pereau, Richard Baron, Stephen Smith, Cédric Philibert, Frank Jotzo, Thierry Bréchet, Wallace Oates, Partha Sen, Alain Haurie, Florent Pratlong, Marzio Galeotti, Asbjorn Aaheim, and Minh Ha Duong. Also the editors thank Dana Andrus and Lianna Kong of The MIT Press for their immensely helpful editorial contributions.

1. The other keynote speakers were Roger Guesnerie (chapter 3), Parkash Chander and Henry Tulkens (chapter 8), and William Pizer (chapter 10).

2. "All is for the best in the best of all worlds."

3. The same remark applies to the simple setting taken by Guesnerie in order to model Bradford's GPGP and other post-Kyoto competitors.

4. See, for instance, Germain et al. (2003).

Reference

Germain, M., Ph. Toint. H. Tulkens, and A. de Zeeuw. 2003. Transfers to sustain dynamic core-theoretic cooperation in international stock pollutant control. *Journal of Economic Dynamics and Control* 28 79–99.

I Design of Climate Institutions

2

Improving on Kyoto: Greenhouse Gas Control as the Purchase of a Global Public Good

David F. Bradford

The main purpose of this paper is to sketch out an alternative approach to controlling greenhouse gas emissions. To give the thing a name, I call it the "global public good purchase" (GPGP) approach. The proposed system has two main advantages, relative to the straight cap and trade approach that is (more or less) embodied in the Kyoto plan and relative to other approaches that involve countries making commitments to policies and measures to control greenhouse gas emissions. First, the enforcement is greatly simplified. As will be seen, countries can freely choose the extent of their emission reductions in pursuit of the global emission objectives, just as suppliers of ordinary goods and services choose the extent of their "contribution" to, for example, national defense. Second, the sharing of the burden of financing the global public good of greenhouse gas control is transparent, rather than buried implicitly in, for example, country-specific emission limits or commitments to policies and measures. Among other things, this property would ease the critically important incorporation of developing countries to a global control regime.

I would emphasize that the issues addressed here concern institutional design. The paper is silent on the policies that would be implemented using the new institution, involving both the allocational questions of the total emissions of greenhouse gases and the assignment of emissions to countries and enterprises, and the matter of the sharing of burdens and benefits from alternative arrangements. The proposed system is conceived of as an alternative to the particular cap and trade plan put forth in the Kyoto Protocol. If that protocol takes effect in the form it has after the 2001 Bonn conference of the parties to the Framework Convention on Climate Control, the proposed system could be taken as the basis for a worldwide system embodied in a successor agreement. The system is also described as a global one. I think it could, however, be deployed by any set of countries wishing to coordinate their climate control efforts.

In setting down these notes, I have taken for granted that readers are broadly familiar with the climate control problem and the development of the international negotiations addressed to it. Most of those readers will also be familiar with the analytical tools of environmental economics on which I depend. I have observed, however, that sometimes fairly basic economic analytical ideas get lost in the policy debates. So I hope readers will not be offended by my recitation of such ideas below. The next section, in particular, is a reminder of the potential power of well-designed property rights to solve problems of coordination, while recognizing the collective nature of the objective to be served. The third section contains the description of the "global public good purchase" system. A fourth section offers some concluding thoughts. An appendix uses some simple economic analytical tools to sketch some of the allocational and distributional properties of the system.

2.1 Reminder of Some Lessons of Basic Economics

2.1.1 *Public Goods*

I would stress two basic economic ideas here. The first is the idea of a public good. It is generally understood that the climate system is a global public good par excellence: For better or for worse (some projected impacts of climate change are positive for some people) a change in the climate affects everyone in the world. What seems to have been less often considered is the usual method by which a public good is acquired. For a classic public good, such as national defense, a national government typically purchases the needed resources, such as weapons systems. The burden of financing the collective purchase is normally determined separately, using a tax system. (Defense is actually a mixed example because, quite often, even in modern economies, the important resource of military manpower is not obtained through the market only, but a significant part is rather conscripted. Although I do not pursue the question, there may be useful lessons to be learned for the extension that I suggest here to the purchase of the public good of climate control from the circumstances that favor conscription in the case of defense.)

An example of an international public good would be peacekeeping operations of the United Nations. In that case national entities do the work of producing the public good, while the financing is shared according to United Nations rules.

The approach to climate control investigated here builds on the analogy of mustering the defense of the world against an approaching asteroid by deflecting it from a collision path. I would think we might cope with this global public good by organizing to purchase the needed resources and, at the same time, figuring out how to spread the cost of paying for them among the countries of the world. The idea is to identify the "needed resources" and work out a method to buy and pay for them.

2.1.2 Property Rights

The second basic idea is that of property rights. The negotiation of the Kyoto Protocol, including, importantly, the COP3bis round in Bonn in 2001, reflects the gradual acceptance of one of the most important but poorly understood ideas in economics: the economizing incentives that derive from private property. It is perhaps not surprising that this idea is unfamiliar to many environmental scientists. But economic theory teaches the power of property rights to serve environmental objectives and the emerging details of the Kyoto Protocol can be understood as exploiting the idea of property rights to certain levels of emissions of greenhouse gases. Some involved in the negotiating process were and are uncomfortable with this idea; some readers will view the concept as raising an ideological red flag. I would urge the economists to stick more firmly to their analytical guns and the others to consider objectively the possibility of exploiting the extraordinary power of private property in the service of global climate policy.

Owners of private property (who might be governments) have strong incentives (economists would say the "right" incentives) to put that property to productive use. Where relevant, owners and potential owners also have appropriate incentives to invest in the creation of substitutes for property through invention. A corollary, one of the important insights of environmental economics, is that environmental problems can be understood as due to the absence of appropriate property rights. Policies to define and enforce new property rights can be one of the most effective lines of attack on environmental problems.

Take as an example the textbook case of companies that purchase inputs of labor and ore, and use them to produce steel. In the process the companies' factories emit smoke with detrimental impact on the health of the surrounding population. In the resulting equilibrium there is a pollution problem: "too much" smoke is used. One way of describing the problem is that the companies are not obliged to pay for the services provided by the surrounding population in contending with the smoke (by accepting the health risk or by incurring expenses to adjust for the presence of smoke in the air). This is in contrast to the companies' use of labor, which also comes at a cost to the population that provides the service. This cost is recognized in the companies' calculations because they are obliged to pay their workers.

A possible approach to correcting the situation is to define a new property right, for example, the right to emit smoke. Suppose the community decided that a certain amount of smoke emitted was acceptable, say, measured in tons per year. Rights to emit would be defined, implemented as allowances; an emitter would be obliged to own an allowance for each unit of smoke emitted. These allowances could be bought and sold, just like units of steel or ore or labor. Companies would then have an incentive to curb the emissions of smoke by a variety of methods, including innovation, just as they have an incentive to control the quantities of other inputs used in production. In the resulting economic equilibrium the cost, measured in goods like steel, of achieving the target level of emissions would be minimized. A large literature attests to the promise of methods like this to improve on "command and control" approaches to controlling pollution of this type.

Some lessons to take away from this homely example:

Fixing the problem requires collective action. This is actually pretty general. Private property itself is a social artifact. Even what we think of as ordinary private property rights (e.g., to real estate) are defined collectively. Often the definitions are complex. For example, the rights of an owner of a piece of real estate property incorporate restrictions due to land use zoning rules. In the jargon of economics, the basic policy problem, which has been pretty well settled in the case of real estate, is to design the property rights in such a way that, in practice, the social payoff is maximized. (A classic exposition of this perspective on property rights is that by Coase, 1960.)

The economizing power of the newly defined property rights is not dependent on who gets them to start with. If the newly defined property is assigned to the (owners of the) companies that were doing the excessive emitting in the first place, those companies will have the same incentives (because the emissions are priced) as they would have if the newly defined property were handed out to some other group in the population or sold by the government. This may seem paradoxical; the key thing to keep in mind is that an entity that uses the rights must pay for them, either by buying them from another owner or by forgoing selling them on the market. Who gets the rights initially is significant mainly as a matter of what we rather sloppily call "income" distribution. Such distributional issues are important, to be sure, but separable from the allocative function of the rights.

The example does not include a "supply curve" of allowances. It simply incorporates a fixed limit on emissions. One could imagine, however, that the community would have a supply curve of allowances. If the price of allowances is very low, the community will not want to supply many allowances but rather enjoy very clean air. If the price is high, they will sell more, enduring dirtier air in return for the other desirable goods and services that can be bought, as indicated by the price. A limiting case of a supply curve for allowances would be an infinitely elastic one at some price. This would duplicate the economic effect of a tax on emissions, although the institutional form, involving priced allowances or permits, might have a different look and feel.

One could, alternatively, imagine a world in which a baseline quantity of emission allowances is set at whatever would have been emitted in the absence of any control regime, with those allowances put in the hands of the companies that, in effect, had the preexisting "right" to emit. The collective decision would then be how many of the allowances to buy and retire. This decision would be made in the light of some system for financing the purchase of allowances. Such an approach might be attractive as a way of getting started and as a way of separating the question of who should pay from the method of implementation. It is the approach I suggest here for controlling the greenhouse gas emissions.

As I have noted, Kyoto can be understood as making a start on defining new property rights: to emission of greenhouse gases. In the spirit of the preceding remark, note one could have adopted other property right concepts, such as the right to add an increment to radiative forcing (at a specified point in time). Focusing on emissions has some problems—importantly, the problem of aggregation of greenhouse gases—but has the advantage of being close to what everyone sees as the essential control problem.

Two things are notable about the property right regime incipient in the Kyoto Protocol. First, there remains considerable controversy about, for example, how much of an annex-B country's emission reductions may be purchased from other countries. The idea of property rights is not fully accepted. Second, the distribution of burdens associated with achieving the reductions in emissions called for in the Kyoto Protocol are mostly implicit in the distribution of implicit property rights, rather than explicit, as I have suggested is more typically the case when public goods are collectively provided. This conflating of the allocational and distributional tasks of the system may have made reaching agreement more difficult that would use of an approach that more clearly distinguished the two. (I should concede that "may" is the right word; it may also be that making the distributional aspects of the system harder to observe contributes to reaching agreement.)

2.2 The Global Public Good Purchase (GPGP) Approach

For purposes of discussion, I have focused on the control of the burning of fossil fuel—in fact I suggest implementing controls to influence the "import" of fossil fuels to countries, whether out of the ground or across the border. (Hereafter I keep the quotation marks on "import" when used in this sense.) In this I am influenced by the discussions and proposals of Hargrave (1998), Lackner et al. (undated) and McKibben and Wilcoxen (1999). Treatment of carbon sinks is straightforward conceptually, if not necessarily as a monitoring problem. Inclusion of other greenhouse gases is a little less straightforward, both as a matter of aggregation and from the point of view of monitoring. The Kyoto Protocol has settled for a rough approximation to a conceptually correct approach to aggregation, which could equally be applied to the GPGP approach. The basic system sketched here could be applied to any aggregation of gases.

The GPGP approach identifies the "needed resources" that have to be diverted from other uses in order to produce climate control, analogous to, say the services of scientists that would have to be diverted from alternative employment to assist in the defense against the extraterrestrial object, as the use of fossil fuel that countries would otherwise choose In other words, to produce climate control, we need to acquire the services of countries' deviations from their BAU emissions trajectories. The GPGP approach is to use the market to acquire those services. It has three elements.

2.2.1 Element 1: BAU Trajectories Assigned

Each participating country, including each LDC, is assigned a BAU trajectory of fossil fuel "imports." The BAU amount for each country would be expressed in terms of "allowances" to bring a unit of CO_2 in the form of fossil fuel into the country. Participating countries would be obliged to have such an allowance for each unit "imported." If such allowances were required of private agents within a country, the aggregate quantity demanded at a price of zero would be the BAU amount—by definition of BAU.

Conceptually the determination of the BAU trajectory is a purely technical matter, not an ideological or value-dependent step. (In a negotiating context this technical matter, like many similar ones, would presumably be contentious. The separation of the financing step, discussed below, could, however, take some of the pressure off.)

It is important to emphasize that the BAU trajectory would not be a simple fixed path, related, for example, to a country's fossil fuel "imports" in some base year. Instead, it would be explicitly contingent on the country's economic performance, as well as on technological developments generally.

2.2.2 Element 2: The International Bank for Emissions Allowance Acquisition

An agency would be created with the sole function of buying and retiring allowances. This retirement would constitute the acquisition of resources needed to produce the global public good of climate control. To be concrete, I denote this agency the International Bank for Emissions Allowance Acquisition (IBEAA). Periodically the COP to the FCCC (or some other entity designated for the purpose) would meet and determine the quantity of (dated) allowances to be purchased and retired. These purchases might be implemented in an active international market with lots of private traders (arguably the setting best situated to "search" for economical emission reductions) or maybe just by putting out tenders to the countries of the world. In the long run, in order to control the concentration of CO_2 in the atmosphere, a very substantial retirement of allowances in the more distant future is going to be required—at some point, to the extent of zero net carbon emissions.

IBEAA would have only one central function: to purchase emission allowances (which would imply ancillary functions, such as monitoring compliance with the allowance regime and, perhaps, issuing debt to finance its purchases). Apart from monitoring the allowance system, the IBEAA would have no role in checking on the activities of countries or businesses. The key operating rule: No verified reduction of "imports" from BAU, no payment from the IBEAA.

2.2.3 Element 3: Cost Sharing

All countries are sellers of allowances in this story. The third element of the system is a procedure for sharing the cost of the allowances purchased by the IBEAA. All participating countries would share in the financing. Just how the costs would be allocated among the participant countries would be determined in the negotiations that set up the GPGP system. The system, per se, is silent on the sharing arrangements. The analogy is the sharing of costs of international peacekeeping. Cost shares might, for example, depend on per capita income or consumption levels and perhaps be responsive to the benefits countries get from protection against climate change. One would expect rich countries would pay most of the bills but "rich" might change over time.

2.3 Issues Raised by and Properties of the GPGP System

2.3.1 All Participating Countries Are Sellers of Allowances

Unlike the Kyoto-style cap and trade system, under the GPGP system all participating countries are sellers of allowances; none are buyers except in the sense that all participate in the collective purchase and retirement of allowances. (I note below a qualification to this assertion.) This is the basis for the compliance advantage of the system: If a participating country chooses not to sell, it is shooting itself in the foot (since, by construction, reductions of emissions from the BAU levels are zero cost) but it does not directly harm the overall emission control effort. I use the modifer "directly" here to recognize that a country's choice of the number of allowances to sell may affect the price obtained by other sellers and paid by the collectivity.

2.3.2 Who Participates?

The GPGP system is predicated on a set of participating countries. Since the objective is a Pareto improvement from the path the global economy would otherwise follow, we know that, in principle, there are arrangements whereby all countries could participate and still be better off than if no controls were put in place. I do not, however, claim to have solved the free-rider problem. Arranging for the provision of this public good requires that countries want to cooperate to do it, just as they do, for example, in military alliances. In the latter case, however, one can see "private" advantages that could be important elements of the story of collaboration. What one can say is that countries should, in principle, be willing to pay something to participate, since only participants are eligible to sell emission allowances to the IBEAA. Beyond that point, getting countries to pitch in poses the same problem in this framework as in any other case of organizing for a collective benefit.

2.3.3 How Does One Determine the BAU Quantities?

There are many conceptual as well as practical problems associated with the idea of BAU quantities of emissions/"imports." Integrated assessment/energy modelers are familiar with many of these. To a first approximation, I think of each country i as having a demand for net "imports" as a function of various determinants, important among them,

- the level of its own economic activity, y_i;
- its population, pop_i;
- the world prices of the different of fossil fuels, p_o, p_g, p_c (oil, gas, and coal; the time path of the prices might need to be made explicit);
- the state of technology/knowledge, k, and time, t, (reflecting, among other elements, the country's capital stock).

Thus a country's BAU would be determined as a function of a number of elements, each of which would need to be expressed in observable terms and each of which would vary through time and depend on contingencies such as technological developments:

$BAU_i(y_i, pop_i, p_o, p_g, p_c, A, k, t)$.

2.4 Estimating BAU "Imports"

Note that conceptually, the BAU amount incorporates a country's policies toward, for example, automobile transportation and industrial subsidies. The idea is to say what a country would have "imported" but for the climate problem. To apply the idea, one would presumably use econometric techniques to estimate such a relationship. One complication is that some countries have already taken steps to contribute to solution of the climate problem. One would need to find a way to allow for this in the data used for estimation.

2.5 BAU_{tr}: "Treaty BAU 'Imports'"

Perhaps one should call the quantities used in a treaty, "treaty BAU" quantities, denoted "BAU_{tr}." Then we could distinguish the conceptual idea, BAU, the econometric estimate, "BAU_{est}," and the amount settled on in a treaty, BAU_{tr}. To allow for estimating errors and in view of the desirablility of having as a BAU quantity a level that a country would not want to exceed, one would presumably choose BAU_{tr} larger than BAU_{est}, perhaps significantly larger. The cost of slack in the context of the GPGP system is the efficiency cost of raising the revenue to "buy it in" to achieve any given climate control target. On the other hand, the BAU allowances themselves are in completely inelastic supply, making them a natural source of revenue with no efficiency cost—neglecting any associated distributional impacts.

From the point of view of the enforcement advantages of the system, it is not necessary that BAU_{tr} be close to BAU. The important thing is that it definitely be larger than BAU, so that a country is very unlikely to want to exceed its BAU_{tr} level. Naturally, in the context of the GPGP system, a country will want the highest possible level of BAU_{tr}. But, since a country's contribution to the cost of the program is negotiated simultaneously with the setting of BAU_{tr} levels, there is an opportunity to design into the system very "generous" BAU_{tr} levels, compensated for by high levels of contributions to the resultingly apparently high cost the financing the allowance purchases. The efficiency and administrative trade-offs involved in this design feature are among the questions meriting closer analysis.

2.6 Updating and the Long Run

An important issue that calls for further investigation is how one would update the BAU_{est} amounts over a longer time period. It is arguably reasonable to contemplate estimating BAU levels at times close to the present, but a system designed to operate for centuries would need the capacity for updating.

The long run also poses challenges that merit further thought. As Jae Edmonds has emphasized to me, to stabilize concentrations of CO_2 in the atmosphere requires, ultimately, zero net emissions. In other words, in the long run one would be looking for the retirement of 100 percent of BAU allowances. The technical problems to one side, the financial magnitudes are daunting, and Edmonds has raised the issue whether the BAU_{tr} levels themselves could be systematically reduced over time (thereby changing their relationship to the conceptual BAU levels that indicate what countries would want to do in the absence of climate considerations). A question would be whether these levels could be cut without losing the enforcement advantage of the system.

2.6.1 What If a Country Exceeds Its BAU Allowance Level?

Conceptually, since the BAU allowance is supposed to indicate what a country would "import" were it not for climate considerations, in theory, there would be no reason to exceed its BAU allowance level. However, as I have suggested, it would make sense to build a substantial cushion in to the BAU_{tr} levels. If nevertheless a country wanted to exceed its BAU allowance, then the logic of the system would call for its buying allowances from the IBEAA. To obtain the enforcement advantages of the system would imply designing it to reduce this possibility to a minimum.

2.6.2 Leakage

The issue of "leakage" arises if the coverage of the GPGP system is less than global. One might think that a nonparticipating country would find it advantageous to specialize in energy-intensive production, thereby exacerbating the climate control problem confronting the participating country. Thus the demand by a country for fossil fuel, as a function of the world prices, will presumably depend on the prices of fossil fuel in effect inside other countries, since that will influence the prices of goods in international trade.

If all countries are participants in the GPGP system, the effective price of fossil fuel "at the border" will be the world price plus a going price for allowances. In that situation variation in the price of fuel inside countries would reflect policy differences, as at present, and hence introduce no new problem.

A country that is not participating in the system would not be eligible to sell allowances to the IBEAA. Furthermore exports from a participating country to a nonparticipating country would presumably not qualify as a deduction from the exporting country's BAU_{tr}. A participating country would, however, be charged with a unit of "imports" for fossil fuel that crosses its borders, whether from a participating or a nonparticipating country. In that situation we would expect the price of a tonne of fossil fuel ruling among nonparticipating countries would be the same as the price of the "package" of the tonne of fuel plus an allowance ruling among participating countries. (Imagine a world in which fossil fuel trades for $150 per tonne between nonparticipating countries and allowances sell for $50 per tonne. A participating country would be indifferent between selling to a participating country for $100 per tonne and to a nonparticipating country for $150 per tonne. So the price ruling in both kinds of countries should be the same.)

2.6.3 Banking

In describing the purchase and retirement of allowances, I was silent on the details as to timing. It is likely to make sense to date allowances. But, if a country contemplates neither using nor selling all of the excess of its BAU allowances over its "imports" in a year (as would be the case for a country that has not yet joined the group of participating countries), there should still be an incentive for it to economize on them. That would suggest that the IBEAA might be provided with a set of equivalences among dated allowances, for example, one 2012 allowance equals 1.05 2011 allowances. It would even be conceptually possible for such a set of equivalences to be set in advance with regard to future "imports" so that the IBEAA could engage in futures trading.

Enforcement considerations—no monitored excess of BAU_{tr} allowances over "imports" in a year, no payment from the IBEAA—would suggest confining IBEAA transactions to allowances for the current or earlier dates.

2.7 Comparison with Kyoto-Style System

I note without commentary differences between the GPGP system and the Kyoto Protocol regime in its current state of evolution:

2.7.1 Worldwide Emissions

In a Kyoto-style system, participating countries are assigned allowable amounts of emissions (specified by required reduction in emissions below a baseline). The total of these allowable amounts, as modulated by the compliance of participants and the behavior of nonparticipants, influenced in some degree by voluntary control efforts, by the Clean Development Mechanism and by leakage, determines the total of emissions. Putting aside questions of compliance and the behavior of non-participating countries, the Kyoto Protocol rigidly specifies worldwide emissions. Note that it would be conceptually possible to use a more flexible mechanism to assign allowable emissions–"safety valve" ceiling on the price of allowances has been suggested, for example. Furthermore the banking provisions of the Kyoto Protocol provide for a degree of smoothing of emissions over the five-year period over which the limits apply.

Flexibility of this sort would be natural aspect of the GPGP system, since the global total of emissions from participating countries would depend jointly on the evolving BAU_{tr} and the collective decision as to how much reduction to purchase from that level. These amounts could be specified in various ways.

For example, the buying agency (labeled in these notes the IBEAA) could agree to buy all amounts submitted at a specific price. Or it could specify an amount of money it was prepared to spend. It could even, in effect, announce a demand schedule for allowances. These approaches would reduce the uncertainty about compliance cost at the price of less certainty about the climate control in any given year or budget period. The agency could, however, be instructed to purchase allowances to meet a fixed total emission target, thus duplicating the Kyoto regime.

2.7.2 Burden Distribution

Arguably, most participants in the policy process think about the allowable emission amounts negotiated in a Kyoto-style system as indicating the emissions that will actually occur within a country that lives up to its obligations under the treaty. If, however, unlimited allowance trading is allowed—as the private property paradigm would imply—and countries take advantage of the opportunity to minimize their costs of compliance, actual emissions from a country will be determined by the market equilibrium. Indeed this is the source of the efficiency advantage that economists see in the allowance trading regime. The division of allowed amounts within a fixed total thus serves not an allocational function but, instead, determines implicitly the distribution of the cost of meeting the climate objective.

The GPGP system does not commit any country to any reductions in emissions, whether directly at home or by purchase from other countries. Instead, every participating country has the opportunity to enter the market to sell emission allowances. It does commit participating countries to share in the cost of the program; a key feature of the GPGP approach is to make the cost-sharing arrangements explicit. The differentiation in financial burdens among countries would be determined according to whatever determines such things in international negotiation. Again, I would invoke international peacekeeping as an analogue. Cost shares could be based on a standard such as equal per capita use of the atmospheric carbon reservoir, an approach often advocated for the long run. I doubt that this would be the outcome of conventional international politics, but nothing rules it out, either. It is not necessary to posit a particular resolution of the world's income distribution problem to deal with climate change.

2.7.3 Enforcement

Apart from monitoring the fossil fuel flows, the system does not involve scrutiny of individual policies of or actions by countries, or of characteristics of projects. This characteristic may be contrasted with the CDM, for example. Conceptually, however, the same property holds for the Kyoto-style cap and trade regime. A participating country is free to meet its reduction target by whatever method it chooses. If it chooses to subsidize its steel industry, that is its affair.

The GPGP system differs significantly, however, in the need for enforcement of a country's overall emission level. The default for each country in the GPGP system is its BAU trajectory. If the BAU levels are appropriately chosen, there is no need to enforce compliance with limits. (To be sure, it would be as critical under the GPGP system as under Kyoto that the system of monitoring be reliable.) Enforcement comes into question in collecting committed contributions. Experience with, for example, the United Nations suggests that this collection function is not trivial. But it is a different sort of problem. One can imagine the IBEAA succeeding in the climate objectives set for it at the same time that the system contends with collecting from participants in arrears. Presumably the IBEAA could set off payments to a country for allowances to be retired against that country's contributions. (One would need to think carefully about the incentives for "dropping out" of such a system.)

Appendix: Incidence

I present in this appendix some highly stylized analyses of the distributional characteristics of alternative climate regimes. Most economists will be very familiar with this material, but perhaps it may serve to clarify some of the issues.

Distributional Effects within Countries

The choice of aggregate GHG levels over time by the COP will have distributive effects via the impact of the price of allowances throughout economies—much like variations in the price of fossil fuel. But the way the national obligations to chip in are spread across taxpayers within countries can be explicitly determined as a matter of domestic policy. For example, one could imagine a tax imposed on emission allowances as one source of revenue. Or some of the allowances could be auctioned by the government.

Distributional Effects Internationally

To illustrate the possibilities, consider, as in figures 2.1, a homely example of a two-country world, AB (a stylized aggregate of the annex-B countries under the Kyoto Protocol) and ROW (rest of the world). The

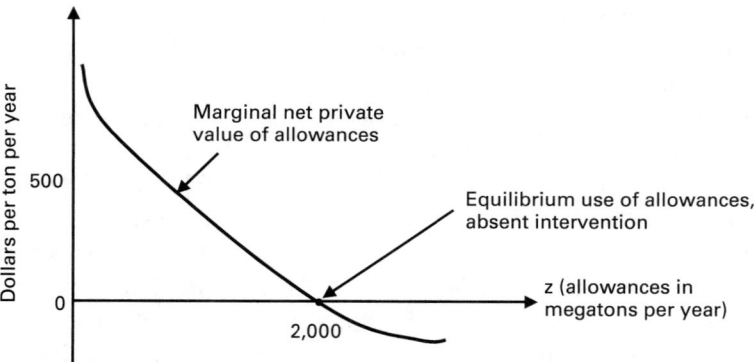

Figure 2.1
Starting point: Illustrative demand for CO_2 allowances in AB

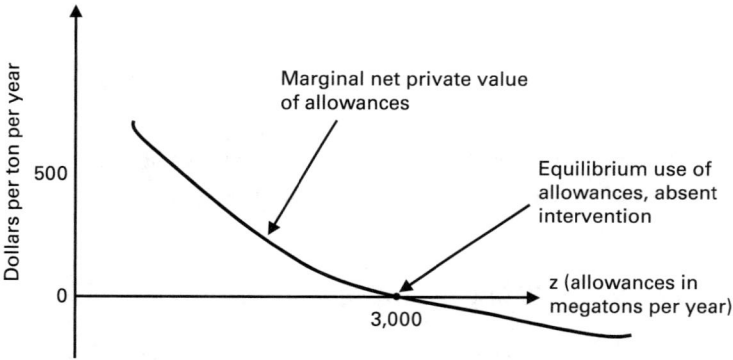

Figure 2.2
Illustrative demand for CO_2 allowances in ROW. In the example, it is cheaper to cut emissions from the initial equilibrium level in ROW than in AB. In the hypothetical scheme, AB would be assigned 2,000 allowances and ROW 3,000 allowances as BAU levels.

figure below sketches a hypothetical demand curve for allowances by AB if they were priced. I take for granted that my readers have an idea how one derives such a thing. The numbers are made up (although one could make a reasonable stab at realistic relationships), and I have completely neglected the time dimension.

Figure 2.2 shows the same thing for ROW. Under the suggested scheme, ROW would be assigned 3,000 million allowances, reflecting its BAU demand. AB would be assigned 2,000 million allowances. In the illustrative BAU scenario, worldwide emissions would be 5,000

Chapter 2 Improving on Kyoto

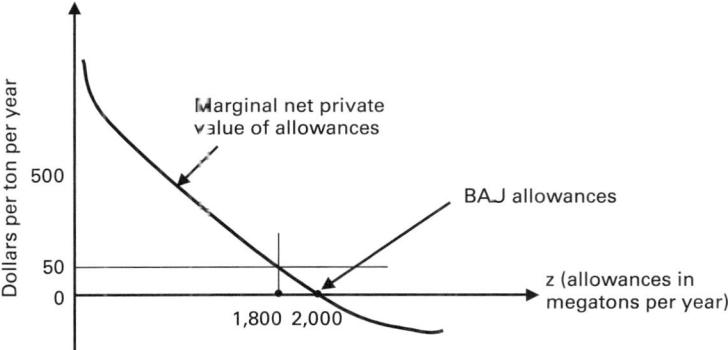

Figure 2.3
Price of allowances that induces a 200 cut in AB. To get AB to sell 200 from its BAU level would call for a price of $50 per tonne.

megatonnes. (In equating allowances with emissions, I neglect the refinement that an allowance might not be used to burn fuel in a particular year.)

Suppose that the international community decides to cut emissions by 20 percent from BAU. (Kyoto is thought to imply about a 30 percent cut for the United States during the first budget period.) Then one looks for the price that will call for that much reduction from BAU from the two countries. For illustrative purposes, suppose that $50 per tonne does the job. (This is probably very high if the cut is not done in a rush.) Figures 2.3 and 2.4 indicates the story for the two countries. The sale of allowances to the IBEAA will earn the countries the amounts in the shaded rectangles in figures 2.5 and 2.6.

The total revenue from the sale of allowances in this illustrative case would be $50 billion (if I got my units right). This total outlay by the IBEAA would be financed by payments to the IBEAA by the two illustrative governments. Just viewed as a matter of national flows, the net result would be a wash for each country in the illustrative case if the financing happened to be divided 20 percent from AB and 80 percent from ROW. More plausible, interpreting ROW as developing countries would be a much larger share of financing from AB. In the extreme case, if AB paid for all of the allowances, the result would be a net transfer of $40 billion from AB to ROW (AB is paid $10 billion for allowances sold; ROW is paid $40 billion for allowances sold; AB contributes $50 billion to IBEAA to finance the purchases).

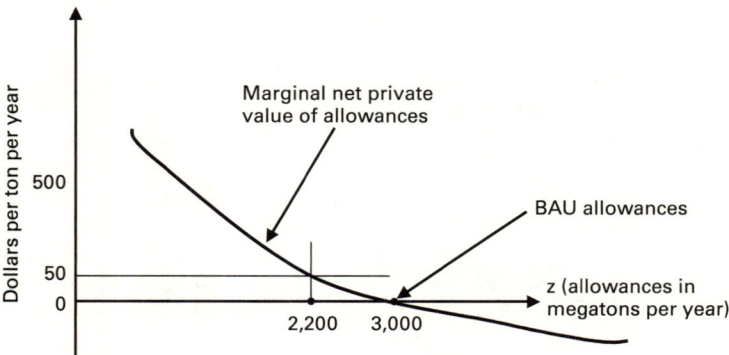

Figure 2.4
Price of allowances that induces an 800 cut in ROW. At a price of $50 per tonne, ROW is prepared to sell 800 megatonnes of allowances.

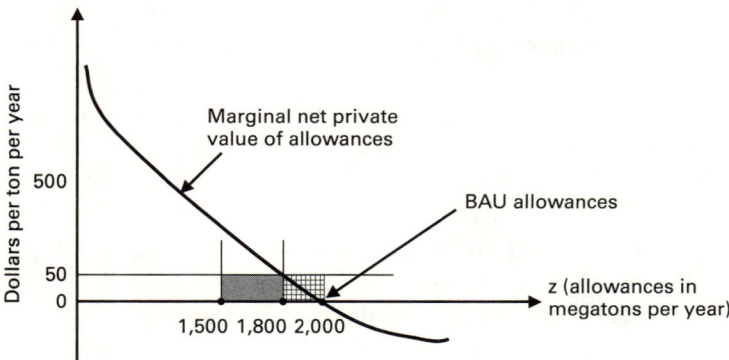

Figure 2.5
Net cost to AB if AB covers cost of retired allowances. Shaded/cross-hatched rectangle shows the revenue from sale by AB of 200 megatonnes of allowances. Solid rectangle shows net payment to ROW. Full cost to AB is the net payment to ROW plus the cross-hatched triangle of real compliance cost.

Chapter 2 Improving on Kyoto 31

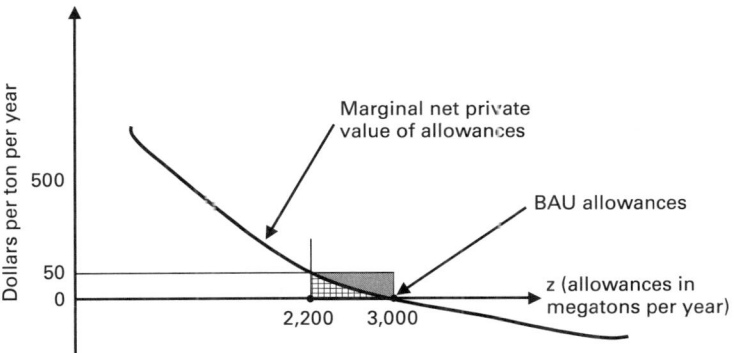

Figure 2.6
Net gain from to ROW if AB finances purchase of allowances. Shaded/cross-hatched rectangle shows the revenue from sale by ROW of 800 megatonnes of allowances. Net gain is the shaded triangle of surplus.

The pictures give some idea of the complexity of the incidence issues that this scheme would raise. (But similar incidence effects are buried in the command and control regimes as well.) For example, who, hypothetically, owns the BAU allowances, which would increase in value from $0 to a large positive value as a result of the illustrative 20 percent retirement and would increase more with a more ambitious intervention? Of course, as the price of allowances rises, so does the payment required from those who would need them in their operations. If all of the BAU allowances were allocated to preexisting users (a version of grandfathered allocation), and they in turn were the ones who needed to buy allowances to operate, then in this very simple story they would, on balance, gain from the policy (as measured by the triangles of surplus above the curves in the rectangles in the graphs above). In this little example it appears that one could finance about half the purchase amount through a tax on allowances and still leave the grandfathered holders in the aggregate a little better off than they were in the pre-regulation situation.

This neglects entirely the benefits from climate control. I would furthermore emphasize again that this is an extremely simple model, just to stimulate thinking. Importantly, it also neglects entirely other price effects of the program—for example, effects on the price of oil. The usual assumption underlying the sorts of demand curves for allowances that I have drawn is that all prices other than allowance prices

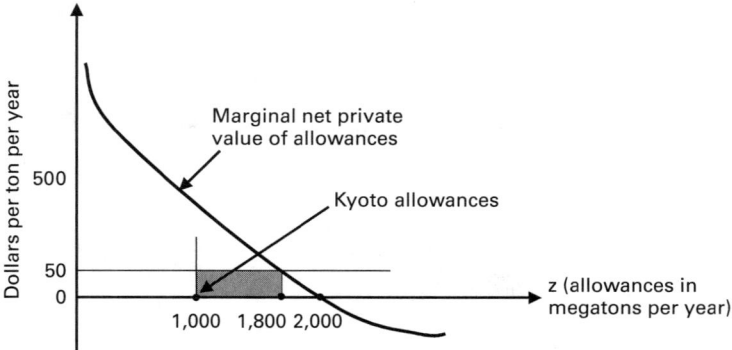

Figure 2.7
Purchase of allowances by AB under hypothetical Kyoto style plan. Shaded rectangle shows the purchase of allowances under Kyoto with full trading, assuming all required reduction is by AB (from BAU of 2,000 to 1,000). The 200 megatonnes of reduction would be carried out "at home."

are fixed. This would clearly be wrong. One can deal with it, but it takes more than our pictures.

Mimicking Kyoto

One can use the simple example to show how an abstract version of the Kyoto regime, one that aimed for the same reduction in worldwide emissions, would implement a particular pattern of financing and implicit purchase by an IBEAA. (For an analysis of the burdens on different countries implicit in the Kyoto Protocol and of the implications of alternative sharing principles for the assignment of Kyoto-style emission limits, see Babiker and Eckaus 2000.) In our stylized world, under the Kyoto arrangements ROW is not obliged to make any reductions but is subject to a cap at its BAU quantity. AB is assigned the full reduction of 1,000 megatonnes. This is equivalent to being given an allowance to emit $2,000 - 1,000 = 1,000$ megatonnes. With full trading, AB will choose to buy 800 megatonnes from ROW, at a cost of $40 billion. Note that giving the usual interpretation of the curves and continuing to ignore climate benefits, AB loses more than $40 billion (by the little triangle under the curve between 2,000 and 1,800 megatonnes, or about $5 billion) and ROW gains from the program (the little triangle above the curve between 3,000 and 2,200, or about $20 billion). Figures 2.7 and 2.8 characterize the "financing arrangements that are

Chapter 2   Improving on Kyoto

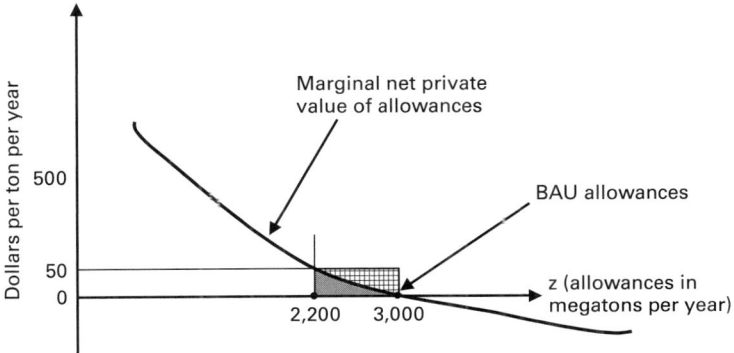

Figure 2.8
Net gain to ROW under hypothetical Kyoto style plan. Shaded rectangle shows the revenue from sale by ROW of 800 megatonnes of allowances. Net gain is the cross-hatched triangle of surplus.

implicit in Kyoto's distributions of obligations to make reductions from business as usual" that I mentioned above.

If there were no international trading at all of allowances, the cost of the Kyoto program to AB would be about $200 billion, and the cost to ROW would be $0 (compared to a gain of about $20 billion under the illustrative plan). The sum of the costs would be $200 billion, instead of $25 billion, a dramatic instance of the sort of gain one can get by relying on a trading regime (figure 2.9). This extra cost (the concept, not the amount) is what the Europeans have been advocating be borne, mainly by the United States and the FSU.

Mimicking a Harmonized Tax

We can use the same graphical tools to illustrate the use of harmonized national tax policies to implement the climate objective, as suggested, for example, by Nordhaus (2001). In this case the required tax would be $50 per tonne. The revenues raised would be as indicated in figures 2.10 and 2.11.

Economists understand that the revenue from the taxes tells us nothing about the effective burdens borne by the two countries. Basically they constitute redistributions among taxpayers within the countries—the revenues would permit reductions in other taxes, if they were not used to compensate the groups on whom the carbon taxes are imposed (as might occur in the political process). Under the simplifying

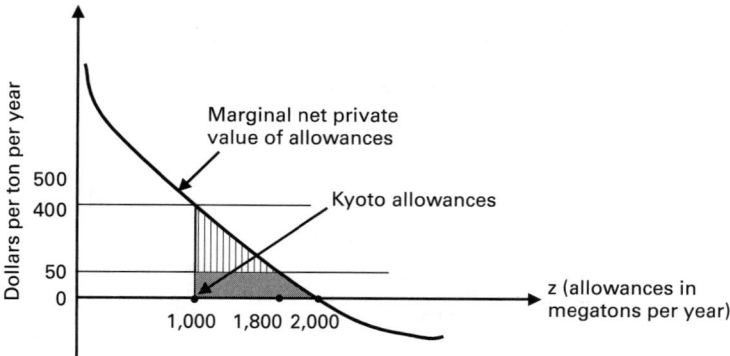

Figure 2.9
Cost to AB under hypothetical Kyoto plan with no trading. Shaded triangle shows the cost of compliance with Kyoto with no trading, assuming all required reduction is by AB. Upper striped area is the extra cost to AB from no trading. In this case, there is no gain to ROW, apart from any climate control benefit.

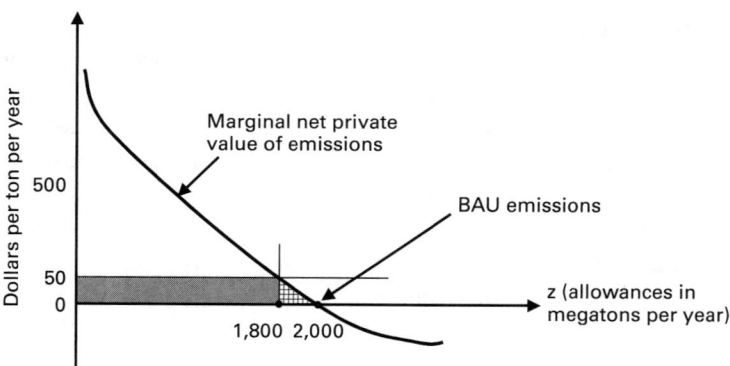

Figure 2.10
AB revenue from harmonized tax. Shaded rectangle shows the AB revenue from a harmonized tax to implement a worldwide reduction of 1,000 megatonnes. No money is transferred abroad. Net cost to AB (neglecting benefit of climate control) is the striped triangle of real compliance cost.

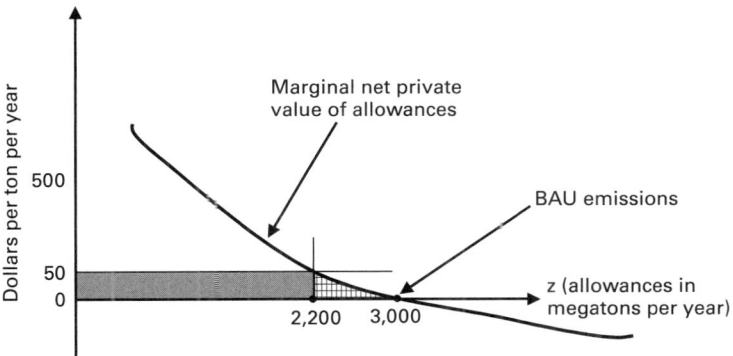

Figure 2.11
ROW revenue from harmonized tax. Shaded rectangle shows revenue raised by harmonized tax. Shaded triangle shows the burden on ROW under the harmonized tax regime before transfers from AB.

assumptions made in constructing the diagrams, the burdens on the two countries are represented by the little triangles under the curves of marginal values of emissions, as shown in the figures. Absent transfers from AB to ROW and neglecting the benefit of controlling the climate, ROW bears a net burden under the illustrative harmonized tax that roughly equals its net gain under the illustrative Kyoto plan with full trading.

Acknowledgments

This paper is based on notes prepared for discussion among colleagues in Princeton's Carbon Mitigation Initiative group. It has had a certain amount of circulation under the title "A No Cap but Trade Approach to Greenhouse Gas Control," a version of which was presented at a conference, "International Climate Policy after COP6," Hamburg, Germany, September 24–26, 2001. Remnants of that terminology may remain in the current draft; I am still looking for a good title. Furthermore comments, suggestions, and references to the literature would be very welcome. For helpful comments and conversations thus far I would like to thank John Deutch, Adele Morris, Michael Oppenheimer, Steve Pacala, Ian Parry, Billy Pizer, David Victor, Mark Witte, and participants in the Hamburg conference, especially Jae Edmonds and Henry Tulkens.

References

Babiker, M. K., and R. S. Eckaus. 2000. Rethinking the Kyoto emissions targets. Report 65. Joint Program on the Science and Policy of Global Change, MIT, Cambridge.

Coase, R. H. 1960. The problem of social cost. *Journal of Law and Economics* 3 (October): 1–44.

Hargrave, T. 1998. US carbon emissions trading: Description of an upstream approach. Washington, DC: Center for Clean Air Policy.

Kopp, R., R. Morgenstern, W. Pizer, and M. Toman. 1999. A proposal for credible early action in U.S. climate policy. Weathervane/Resources for the Future. Available at ⟨http://www.weathervane.rff.org/features/feature060.html⟩.

Lackner, K. S., R. Wilson, and H.-J. Ziock. ND. Free market approaches to controlling carbon dioxide emissions to the atmosphere: A discussion of the scientific basis. Unpublished manuscript.

McKibben, W. J., and P. J. Wilcoxen. 1999. Permit trading under Kyoto and beyond. *Brookings Discussion Papers in International Economics*. Washington, DC: Brookings Institution.

Nordhaus, W. D. 2000. Alternative mechanisms to control global warming. Paper presented at the 20th anniversary meeting of the International Energy Workshop, Laxemburg, Austria, June 21.

Victor, D. G. 2001. *The Collapse of the Kyoto Protocol and the Struggle to Slow Global Warming*. Princeton: Princeton University Press.

3 The Design of Post-Kyoto Climate Schemes: Selected Questions in Analytical Perspective

Roger Guesnerie

Climate policies raise many issues. These issues can effectively be addressed by grouping them into two categories: as questions on the desirable pace of action or as questions on the means of action. Although these two categories do overlap, separate assessments are often appropriate.

The analytical and conceptual problems feeding the thinking on the desirable pace of action are more difficult to assess. They have emerged, again, at the forefront of the climate change debate with the recent publication of the *Stern Review*. In particular, discussion has revived the controversy over long-run discount rates—and this was in fact the topic I had initially chosen to treat at this conference. My intent was to present and develop an argument I had sketched out earlier (e.g., see Guesnerie 2004). My argument relied on more standard views of discounting (relative prices) than those advocated by *Stern*, although it did support more radical environmental friendly conclusions.

The second broad set of issues, the analysis of the means of action, and especially of the design of international action, is the subject of David Bradford's contribution. This is the keynote lecture reprinted as chapter 2 of this book, and in it David makes the striking proposition of GPGP, or global public good purchase. Because David is no longer with us to expand his views on this topic, I suggested, and Henry Tulkens agreed, that it would be appropriate to redefine my intervention as a kind of complement to the presentation read at the opening session of the conference. Thus the present chapter may be viewed as a broadening of a discussion on climate policies that starts from, and plays freely on, David Bradford's concept of GPGP.

It should be obvious at the outset that an assessment of GPGP cannot be made in a vacuum. In particular, such a proposed scheme calls for a comparison with the competing arrangements of the Kyoto

Protocol. And by this agenda, the emphasis cannot be put on the strict version of the Kyoto Protocol but, for reasons that will be explained shortly, on a variant that is both open and flexible. Hence, in this chapter, I start by setting the discussion of GPGP in the broader perspective of its comparison with what I call "Kyoto-compatible" schemes. I do not, however, attempt a comparison with Kyoto that is exhaustive; my emphasis is rather on two important dimensions of the design of climate schemes: the question of participation and the type of climate policy target. By this perspective I will enlarge last the scope of comparison of climate policies to bring into the discussion a tax harmonization scheme.

3.1 Comparing GPGP and Kyoto-like Arrangements

3.1.1 GPGP, a Brief Recap

The GPGP, as presented in chapter 2 of this book, can be summarized as follows (Bradford 2001):[1] A number of countries provide voluntary initial contributions. The set of contributing countries, their characteristics, and the amounts of their contributions are treated as exogenous. A business as usual (BAU) level of emissions allowance is attributed to "all" countries (the allowances can be more generous for noncontributing countries). An agency spends the contributed money and buys, at a market price, any reduction from the BAU level by any country in the world; this agency is called the International Bank for Emissions Allowance Acquisition (IBEAA).[2]

The GPGP scheme is usefully illustrated in a stylized static and certain world. It is assumed that there is only one-period, one single greenhouse gas, and that abatement costs are certain. To put this more formally, countries are indexed by $i \in I$. The costs of q_i units of emissions abatement, measured from a BAU basis, are given by a convex cost function C_i that is defined for every q_i. The utility generated in each country i, from total reductions Q, is $U_i(Q)$, that is, $U_i(Q_{-i} + q_i)$, with $Q_{-i} = \sum_{j \neq i} q_j$. The function U_i is then assumed to be concave.

Metaphorically, the two groups of countries are called, respectively,[3] annex B, for $i \in B$, where $dU_i/dQ > 0$, in the whole range under consideration, and non–annex B, for $i \in NB$, which are countries that are presumed to be unconcerned[4] about climate policy, so that

$$\frac{dU_i}{dQ} = 0 \quad \forall Q, U_i = C_{ste}(=0).$$

Chapter 3 Design of Post-Kyoto Climate Schemes

Let us consider a GPGP equilibrium associated with exogenously fixed contributions $\bar{T}_i, i \in B$

DEFINITION 1 A GPGP equilibrium, associated with fixed contributions $\bar{T}_i$, $i \in B$, consists of abatements $(q_i^* \geq \bar{\bar{q}}_i, i \in B; q_i^* \geq 0, i \in NB)$, a positive carbon price t^*, such that:

1. q_i^* is a solution of $\max\{U_i(Q_{-i}^* + q_i) + t^*(q_i - \bar{\bar{q}}_i) - C_i(q_i)\}$, $q_i \geq \bar{\bar{q}}_i$, $i \in B$, with $Q_{-i}^* = \sum_{j \neq i} q_j^*$;
2. q_i^* is a solution of $\max\{t^* q_i - C_i(q_i)\}$, $i \in NB$;
3. $\sum_{i \in B} t^*(q_i^* - \bar{\bar{q}}_i) + \sum_{i \in NB} t^* q_i^* = \sum_{i \in B} \bar{T}_i$.

The GPGP equilibrium is said to be individually rational iif:

4. $\{U_i(Q^*) + t^*(q_i^* - \bar{\bar{q}}_i) - C_i(q_i^*) - \bar{T}_i\} > U_i(\sum_i \bar{q}_i) - C_i(\bar{q}_i)$, $i \in B$, with $Q^* = \sum_i q_i^*$, and where $\bar{q}_i$ is an a priori given reference level.

Put more informally, we have the annex-B countries contributing an exogenously fixed amount $\bar{T}_i$; t^* is the carbon price paid by the central agency. The BAU abatements are q_i^* in the annex-B countries and zero in the non–annex-B countries. So country i's abatements $i \in B$ are equal to $\max(0, q_i^* - \bar{\bar{q}}_i)$ or equal to q_i^*, $i \in NB$. The sum of payments associated with abatements beyond the BAU level is given by the left-hand side of condition 3, and it equals the right-hand side $\sum_{i \in B} \bar{T}_i$. Conditions 1 and 2 define the optimal decisions, respectively, of annex-B and non–annex-B countries.

The viability condition—which is equivalently an individual rationality condition—expresses the fact that an annex-B country is better off than in the initial reference situation. This condition is necessarily satisfied for a non–annex-B country, but not necessarily for an annex-B country if its initial contribution (or its $\bar{\bar{q}}_i$)[5] $\bar{T}_i$ is too high. Here, as later in the appendix, I will assume that the reference situation is an interior noncooperative Nash equilibrium of abatements.[5] In all of the discussion that follows I will be assuming that $\bar{\bar{q}}_i = \bar{q}_i$.[6]

My evaluation of the GPGP scheme in terms of Kyoto-like arrangements will begin with the free-riding issue.

3.1.2 The Participation Problem

Free-Riding

There has been much debate about climate policy arrangements, and this conference has a number of contributions to add to this debate.

Indeed the Kyoto arrangements are vulnerable to free-riding in some form or another. As is by now understood, a cooperative policy can be expected to be stable only in a weak sense. Carraro (1999) provides a review of existing results, and a number of papers in this conference are concerned with this issue.[7]

One good argument for the GPGP scheme is that it solves, or at least helps solve, a problem coming under the rubric of free-riding, the so-called participation problem. If the BAU cap is computed correctly, then it is in the unambiguous interest of a non–annex-B country to participate in the abatement process. It cannot lose and, by selling allowances, can only win. As I assume, in definition 1, that the arrangement is viable for annex-B countries (this is the viability constraint 4), all countries should be willing to participate in the abatement policy. This assertion is trivially true in the above stylized model and would remain true in an uncertain world if the proposed cap were contingent on sufficiently verifiable events. Hence, as David Bradford suggested, the allowances for contingencies used to determine BAU levels have to be accurate and, if not, generous enough. Hence, although the contributions to the financing of GPGP may be optional and made on voluntary basis, all nations are expected to gain from participating in the abatement scheme. This is a most welcome achievement of the GPGP.

A scheme that leaves out of its arrangement a significant part of the world suffers from serious drawbacks. One such drawback is captured by the curve shown in figure 3.1, with the effort of the climate coalition (e.g., as measured by the carbon price) represented along the horizon-

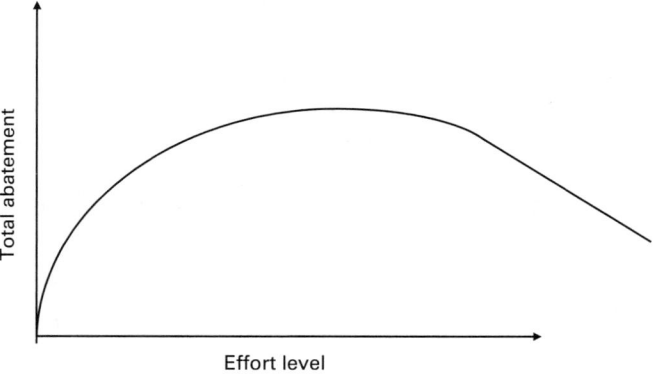

Figure 3.1
Reduction of carbon emission as a function of increased effort by a climate coalition

tal axis by and the reduction of world emissions along the vertical axis. By the hump shaped curve, we see that at a low level of effort by the coalition, emissions do decrease, but as the effort increases, the amount of reductions reach a maximum. The explanation here is that polluting industries of the coalition lose market shares and/or migrate outside of the coalition borders. This phenomenon is exacerbated as the effort levels of the coalition increase.

Possibly, under the same assumptions on the "out-of-coalition standards," higher effort would yield lower performance. A sketchy but more formal presentation of the "leakage" argument is provided in appendix A, and puts emphasis on the elasticity of substitution between "de-carbonated" and "carbonated" goods. Although the present estimates show such a phenomenon unlikely to occur in the aggregate, unless the climate coalition level of effort becomes very high, a bold empirical assessment might suggest that the phenomenon ("carbon leakage" over 100 percent) is plausible at the industry level. So increasing the effort by virtuous countries might increase the world emissions within a given industry, provided that the coalition level of effort is significant (for an empirical assessment, see chapter 16 by Demailly and Quirion in this book).

To see the advantage of the GPGP scheme over the Kyoto-like variants (denoted below by K), it may be useful to start with a formal definition of a flexible Kyoto arrangement ("flexible" in the sense that quotas are tradable on an international market for permits, the same as they are in the Kyoto Protocol).

DEFINITION 2 A flexible Kyoto equilibrium consists of (exogenously given) quotas $\hat{q}_i$, $i \in B$, abatements $(q_i^{*\prime})$, $i \in B$, and a positive carbon price t^* such that:

1. $q_i^{*\prime}$ is a solution of $\max\{t^*(q_i - \hat{q}_i) - C_i(q_i)\}$, $i \in B$,
2. $\sum_{i \in B} q_i^{*\prime} = \sum_{i \in B} \hat{q}_i = Q^{*\prime}$.

The equilibrium is individually viable if:

3. $\{U_i(Q^{*\prime})) + t^*(q_i^{*\prime} - \hat{q}_i) - C_i(q_i^{*\prime})\} > U_i(\sum_i \bar{q}_i) - C_i(\bar{q}_i)$.

In a flexible Kyoto equilibrium the sum of abatements equals the sum of the initially attributed quotas (non–annex-B countries are not included). By condition 1, this is taken as given in a country's optimization confronting the costs and market price of abatements. But quotas are traded, and the equilibrium price of the market for permits is t^*. In the definition the quotas are fixed even if they have been ex ante bargained in order to spread the burden across participating countries.[8]

Of course, a flexible Kyoto (FK) scheme does not solve the participation problem of developing countries (as is clear in the stylized model above, where non–annex-B countries are assumed not to participate).[9] If the FK scheme fails to attract participation in a part of the world where emissions are likely to become sizable, it is bound to become much less effective than a GPGP scheme. This is the participation argument that is at the heart of David Bradford's proposition. I would like to propose instead that the GPGP be compared not necessarily with the formal Kyoto Protocol but with variants of Kyoto that would mimic the GPGP solution in order to solve the so-called participation problem.

It is with this aim that I introduce an open flexible Kyoto variant (denoted OFK). Countries that are noncontributing in the GPGP scheme would be given, within the OFK scheme, the same BAU targets that they have in the GPGP scheme and would be induced to participate in much the same way. By the OFK scheme the targets would be nonbinding, or if worded slightly differently, the targets would be no-loss or one-way targets so that developing countries could be persuaded to join in the scheme (e.g., see Philibert 2000; Philibert and Pershing 2001). That is,

DEFINITION 3 An open, flexible Kyoto equilibrium consists of (exogenous) quotas $\hat{q}_i$, $i \in B$, abatements (q_i^{**}), $i \in I$, and a positive carbon price t^{**}, such that:

1. q_i^{**} is a solution of $\max\{t^{**}(q_i - \hat{q}_i) - C_i(q_i)\}$, $i \in B$;
2. q_i^{**} is a solution of $\max\{t^{**}q_i - C_i(q_i)\}$, $i \in NB$;
3. $Q^{**} = \sum_i q_i^{**} = \sum_{i \in B} q_i^{**} + \sum_{i \in NB} q_i^{**} = \sum_{i \in B} \hat{q}_i$;
4. if $\forall i \in B$, $\{U_i(Q^{**}) + t^{**}(q_i^{**} - \hat{q}_i) - C_i(q_i^{**})\} > U_i(\sum_i \bar{q}_i) - C_i(\bar{q}_i)$, with $Q^{**} = \sum_i q_i^{**}$, then it is individually viable.

In an OFK equilibrium the environmental performance is still determined by the quotas for the annex-B countries. But non–annex-B countries, which are given nonbinding commitments, do participate and, as in the GPGP scheme, benefit from participation while helping to decrease the cost burden of annex-B countries. Appendix B elaborates on this formal definition of the GPGP/OFK schemes by introducing other references, such as Nash equilibrium, in a more comprehensive analysis. As the appendix shows, for fixed contributions there exists a unique GPGP equilibrium. This is a convenient, though not obvious, statement. Proposition 6 asserts that if one omits the viability constraint 4, the abatement targets attainable through a GPGP scheme when varying contributions are identical to the abatement targets attainable in an OFK scheme when varying quotas.[10] The analysis also suggests, although it is not always true, that the contributing countries, let us call them annex-B countries, using the Kyoto vocabulary are likely to be better off in a GPGP scheme compared with an OFK scheme that leads to the same ecological performance. Hence participation constraints for annex-B countries are more likely to be satisfied with GPGP.

The reason may be surprising.[11] In a GPGP scheme annex-B countries faced with a given carbon price tend to have a more ambitious abatement policy than if they face the same carbon price in an OFK scheme, and hence may pay a cheaper bill to attract non–annex-B countries; see appendix B, corollary 7. Here the GPGP's advantage in ensuring participation, as suggested, is not strong, and it would lessen as the size of annex B increases. Further, although the effort among annex-B countries may be more straightforward, negotiations are not necessarily easier. This more general issue relates to coalitional stability.

The GPGP's transparency of costs was argued by David Bradford to be an advantage for real world negotiations. With OFK, such transparency of the costs would only be guaranteed in a world without uncertainty. Hence the differences in the stability assessments of GPGP and OFK would reflect real effects, as those stressed in the context of certainty as well as those related to uncertainty.[12] The differences might also reflect "framing" effects whereby two equivalent games are played differently.

By this point the advantage of GPGP over the actual Kyoto (K) arrangements, in terms of participation cost, should be clear. But this is not a decisive advantage, since variants of Kyoto can mimick the central feature of GPGP: voluntary costly participation could, in principle, be triggered, as well as voluntary noncostly participation, by adopting an open flexible Kyoto (OFK) scheme. Hence, as just argued, and more fully in appendix B, the suggested advantage of GPGP over OFK remains at this point inconclusive.

3.1.3 The Ratchet Effect

David Bradford was, of course, fully aware that the participation decisions of countries that consider joining these schemes are not as transparent as suggested by the preceding analysis. Indeed the argument above, as well as the stylized one made in David Bradford's chapter 2, take place in a static world. In a dynamic world the initial BAU quotas would be followed in the next period by adjusted BAU quotas. A participation scheme would then face what is known as a "ratchet effect." The ratchet effect is what plagued the Soviet economy: firms were reluctant to meet current quotas because of the expectation that the current quotas would be raised tomorrow as a consequence. Asymmetric information between central agency and the firms, and the absence of commitment from the central agency, are at the heart of the ratchet effect. Optimal bonuses can be designed if such asymmetry of information and noncommitment are accounted for (see Freixas, Guesnerie, and Tirole 1985).

In the GPGP scheme, the ratchet effect could take a simple form: a country considering whether to sell its allowances would take into account not only the benefits from its present sale but also its likely loss due to the effect of the decision on the next allocation of BAU allowances. It is not difficult to come up with situations where the participation decision, particularly of noncontributing countries (non–annex B, in this case), would be postponed. Even if such were not the case, the incentive to reduce emissions would be lower than the apparent monetary incentive (x dollars per ton of carbon on the market for allowances). Hence the main attractive feature of the GPGP, that it solves, the developing countries participation problem, is not as convincing as it may at first appear. Naturally the same can be said about a related feature of the OFK scheme (with nonbinding, one-sided constraints).[13]

The way around the ratchet effect, whether it be in an OFK scheme or a GPGP scheme, is to ensure long-term commitment. Strict commitment would mean that the BAU allowances of the GPGP are fixed at the outset for a very long period of time but account for contingencies that might cause them to become unrealistically high. An alternative might be to agree on long-term targets set rather uniformly across countries but probably based on per capita emissions corrected by a production index and climate damage exposure index. This approach might not suppress opportunistic considerations in abatement decisions but could limit their scope.

3.2 Suggestions for Improvement

Basically the GPGP scheme is a polar solution to the classic problem of financing a public good, not through taxes but through voluntary contributions. The GPGP scheme further permits that countries not able to contribute financially be given allowances that induce them nevertheless to participate in the implementation of climate policy. As I have just argued, while the GPGP scheme consequently favors participation in a way that makes it superior to the Kyoto Protocol, it is not necessarily superior to the suggested variant of the Kyoto scheme. The main drawback, which is the same for the improved Kyoto scheme, is that the GPGP scheme is susceptible to ratchet-like effects.

Taxation has proved historically to be successful in financing a public good. Within the GPGP framework, let me suggest that the financing rules might be changed to include, to some extent, an emission tax that would be paid by countries each period, based on the amount of their previous period emissions. In principle, the taxation scheme would remain combined with a voluntary contribution (the a priori weight given to the taxation component of the resources of the agency would need to be thought out carefully). This way every country would have part of its contribution, at least in future periods, determined by a tax on its emissions of the previous period.

One could, of course, object that taxation suggests a regalian power and that there is no such thing as a world government, not today nor in the foreseeable future. However, in the Kyoto variant that I propose in this chapter, voluntary participation might still be triggered, probably at total contributing burdens close to the burden level of contributions of the GPGP scheme. The idea here may be better illustrated by treating this scheme formally as a type of GPGP variant: a GPGPPT (global public good purchase plus taxation) scheme.

DEFINITION 4 A GPGPPT equilibrium, associated with (exogenous) contributions $\bar{T}_i$, $i \in B$, consists of abatements ($q_i^* \geq \bar{\bar{q}}_i$, $i \in B$, $q_i^* > 0$, $i \in NB$), and a positive carbon price t^*, such that:

1. q_i^* is a solution of $\max\{U_i(Q_{-i}^* + q_i) + t^*(q_i - \bar{\bar{q}}_i) - C_i(q_i)\}$, $q_i \geq \bar{\bar{q}}_i$, $i \in B$, with $Q_{-i}^* = \sum_{j \neq i} q_j^*$;
2. q_i^* is a solution of $\max\{t^* q_i - C_i(q_i)\}$, $i \in NB$;
3. $\bar{T}_i = \bar{t} E_i^{-1} + e_i$, where $\bar{t}$ is a tax rate and E_i^{-1} is the emissions of country i at period -1;
4. $\sum_{i \in B} t^*(q_i^* - \bar{\bar{q}}_i) + \sum_{i \in NB} t^* q_i^* = \sum_{i \in B} \bar{T}_i$.

As is evident, in the short run the burden remains the same as in the original GPGP scheme for the voluntary contributors to the GPGPPT scheme (see also appendix D). Going a bit further, one could make the tax nonlinear. Introducing an exemption level would leave the developing countries temporarily out of the business of financing the scheme, providing another rationale for the separation of annex-B countries from non–annex-B countries.

There are two advantages to combining voluntary contribution with taxation in an amended GPGP framework. First, the tax would not only serve as a financial instrument but also leverage the incentive to contribute to the scheme. Provided that in a steady state the price of carbon on the market managed by the IBEA agency is y dollars per ton and the (marginally flat) tax on carbon emissions is x dollars per ton, then, without discounting, the incentives to reduce emissions of one unit in the present period would be $x + y$ (instead of just y). Second, the system would likely work to limit a ratchet effect. It may be the sufficient inducement for countries that currently do not contribute, and are not taxed, to consider how their present effort might lessen the burden of a possible future taxation. The perspicacious reader will recognize a hidden hypothesis here. What is behind the argued improvement in performance of the GPGPPT scheme is not that it involves some *quantity commitment* in the future but rather a *procedural commitment* in that the taxation of past emissions will be on the negotiation table at the next stage of the game. Whether this scheme will prove to be politically viable remains in question. It seems, however, to provide a better framework for negotiations than the initial GPGP or OFK (although OFK could be amended by an analogous a tax arrangement).

From introducing taxation into the GPGP scheme my discussion will proceed to some thoughts on some other possible climate policies. I will compare the policies of the GPGP scheme and the variant just suggested—global public good purchase plus taxation, GPGPPT—and also the OFK to an alternative scheme, a scheme suggested[14] early in the climate negotiations and labeled "harmonized taxation."

3.3 Flexible or Rigid Targets: GPGP, Kyoto-like Schemes, and Tax Harmonization

Harmonized taxation is often viewed as a prototype price-style policy, as opposed to the Kyoto-like quantity-style policy. This distinction will be assessed in this section. However, before embarking on a general discussion, I will remark on the conventional wisdom regarding price versus quantity within the context of climate policy.

3.3.1 Price versus Quantity

The Kyoto Protocol implements a quantity policy in the sense that it sets global emissions objectives for the participating partners. National performances may differ from national quotas because the quotas can be exchanged on an international market. But the total authorized amount of emissions is left unchanged. This basic mechanism of the Kyoto Protocol has been criticized as not being flexible enough given the present uncertainty about the costs of abatement. It can lead to too little reduction if it turns out that the (marginal) cost of abatement is smaller than expected; it can lead to too much reduction in the opposite instance. Hence the performance of a policy based on quantitative objectives is suboptimal in the presence of uncertainty. Use of a price mechanism (implicitly associated with a uniform tax on carbon emissions) can also lead to suboptimal performance, although in a different way: fixed pricing of emissions introduces a discrepancy between the actual and the optimal effort that increases with the slope of the marginal benefit curve. The comparison between the two policies, price and quantity (which are generally both suboptimal), is usually made along the lines of the argument just briefly suggested and initially made by Weitzman (1974).

The transposition of the argument to the greenhouse problem requires assumptions to be made about the shape of the marginal benefit curve: the curve is generally assumed to be flat, since the amount of emissions reduction under consideration has to be small compared to the stock of carbon dioxide in the atmosphere. By this assumption, a price policy is generally viewed as superior to a quantity policy. The argument of superiority of a "price policy" over a "quantity policy" in a climate context has already been revisited in a dynamic framework and confirmed with operationally plausible calibrations (see the work of Newell and Pizer 2000; Pizer 2001). Nevertheless, a number of well-reasoned objections to the price policy argument have been made (see Hoel and Karp 2001, 2002; Karp and Zhang 2005; Victor 2001). These objections will not be discussed here, but later in this chapter I will raise a different set of objections.

Here I mean only to remark that a price policy approach should not be limited to partial tax harmonization (by which the additional carbon tax is small and equal for all countries). A larger tax harmonization scheme would wipe out the present huge differences in total carbon taxation among countries. It remains to be seen whether it is possible to set a more ambitious goal, such as stabilizing the total carbon price (which includes not only the carbon tax but also the price of the carbon implicit to the price of the fossil fuel where it comes from), a point I will return to later. In fact, by the standard price–quantity argument, a GPGP performs better than a K scheme. In the price–quantity space (as detailed in appendix C) the suboptimal outcome of a GPGP scheme obtains at the intersection of a hyperbola corresponding to the fixed expenditure (the price multiplied by emission quantity) and the marginal cost curve. The quantity abatement in that case fluctuates less than is optimal, but fluctuates more than under a pure quantity policy with comparable expected costs.

3.3.2 Price versus Quantity: Should We Follow Conventional Wisdom?

I come now to my criticism of the standard assessment associated with the conventional wisdom. My critique may be tentative on several points, but my aim is to put the comparison of quantity policies (Kyoto), price policy (tax and price harmonization), and mixed price–quantity policies (GPGP) into a better perspective.

First, there is a technical objection to the standard argument that does not get much attention in the literature I quoted earlier. This relates to time aggregation and separability of costs. The sequential model under scrutiny is one that uses a marginal cost curve that in each period is increasing. The rising marginal cost implies that the first action at time $t+1$ is (much) less costly than the last action at time t. This modeling artifact does bias the argument against the quantity policy.

Second, an analysis of a pure taxation policy, absent any quantity constraints, should take into account the effect of taxation on fossil fuel prices. This is not an easy task mainly because, on the one hand, the short-run and the long-run incidence effects have to be disentangled and, on the other, the oligopolistic dimensions of the problem make the actual pricing of fossil fuels much more difficult to assess than in the simple reference competitive model.

An insight from the static competitive model is that in the exploitation of an exhaustible resource, a tax has no effect on the producer price; it transfers rent from the producers to the taxing authority. This key insight from the static model has a dynamic counterpart. In a dynamic setting a tax increase induces offsetting movements of producer prices, even if only in rare circumstances do tax changes leave final prices and quantities unchanged. Taxes would likewise transfer rent from owners of an exhaustible resource to consuming countries. Note also that an appropriate modulation of taxes should, over time, delay the exhaustion of that resource (see appendix H).

Although there has been in the past a significant amount of work on the price of fossil fuels (see the survey of Karp and Newbery 1993), the climate policy dimension of the problem does not seem to have attracted enough interest. The exception are Chakrovorty, Magné, and Moreaux (2003) and Magné and Moreaux (2002); these are papers that take into account neither uncertainty nor oligopolistic pricing, but they convey messages that are relevant for a discussion of policy design.

3.4 Tax Harmonization versus Kyoto-Compatible Schemes

Harmonized taxation has been often presented as a better arrangement than any of the Kyoto-like schemes. I attempt in this section a reassessment of the issue in terms of both the GPGP and OFK schemes.

The formal definition of a harmonized tax equilibrium (within the simple framework of appendix B) ensures that the price of carbon is controlled,[15] and it is generally associated with full tax harmonization, although I will challenge this interpretation later.

DEFINITION 5 A harmonized tax equilibrium consists of abatements (q_i^{**}), $i \in I$, a positive "carbon tax" or "carbon price" t^{**}, such that:

1. q_i^{**} is a solution of $\min\{t^{**} q_i - C_i(q_i)\}$, $i \in I$;
2. $Q^{**} = \sum_i q_i^{**} = \sum_i R_i(t^{**})$, where $R_i(t)$ is the solution of $\max\{tq_i - C_i(q_i)\}$;
3. the welfare of country i is $\{U_i(Q^{**}) - C_i(R_i(t^{**}))\}$.

This definition calls for several comments. Tax harmonization does not allow for any freedom in the distribution of burdens (as in Bradford 2002, app.). Its outcome can be mimicked, again in the simple world (of appendix B) considered here, by a Kyoto-like arrangement where quotas are such that there is no trade on the world permit market. Placing severe limits on any income transfers may be viewed as an advantage by those who are suspicious of income transfers from developed countries to developing countries.[16] Because there is no strict connection between present and future targets, there is an added advantage of the harmonized taxation scheme being, in principle, less vulnerable to the ratchet effect described earlier. However, the absence of income transfers may be problematic, since such transfers allow for flexibility in the search for a stable arrangement. In particular, absent flexibility, voluntary participation is likely to be triggered only at low levels of effort.

The complex relationship between fossil fuels prices and taxes suggests that full tax harmonization, strictly speaking, would create uncertainty about the final price of fossil fuels. Hence the model policy might rather be described as a carbon price control policy and not harmonized taxation.[17] In fact the quantity performance associated with the harmonized taxation policy may be grossly suboptimal compared with the carbon price control policy embedded in the definition above. One explanation can be found in the model of a simple world with a single exhaustible resource and competitive owners who have perfect foresight. As was argued above, the price responses would then partly offset the effect of taxes. Another explanation is ordinary skepticism about the long-run outcome of tax harmonization schemes (and pure price policies) that cannot be fully committed. For this reason the time profile of taxes should be announced in a credible way at the outset: if not, the final price reaction is most difficult to assess. Last, the quantity reaction of providers of the exhaustible resources in response to taxes may be plagued by "eductive" instability.[18]

Appendix F shows graphically why the ranking of a pure Kyoto quantity policy, of a GPGP arrangement, or of a tax harmonization with uncertain effects is ambiguous, even when the marginal benefit curve is flat. On the whole, the reassessment of the price versus quantity debate, only attempted here, gives more credence to the pure quantity policies of the Kyoto-like schemes. It introduces another and yet unexplored dimension of comparison between GPGP and OFK, a dimension that may weaken the case for GPGP. More thoughts on the effects of carbon tax schemes on fossil fuels pricing would be most welcomed.

3.5 Conclusion

I have compared GPGP and Kyoto-like schemes in this chapter and have argued that the power of the GPGP scheme in eliciting universal participation can be duplicated by a more open and flexible variant scheme of the Kyoto Protocol, which I call the OFK. My comparison of the OFK scheme with that of the GPGP is somewhat inconclusive. The GPGP scheme has three important strengths. It may be (slightly) superior in prompting world participation. It can be modified to include a form of emissions taxation that can attenuate the ratchet effect. It has more flexible targets than the Kyoto Protocol and hence is less vulnerable to the standard criticism against quantity policy. However, as I argued last, conventional wisdom on the relative virtue of price versus quantity has to be carefully analyzed in terms of the response of fossil fuel pricing to carbon taxes. When this dimension is accounted for, it may be that the merit of harmonized taxation will have to be reassessed downward. It follows that the relative evaluation of GPGP and OFK in terms of flexibility may be significantly affected in a way that reopens the debate on their the relative merits.

Appendix A: Effort Level of Annex-B Countries and Global Reductions

Let us assume that the effort level in annex-B countries is measured by a normalized price of carbon, denoted t, so that the price of the carbon-intensive good is $p + t$. Consequently the level of reductions per unit of the carbon-intensive good is $(\partial C)_{-1}(t)$, and the level of reductions is $D(.)(\partial C)_{-1}(t)$, where $D(.)$ is the demand for the good. In non–annex-B countries the price is p, but no reduction is implemented, and there is at the outset an excess emission of u per unit of production.

Calling respectively $D(B,.,.)$ and $D(NB,.,.)$ the demand functions for the goods produced in the two areas, the total level of reductions, as a function of t, is $\Delta(t) = I[D(B, p, p) - D(B, p + t, p)] + (\partial C)_{-1}(t) D(B, p + t, p) - (I + u - \epsilon)[D(NB, p + t, p)) - D(NB, p, p)]$, where I is the initial carbon content of annex-B goods and where ϵ may (here exogenously) take into account the fact that the new demand in annex-B countries is met both by firms using the old non–annex-B technology and by migrating firms using the initial annex-B technology. So we have

$$\left(\frac{d\Delta}{dt}\right) = -\partial_1 D(B,.,.)\{I - (\partial C)_{-1}(t)\} + ((\partial C)'_{-1}(t))D(B,.,.)$$
$$- (I + u - \epsilon)[\partial_1 D(NB,.,.)].$$

The shape of $\Delta(t)$ depends on the form of the demand function. For example, with CES demand functions, when the elasticity of substitution between the de-carbonated good and the carbonated good is greater than one, the effect of an increase in the carbon tax in the annex-B countries on total emissions becomes ambiguous: the two first (positive) terms may become dominated by the last term (which is negative). Note, however, that the required elasticity of substitution is more plausible at the industry level (e.g., steel) than at the aggregate level.

Appendix B: Comparison of GPGP and OFK

Countries are indexed by $i \in I$. The cost of emission reductions of amount q_i, measured from a business as usual (BAU) basis, are associated with a convex cost function C_i, defined for every q_i. The utility generated in country i, from total reductions Q, is $U_i(Q)$. That is to say, $U_i(Q_{-i} + q_i)$, with $Q_{-i} = \sum_{j \neq i} q_j$. The function U is concave.

There are two groups of countries, let us say annex-B countries, $(i \in B)$, where $dU_i/dQ > 0$, in the whole range under consideration, and non–annex-B countries $(i \in NB)$ that are associated with $dU_i/dQ = 0$, $\forall Q$, namely $U_i = C_{ste}(= 0)$, and are not affected[19] by the policy.

We can define this relationship formally as follows:

DEFINITION 6 A noncooperative Nash equilibrium is a set of abatement levels $\bar{q}_i$ subject to $\bar{q}_i$, which is a solution of the problem $\max\{U_i(\sum_{j \neq i} \bar{q}_j + q_i) - C_i(q_i)\}$.

In a Nash equilibrium $\bar{q}_i > 0$, $i \in B$, $\bar{q}_i = 0$, $\forall i \in NB$. Note that existence of a Nash equilibrium follows from very weak assumptions.[20] The uniqueness problem is, more complex. However, as it is formally analogous to the problem of existence in a Cournot oligopoly of firms having a concave maximand we can import the classical results from oligopoly theory:

PROPOSITION 7 When all annex-B countries are similar (a) there exists an equilibrium, and (b) if the utility function is concave, there exists a unique equilibrium.

Proof (Sketch) For part a, it is enough to note the formal similarity of this problem to the Cournot problem. Check that as in a Cournot problem, the discontinuities of the best reply are upward, and transpose the proof of Bamon and Fraysse (1985) or Novshek (1985).

For part b, it is enough to check the continuity of the best-reply functions and to show that the Selten trick applies. As the Selten trick will be explained later, the reader is advised to check my suggestion after doing some further reading.

Chapter 2 Design of Post-Kyoto Climate Schemes

Now we can assume, with the slight loss of generality suggested above, that the equilibrium is unique and, without loss of generality, that $\bar{\bar{q}}_i = 0$, $\forall i$ (however, we may need to redefine $C_i(q_i)$, $i \in B$). Let us consider a GPGP equilibrium associated with fixed contributions $\bar{T}_i$, $i \in B$ (i.e., we do assume that annex-B countries contribute the, now exogenously, fixed amount $\bar{T}_i$).

DEFINITION 8 A GPGP equilibrium associated with fixed contributions $\bar{T}_i$, $i \in B$, consists of abatements $(q_i^* \geq \bar{\bar{q}}_i, i \in B, q_i^* > 0, i \in NB)$, and a positive "carbon price" t^*, such that:

1. q_i^* is a solution of $\max\{U_i(Q_{-i}^* + q_i) - t^*(q_i - \bar{\bar{q}}_i) - C_i(q_i)\}$, $q_i \geq \bar{\bar{q}}_i$, $i \in B$, with $Q_{-i}^* = \sum_{j \neq i} q_j^*$;
2. q_i^* is a solution of $\max\{t^* q_i - C_i(q_i)\}$, $i \in NB$;
3. $\sum_{i \in B} t^*(q_i^* - \bar{\bar{q}}_i) + \sum_{i \in NB} t^* q_i^* = \sum_{i \in B} \bar{T}_i$.

The GPGP equilibrium is said to be individually rational iif:

4. $\{U_i(Q^*) + t^*(q_i^* - \bar{\bar{q}}_i) - C_i(q_i^*) - \bar{T}_i\} > U_i(\sum_i \bar{\bar{q}}_i) - C_i(\bar{\bar{q}}_i)$, $i \in B$, with $Q^* = \sum_i q_i^*$.

The existence problem is again not straigthforward. The following proposition must be proved:

PROPOSITION 9 If the utility function is concave, there exists for a fixed set of T_i a unique GPGP equilibrium.

Proof The proof has several steps.

1. We start by defining a pseudo t-GPGP equilibrium, the "fixed" carbon price t being positive, as consisting of abatements $(q_i^0 > 0, i \in B \cup NB)$, such that:
a. q_i^0 is a solution of $\max\{U_i(Q_{-i}^0 + q_i) + tc_i - C_i(q_i)\}$, $i \in B$, with $Q_{-i}^0 = \sum_{j=i} q_j^0$;
b. q_i^0 is a solution of $\max\{tq_i - C_i(q_i)\}$, $i \in NB$.
We call $r_i(Q_{-i}, t)$ the best-reply function in part a (clearly continuous in Q_{-i} and t) and write (using Selten's trick) $f_i(Q,t) = r_i(Q - f_i(Q,t), t)$: $f_i(Q,t))$ is the pseudo–best reply of i to the total level of public good Q, for fixed t. For $i \in NB$, we have the best reply $R_i(t)$, independant of Q_{-i}.

2. We next prove the existence and uniqueness of a pseudo-t GPGP equilibrium. Such an equilibrium is now associated with $Q^0(t)$ subject to

$$Q^0(t) = \sum_{i \in B} f_i(Q^0(t), t) + \sum_{i \in NB} R_i(t).$$

We then show that $df_i/dQ = (dr_i/dQ_{-i})(1 + dr_i/dQ_{-i})$, and from the inspection and differentiation of the first-order conditions of part a (left to the reader) that the denominator is positive and the numerator negative. The left-hand side of the equilibrium equation is at fixed t, decreasing in Q. Equilibrium and uniqueness follows.

3. We now prove that the left-hand side $Q^0(t)$ is increasing in t (and equal to zero—Nash—when $t = 0$). It follows that $tQ^0(t)$ is increasing in t, starting from zero and going to infinity. It necessarily reaches $\sum \bar{T}_i$, at some and unique t. □

A corollary obtains:

COROLLARY 10 The equilibrium carbon price of a GPGP equilibrium is increasing with $\sum \bar{T}_i$.

Now

DEFINITION 11 A flexible Kyoto equilibrium consists of (exogenously given) quotas $\hat{q}_i$, $i \in B$, abatements $(q_i^{*\prime})$, $i \in B$, a positive "carbon price" t^*, such that:

1. $q_i^{*\prime}$ is a solution of $\max\{t^*(q_i - \hat{q}_i) - C_i(q_i)\}$, $i \in B$, and
2. $\sum_{i \in B} q_i^{*\prime} = \sum_{i \in B} \hat{q}_i = Q^{*\prime}$.

A flexible Kyoto equilibrium is individually viable if

3. $\{U_i(Q^{*\prime})) + t^*(q_i^{*\prime} - \hat{q}_i) - C_i(q_i^{*\prime})\} > U_i(\sum_i \bar{q}_i) - C_i(\bar{q}_i)$.

In a flexible Kyoto equilibrium the sum of abatements equals the sum of initial quotas (non–annex-B countries are not concerned), and this is taken as given in individual optimizations that take account of the cost and the market price of abatements (condition 1). The trade of quotas is determined from such optimizations.

The flexible equilibrium is unique since the demand for abatement is an increasing function of t, and it is Pareto superior[21] to a "rigid" Kyoto equilibrium, which disallows the trade of quotas. Hence the flexible equilibrium is viable for individual countries whenever the inflexible equilibrium is. Note that in our formalism, the Kyoto Protocol scheme may be viewed as a "flexible" arrangement in which the quotas are fixed by the consideration of a basis year. Now we come to the open, flexible scheme proposed in the main text:

Chapter 3 Design of Post-Kyoto Climate Schemes

DEFINITION 12 An open, flexible Kyoto equilibrium consists of (exogenous) quotas $\hat{q}_i$, $i \in B$, abatements (q_i^{**}), $i \in I$, a positive "carbon price" t^{**}, such that:

1. q_i^{**} is a solution of $\max\{t^{**}(q_i - \hat{q}_i) - C_i(q_i)\}$, $i \in B$;
2. q_i^{**} is a solution of $\max\{t^{**} q_i - C_i(q_i)\}$, $i \in NB$;
3. $Q^{**} = \sum_i q_i^{**} = \sum_{i \in B} q_i^{**} + \sum_{i \in NB} q_i^{**} = \sum_{i \in B} \hat{q}_i$;
4. It is individually viable if $\forall i \in B$:

$$\{U_i(Q^{**}) + t^{**}(q_i^{**} - \hat{q}_i) - C_i(q_i^{**})\} > U_i\left(\sum_i \bar{q}_i\right) - C_i(\bar{q}_i),$$

with $Q^{**} = \sum_i q_i^{**}$.

In an open, flexible equilibrium the environnemental performance is still determined from the quotas of the annex-B countries, but the non–annex-B countries, which are given nonbinding commitments, participate and benefit from participation, while decreasing the cost of annex-B countries. It is left to the reader to show that a unique OFK equilibrium exists and involves a Pareto-improvement as compared to the FK equilibrium with the same $\hat{q}_i$.

The next proposition stresses the less obvious comparative features:

PROPOSITION 13

a. Given an open and flexible Kyoto equilibrium, there exists a family of GPGP schemes, all with the same total contributions $\sum F_i$, of annex-B countries and all inducing the same total abatements as the open flexible Kyoto equilibrium. In the change the welfare level of non–annex-B countries decreases, but the change of welfare of annex-B countries is ambiguous. It is nevertheless positive by the conditions set out in corollary 7: in this case, the participation constraints are met in the GPGP scheme if they are in the "generating" OFK scheme.

b. Given a GPGP equilibrium, there exists a family of OFK equilibria, each leading to the same level of total abatements as the GPGP equilibrium. In the change the welfare level of non–annex-B countries increases.

Proof For part a, take an OFK equilibrium, t^{**}, q_i^{**}. Using the previous notation, write

$$\sum_{i \in I} R_i(t^{**}) = \sum_{i \in I} q_i^{**} = \sum_{i \in B} \hat{q}_i = \hat{Q}.$$

Consider the (unique) t^{**}-GPGP equilibrium, and write its global abatement as

$$Q^* = \sum_{i \in B} f_i(Q^*, t^{**}) + \sum_{i \in NB} R_i(t^{**}) = \sum_{i \in B} R_i(t^{**} + a_i^*) + \sum_{i \in NB} R_i(t^{**}),$$

with $a_i^* = (\partial U_i/\partial Q)(Q^*)$. It is straightforward that $Q^* > \hat{Q}$. Hence, by the previous corollary, there exists a t' GPGP equilibrium, $t' < t^{**}$, with global abatment $\hat{Q}$, so that the welfare of non–annex-B countries is lowered. Then we have $\sum_{i \in B} R_i(t' + a_i) + \sum_{i \in NB} R_i(t') = \hat{Q}$, with $a_i = (\partial U_i/\partial Q)_{\hat{Q}}$.

The difference of welfare of one annex-B country, measured in "money," is

$$t'R_i(t' + a_i) - t^{**}(R_i(t^{**}) - \hat{q}_i) - C_i(R_i(t' + a_i)) + C_i(R_i(t^{**})).$$

In words, this country would be equally well off if it were subject to a positive T_i transfer such that

$$t'R_i(t' + a_i) - t^{**}(R_i(t^{**}) - \hat{q}_i) - C_i(R_i(t' + a_i)) + C_i(R_i(t^{**})) = T_i.$$

The sum of the left-hand side over $i \in B$ is, by means of elementary computation,

$$t' \sum_{i \in B} R_i(t' + a_i) + t^{**} \sum_{i \in NB} R_i(t^{**}) + \sum_{i \in B} C_i(R_i(t^{**})) - \sum_{i \in B} C_i(R_i(t' + a_i)).$$

The question is whether the transfers can finance the abatement payment, which takes the form of

$$\sum_{i \in B} t'R_i(t' + a_i) + \sum_{i \in NB} t'R_i(t').$$

Hence this expression must be compared with

$$\sum_{i \in NB} t'R_i(t') \quad \text{and} \quad t^{**} \sum_{i \in NB} R_i(t^{**}) + \sum_{i \in B} (C_i(R_i(t^{**})) - C_i(R_i(t' + a_i))).$$

In the case where

$$\sum_{i \in NB} (t^{**}R_i(t^{**}) - (t'R_i(t'))) > \sum_{i \in B} (C_i(R_i(t' + a_i))) - (C_i(R_i(t^{**}))),$$

there exists a GPGP equilibrium where annex-B countries are better off. The right-hand side equals

Chapter 3 Design of Post-Kyoto Climate Schemes

$$\sum_{i \in NB} (t^{**} - t') R_i(t^{**}) + t'(R_i(t^{**}) - R_i(t')).$$

The left-hand side is smaller than

$$\sum_{i \in B} (t' + c_i)[R_i(t' - a_i)) - R_i(t^{**})].$$

Hence a sufficient condition for the property is having

$$\sum_{i \in NB} (t^{**} - t') R_i(t^{**}) + t'(R_i(t^{**}) - R_i(t'))$$

$$> \sum_{i \in B} (t' + a_i)[R_i(t' + a_i)) - R_i(t^{**})].$$

Put differently, because the total abatement is the same in both situations,

$$\sum_{i \in NB} (t^{**} - t') R_i(t^{**}) > \sum_{i \in B} a_i [R_i(t' + a_i)) - R_i(t^{**})].$$

Although, in general, this inequality may not be satisfied, it is still plausible.

For part b, let a GPGP equilibrium be associated with t^*. For any OFK associated with t', with straightforward notation,

$$\sum_{i \in B \cup NB} R_i(t^*) < \sum_{i \in B} R_i(t^* + a_i^*) + \sum_{i \in NB} R_i(t^*),$$

meaning the global abatement is smaller. However, $t^{**} > t^*$ can be chosen such that

$$\sum_{i \in B \cup NB} R_i(t^{**}) = \sum_{i \in B} R_i(t^* + a_i^*) + \sum_{i \in NB} R_i(t^*).$$

The conclusion readily follows that the welfare level of non–annex-B countries is higher. □

The question of the difference of welfare for annex-B countries for the GPGP and OFK equilibria with the same level of abatement can easily be settled in some special cases. The next statement is a corollary that follows easily from the inequality derived in the preceding proof.

COROLLARY 15 If all $m + n$ countries have a similar quadratic cost function $C_i(q) = q^2$, and if all m annex-B countries have similar preferences, then in part a of proposition 13, all annex-B countries are better off provided that the carbon price in the initial OFK equilibrium is not smaller than the marginal willingness to pay by a single annex-B country.

Proof In this case, the OFK and GPGP equilibria are such that

$$(m+n)(t^{**}) = m[t' + a] + n[t'].$$

It follows that $(t^{**} - t') = [m/(m+n)]a$. The inequality

$$\sum_{i \in NB}(t^{**} - t')R_i(t^{**}) > \sum_{i \in B} a_i[R_i(t' + a_i)) - R_i(t^{**})]$$

becomes

$$\sum_{i \in NB}(t^{**} - t')t^{**} + \sum_{i \in B}(t^{**} - t')a > ma^2,$$

$$\left[\frac{mn}{(m+n)}\right]at^{**} + \left[\frac{m^2}{(m+n)}\right]a^2 > ma^2.$$

This inequality remains as long as $t^{**} > a$. □

Appendix C: GPGP, and the Price versus Quantity Policy Debate

Where does the GPGP scheme stand in the standard price versus quantity debate? Figure 3.2 provides a diagrammatic sketch of the formal argument in terms of the standard Weitzman framework. On the hori-

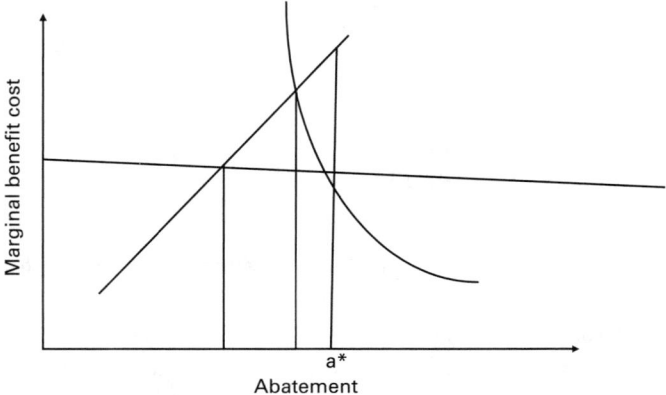

Figure 3.2
Welfare loss (shaded area) due to (optimal) quantity policy

zontal axis is the abatement quantity, and on the horizontal axis is a cost or benefit measure. In conformity with the conventional view of the greenhouse effect, the benefit curve is assumed to be almost flat. Cost is unknown, and there are a priori two different marginal cost curves. The GPGP involves constant spending, and the abatements obtained in the figure are for an (approximately) optimal GPGP. The GPGP scheme shown would then mediate between a price policy and a quantity policy.

Appendix D: GPGPPT and Tax Harmonization

Recalled here is the simple model of appendix B, the definition of a GPGP plus a taxation equilibrium and a definition of the harmonized tax equilibrium.

DEFINITION 16   A GPGPPT equilibrium, associated contributions $\bar{T}_i$, $i \in B$, consists of abatements $(q_i^* \geq \bar{\bar{q}}_i, i \in B, q_i^* > 0, i \in NB)$, a positive "carbon price" t^*, such that:

1. q_i^* is a solution of $\max\{U_i(Q_{-i}^* + q_i) + t^*(q_i - \bar{\bar{q}}_i) - C_i(q_i)\}$, $q_i \geq -\bar{\bar{q}}_i$, $i \in B$, with $Q_{-i}^* = \sum_{j \neq i} q_j^*$;
2. q_i^* is a solution of $\max\{t^*q_i - C_i(q_i)\}$, $i \in NB$;
3. $\bar{T}_i = \bar{t}E_{-1i} + e_i$, where t is a tax rate and E_{-1i} is the emissions of country i at period -1;
4. $\sum_{i \in B} t^*(q_i^* - \bar{\bar{q}}_i) + \sum_{i \in NB} t^*q_i^* = \sum_{i \in B} \bar{T}_i$.

It is easy to check that choosing $t = \min(\bar{T}_i/E_{-1i})$ allows the GPGP contributions to be adjusted to any a priori profile of contributions.

DEFINITION 17   A harmonized tax equilibrium consists of abatements (q_i^{**}), $i \in I$, a positive "carbon tax" t^{**}, such that:

1. q_i^{**} is a solution of $\min\{t^{**}(q_i) - C_i(q_i)\}$, $i \in I$;
2. $Q^{**} = \sum_i q_i^{**} = \sum_i R_i(t^{**})$;
3. the welfare of country i is $\{U_i(Q^{**}) - C_i(R_i(t^{**}))\}$.

The reader will note that a tax equilibrium is an OFK with $\hat{q}_i = R_i(t^{**})$.

Appendix E: Tax Incidence for an Exhaustible Resource

We refer to the model sketched in the text of pure competition, perfect foresight, a constant discount rate, a time-invariant demand function, and different assumptions on the boundary behavior of prices.

- *Case 1: The resource is exhausted at time T before it gets a substitute.* The producer price is denoted p, the total price is denoted P, and the difference is the tax τ. Necessarily $p(t) = p(0)\exp(rt)$, $P(t) = p(t) + \tau(t)$. A tax increase from $\tau(t)$ to $\tau'(t) = \tau(t) + \epsilon$, $\bar{t} \leq t \leq \bar{t} + \epsilon$ ($\tau(t) = \tau'(t)$ elsewhere), reduces the producer price from $p(0)\exp(rt)$ to $p_0(0)\exp(rt)$ subject to

$$\int_0^T D(p'(0)\exp(rt) + \tau'(t))\,dt = \int_0^T D(p(0)\exp(rt) + \tau(t))\,dt.$$

A temporary increase in the tax depresses the producer price over the entire period and depresses the total price over the entire period starting at the time when the tax is increased.

- *Case 2: There is a (perfect) substitute at time T' for price v.* After time T the final price of the resource is v, and the demand is shared between the resource and its substitute in proportions α and $1 - \alpha$. The resource is exhausted at time T. Necessarily $p(t) = p(0)\exp(rt)$, $P(t) = p(t) + \tau(t)$, $P(t) = v$, $t \geq T'$. A temporary tax increase from $\tau(t)$ to $\tau'(t) = \tau(t) + \epsilon$, $\bar{t} \leq t \leq \bar{t} + \epsilon < T'$, reduces the producer price from $p(0)\exp(rt)$ to $p'(0)\exp(rt)$ and increase the exhaustion time to T subject to

$$\int_0^{T'} D(p'(0)\exp(rt) + \tau'(t))\,dt + \int_{T'}^T \alpha D(v)\,dt$$

$$= \int_0^{T'} D(p(0)\exp(rt) + \tau(t))\,dt + \int_{T'}^T \alpha D(v)\,dt.$$

Appendix F: Uncertainty on Tax Incidence: An Example Where a Quantity Policy Is Better Than a Price Policy

Shown in figure 3.3 are two probable outcomes of a harmonized tax. The carbon price is in fact uncertain in the medium run for the following reasons:

- Uncertainty about the future tax
- Absence of commitment

Chapter 3 Design of Post-Kyoto Climate Schemes

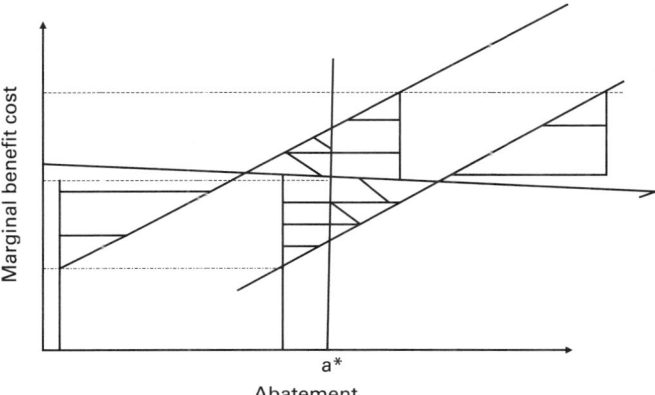

Figure 3.3
Effects of carbon pricing that includes a harmonized tax

- Difficulty in predicting for the medium run the producers' price
- Difficulty in predicting coordination ("eductive" instability)

For a given tax decision there are two possible values for carbon prices: above the marginal benefit or below, as figure 3.3 illustrates. It shows, in the framework of appendix C, a case where the pure quantity (Kyoto-like) policy is superior to the price policy.

Notes

I thank H. Tulkens for his constructive advice, O. Compte for useful criticisms on a previous version of this chapter, and L. Karp and C. Philibert for their detailed comments and suggestions.

1. The proposition was actually circulated both as "no cap but trade" (NBCT) and as "global public good purchase" (GPGP). Here I use the second form.

2. The GPGP scheme is original in the system of control that it advocates, which avoids measurements of emissions. The Kyoto Protocol calls for measurement of emissions of participating countries. The GPGP solution, measuring imports of carbon in fossil fuels, is simpler. Imports may come from "extraction" from the ground within the borders or shipments of the same fossil fuels from another country (exports being treated as negative imports). Naturally some discrepancy may arise between a measured quantity and carbon emissions, but the form of measurement is rather simple. Accurate measurement is not an easy problem; for example, see Hargrave (1998) and McKibbin and Wilcoxen, (2000).

3. The terminology suggests that the GPGP contributing countries would be the annex-B countries of the Kyoto Protocol.

4. This is much too extreme an assumption but a reasonable modeling option in order to separate annex-B and non–annex-B countries.

5. Note that a change in q_i can be matched by a change in T_i that leaves the equilibrium unchanged. Treating the contribution formally as a money contribution plus an abatement contribution allows, without loss of generality, the quota associated with an annex-B country, q_i, to be taken as 0. Naturally the cost function would then have to be redefined accordingly.

6. The BAU levels are assumed to be zero for *NB* countries and nonzero for *B* countries to reflect the fact that in Nash equilibrium (defined in appendix B) annex-B countries would abate whereas non–annex-B countries would not. As in notes 4 and 5, q_i might be taken as equal to zero. However, the interpretation of C_i, and in particular of $C_i(0)$, is affected.

7. Note that we have here some force for cohesion in the sense that free-riding, in the most formal definition of the word, is not likely to be a "dominant strategy" for any country, but cohesion effects are limited in the sense that defecting from the arrangement, at least if others can be assumed to stay, is often an optimal strategy.

8. Contrary to the Kyoto scheme, where the quotas are fixed in reference to the countries' levels of emissions at a given year (1990), with little possibility of bargaining (but on the reference year itself) around this reference arrangement. It seems, however, that increasing the degrees of freedom in the arrangement by playing in a more flexible way on the repartition of national quotas might favor stability even though the negotiations might be trickier. As an illustration of this idea, the ex post favorable treatment of Russia turned out to be conducive to the (incomplete) stability of the Kyoto arrangements even though this was due more to "luck" than to the design of the negotiations.

9. In the real world, *NB* countries are induced to participate via the CDM mechanisms, which seems to provide only limited incentives.

10. This may justify viewing GPGP as a "Kyoto-compatible" scheme as in Guesnerie (2003).

11. Particularly, in view of the fact that the former might induce Pareto optimality whereas the latter cannot.

12. For OFK, without uncertainty, the outcome is, in principle, "transparent."

13. Also, in the present Kyoto Protocol, the ratchet effect is likely to appear at the renegotiation stage, despite some existing dispositions that penalize nonattainment of the objectives. Countries that have performed poorly are likely to find it a reason for lowering their future objectives.

14. Originally this scheme suggested by the Europeans was rejected by the United States but the positions have now reversed to some extent.

15. Again, it is the sum of the carbon tax plus the price of carbon implicit in the price of the fossil fuel from which it comes.

16. Suspicion about income transfers is implicit in the US position in the negotiations over the Kyoto Protocol.

17. A similar objection could be made to the GPGP model but less to the OFK model.

18. For a view on expectational coordination, see Guesnerie (2005). The fact that price coordination is, in this context, not stable is straightforward in the simple models discussed earlier.

19. This assumption is extreme but a reasonable modeling option for separating annex-B and non–annex-B countries.

20. For example, Nash follows from the strict concavity of cost or the boundedness of the utility function.

21. This is straightforward.

References

Aldy, J. E., P. R. Orszag, and J. E. Stiglitz. 2001. Climate change: An agenda for global collective action. Prepared for the conference on Timing of Climate Change Policies. Pew Center on Global Climate Change, October.

Barrett, S. 2001. Towards a better climate treaty. Policy matters. Working paper 01-29. Brookings Institution, Washington, DC.

Bamon, R., and J. Fraysse. 1985. Existence of Cournot equilibrium in large markets. *Econometrica* 53 (3): 587–97.

Chakrovorty, U., B. Magné, and M. Moreaux. 2003. Energy resource substitution and carbon concentration targets with nonstationary needs. Leerna 31. University of Toulouse.

Cooper, R. 1998. Toward a real global warming treaty. *Foreign Affairs* 77 (2).

Carraro, C. 1999. The structure of international agreements on climate change. In Carraro, C., ed. *International Environmental Agreements on Climate Change*. Dordrecht: Kluwer Academic.

Chander, N. L., and H. Tulkens. 1998. The core of an economy with multilateral environmental externalities. *International Journal of Game Theory* 26: 379–401.

Freixas, X., R. Guesnerie, and J. Tirole. 1985. Planning under incomplete information and the ratchet effect. *Review of Economic Studies* 52: 173–91.

Guesnerie, R. 2003. Les enjeux économiques de l'effet de serre. In R. Guesnerie, ed., *Kyoto et l'économie de l'effet de serre*. Paris: La Documentation Française.

Guesnerie, R. 2004. Calcul économique et développement durable. *Revue Economique*: 363–82.

Guesnerie, R. 2005. *Assessing Rational Expectations. 2: "Eductive" Stability in Economics*. Cambridge: MIT Press.

Ha-Duong, M., M. Grubb, and J. C. Hourcade. 1997. Influence of socio-economic inertia and uncertainty on optimal CO_2-emissions abatement *Nature* 390.

Hargrave, T. 1998. US carbon emissions trading: Description of an upstream approach. Mimeo. Center for Clean Air Policy, Washington, DC.

Hoel, M., and L. S. Karp. 2001. Taxes and quotas for a stock pollutant with multiplicative uncertainty. *Journal of Public Economics* 82: 91–114.

Hoel, M., and L. S. Karp. 2002. Taxes versus quotas for a stock pollutant. *Resource and Energy Economics* 24: 367–84.

Karp, L. S., and J. Zhang. 2005. Regulation of stock externalities with correlated abatement costs. *Environmental and Resource Economics* 32: 273–300.

Karp, L. S., and D. M. Newbery. 1993. Intertemporal consistency issues in depletable resources. In A. Kneese and J. Sweeny, eds., *Handbook of Natural Resources*, vol. 3. Amsterdam: North Holland, pp. 881–930.

McKibbin, W. J., and P. J. Wilcoxen. 2000. Designing a realistic climate change policy that includes developing countries. Paper for the 2000 Conference of Economists, the Gold Coast, July 3–5.

Magné, B., and M. Moreaux. 2002. Long run energy trajectories: Assessing the nuclear option in response to global warming. Leerna DP 0226101. University of Toulouse 1.

Neary, P. 1999. International trade and the environment: Theoretical and policy linkages. Mimeo.

Newell, R. G., and W. A. Pizer. 2000. Regulating stock externalities under uncertainty. Discussion paper 99-10. Resources for the Future, Washington, DC.

Nordhaus, W. D. 2002. After Kyoto: Alternative mechanisms to control global warming. Paper prepared for the meetings of the American Economic Association and the Association of IEA/SLT(2002)28.

Novshek, W. 1985. On the existence of Cournot equilibrium. *Review of Economic Studies* 52: 85–88.

Philibert, C. 2000. How could emissions trading benefit developing countries. *Energy Policy* 28 (13).

Philibert, C., and J. Pershing. 2001. Des objectifs climatiques pour tous les pays: Les options. *Revue de l'énergie* 524.

Pizer, W. A. 2001. Combining price and quantity control to mitigate global climate change. *Journal of Public Economics* 85(3): 409–34.

Victor, D. 2001. *The Collapse of the Kyoto Protocol and the Struggle to Slow Global Warming*. Princeton: Princeton University Press.

Rieu, J. 2002. Politiques nationales de lutte contre le changement climatique et réglementation de la concurrence: Le cas de la fiscalité. Mimeo.

Weitzman, M. L. 1974. Prices vs. quantities. *Review of Economic Studies* 41.

4 Design of Climate Change Policies: A Discussion of the GPGP Approach of Bradford and Guesnerie

Sushama Murty

In this discussion of chapters 2 and 3 by Bradford and Guesnerie, I focus on the modeling issues involved in designing the institutional structure underlying the global public good purchase (GPGP) scheme. I find that the GPGP approach, as presented by Bradford and Guesnerie, is reminiscent of the notion of a tax/public competitive equilibrium defined by Foley (1967, 1970) for economies with public goods and that elements of the Foley model can be used to ensure effective participation of nations in the GPGP scheme.

I assume throughout that there are I countries indexed by i and K private goods indexed by k. Contributions of country i, toward the purchase of a level z of the global public good, is denoted by R_i. The market price of the global public good is t, and the price vector for the private goods is denoted by p (I assume free trade). The production and sale of abatement by country i in the world market is denoted by a_i. There is one aggregate consumer per country, and the level of the consumption of private good in i is denoted by x_i. The output vector of the aggregate producer in i is denoted by y_i and the aggregate production function is $f_i(y_i, a_i)$. For all $i = 1, \ldots, I$, the utility level made possible at an initial noncooperative Nash equilibrium is denoted by $\overset{*}{u}_i$.

By this framework, it is possible to define Guesnerie's GPGP equilibrium associated with fixed contributions as follows:

DEFINITION A noncooperative GPGP equilibrium associated with fixed contributions $(\bar{R}_1, \ldots, \bar{R}_I)$ is a set of levels of supply of abatement, private consumption, and production of individual countries $\bar{a}_i, \bar{x}_i, \bar{y}_i$ for all i, a market price of abatement $\bar{t}$, and a price vector for private goods $\bar{p}$, such that for all i, $\bar{a}_i, \bar{x}_i, \bar{y}_i$ solve

$$\max_{x_i, y_i, a_i} U_i\left(x_i, \sum_{j \neq i} \bar{a}_j + a_i\right)$$

subject to

$$\bar{p} \cdot x_i \leq \bar{p} \cdot y_i + \bar{t} a_i - \bar{R}_i, \qquad f_i(y_i, a_i) \leq 0, \tag{4.1}$$

and

$$\sum_i \bar{x}_i = \sum_i \bar{y}_i,$$

$$\sum_i \bar{R}_i = \bar{t} \sum_i \bar{a}_i,$$

$$U_i\left(\bar{x}_i, \sum_j \bar{a}_j\right) > \overset{*}{u}_i, \qquad \forall i = 1, \ldots, I. \tag{4.2}$$

4.1 Foley's Model of Collective Action Applied to Climate Change

I now present an adaptation of Foley's model to the climate change problem. In the next section I will show how it relates to the notion of collective action and a GPGP equilibrium discussed by Bradford and Guesnerie. In Foley's model the consumption sectors across countries form a collective action group that transacts with the producers of the public good of abatement. The collective consumption demand for abatement is obtained at every price system by a process of aggregation of preferences of all consumers of climate change through the adoption of some political process by the collective action group. The production sectors across all countries are fully decentralized.

4.1.1 Demand for Private and Collective Consumption

For the aggregate consumer in each country i who takes the level of the public good z, the price system (p, t), the profit income from the production sector $\Pi_i(p, t)$, and his contribution R_i as given, define the indirect utility function as

$$V_i(p, t, z, R_i) := \max_{x_i} U_i(x_i, z)$$

subject to

$$p \cdot x_i \leq \Pi_i(p, t) - R_i. \tag{4.3}$$

Suppose that the solution to this problem is, for all i,

$$x_i = \overset{*}{x}_i(p, t, z, R_i). \tag{4.4}$$

The demand for collective consumption of the public good by the collective action group as well as the optimal contributions corresponding to a given price system (p, t), are obtained as the solution of the following problem performed by the new organization. The problem defines a utility possibility frontier corresponding to the given price system (p, t).

$$\hat{U}_1(u_2, \ldots, u_I, p, t) := \max_{z, R_1, \ldots, R_I} V_1(p, t, z, R_1)$$

subject to

$$V_i(p, t, z, R_i) \geq u_i \quad \forall i \neq 1,$$

$$tz = \sum_i R_i. \tag{4.5}$$

Suppose that the solution to this problem is given by the following collective demand and contribution functions:

$$z = \overset{*}{z}(u_2, \ldots, u_I, p, t),$$

$$R_i = \overset{*}{R}_i(u_2, \ldots, u_I, p, t) \quad \forall i. \tag{4.6}$$

DEFINITION A *budget proposal* at a price system (p, t) is a combination of levels of collective consumption of public good z and contributions $(R_1, \ldots, R_I)$, such that $tz = \sum_i R_i$.

DEFINITION A budget proposal $(z, R_1, \ldots, R_I)$ is *unanimously rejected* at price system (p, t) if there exists another budget proposal $(\bar{z}, \bar{R}_1, \ldots, \bar{R}_I)$ such that for all i, we have $V_i(p, t, \bar{z}, \bar{R}_i) > V_i(p, t, z, R_i)$.

For a given p and t, the set of budget proposals that will not be unanimously rejected is defined by

$$G(p, t) := \{(z, R_1, \ldots, R_I) \mid z = \overset{*}{z}(u_2, \ldots, u_I, t, p)$$

and

$$R_i = \overset{*}{R}_i(u_2, \ldots, u_I, t, p) \; \forall i \text{ for some } u_2, \ldots, u_I\}. \tag{4.7}$$

All these proposals will result in points chosen on the frontier $u_1 = \widehat{U}_1(u_2, \ldots, u_I, p, t)$. Then, depending on the (Pareto criteria incorporating) political mechanism adopted by the collective action group, the choice of collective consumption and the distribution of contributions gets further restricted to those corresponding to particular subsets of the frontier $u_1 = \widehat{U}_1(u_2, \ldots, u_I, p, t)$.

4.1.2 Decentralized Production

Taking the prices (t, p) as given, the aggregate supply by profit-maximizing producers in country i is obtained as the solution to the following problem, which defines the aggregate profit function of country i:

$$\Pi_i(p, t) := \max_{y_i, a_i} \ p \cdot y_i + t a_i$$

subject to

$$f_i(y_i, a_i) \leq 0. \tag{4.8}$$

Suppose that the solution to this problem is, for all i,

$$y_i = \overset{*}{y}_i(p, t). \tag{4.9}$$

4.1.3 Equilibrium

DEFINITION A *competitive collective consumption with contributions (CCCC) equilibrium*[1] is defined by levels of supply of abatement, contributions, private consumption and production of individual countries $\bar{a}_i$, $\bar{R}_i$, $\bar{x}_i$, $\bar{y}_i$ for all i, a level of collective demand for abatement $\bar{z}$, a market price of abatement $\bar{t}$, and a price vector for private goods $\bar{p}$, such that

$$(\bar{z}, \bar{R}_1, \ldots, \bar{R}_I) \in G(\bar{p}, \bar{t}), \quad \sum_i \overset{*}{x}_i(\bar{p}, \bar{t}, \bar{z}, \bar{R}_i) = \sum_i \overset{*}{y}_i(\bar{p}, \bar{t}). \tag{4.10}$$

Note that the Walras law will ensure that at a CCCC equilibrium, the market for abatement clears, that is,

$$\bar{z} = \sum_i \bar{a}_i. \tag{4.11}$$

The welfare properties of a CCCC equilibrium are summarized in the following theorem:

THEOREM (adapted from Foley) A competitive collective consumption with contributions equilibrium is a world Pareto optimum. Furthermore, for any distribution of global resources over countries, for any world Pareto optimum, there exists a price system and a distribution of contributions such that the given Pareto optimum can be obtained as a competitive collective consumption with contributions equilibrium.

4.2 A CCCC Equilibrium and a Noncooperative GPGP Equilibrium with Fixed Contributions

The collective demand concept, which is part of the definition of the CCCC equilibrium, is resonated in the GPGP scheme discussed by Bradford and Guesnerie. A dependence on this concept is suggested if the following excerpt from Bradford's chapter is compared with the model of Foley, as adapted to the context of climate change:

> One could, alternatively, imagine a world in which a baseline quantity of emission allowances is set at whatever would have been emitted in the absence of any control regime, with those allowances put in the hands of the companies that, in effect, had the preexisting right to emit. The *collective* decision would then be how many of the allowances to buy and retire. This decision would be made in the light of some system of financing the purchase of allowances. Such an approach might be attractive as a way of getting started and as a way of separating the question of who should pay from the method of implementation. It is this approach I suggest here for controlling the greenhouse gas emissions.

This is essentially Bradford's institutional structure, which involves the creation of a new organization (IBEAA plus the COP to the FCCC) through the combined efforts of countries, whose charge is to decide the level of abatement of green house gases to be bought in a world market with its purchases being financed by contributions from voluntary member countries:

> An agency would be created with the sole function of buying and retiring allowances. This retirement would constitute the acquisition of resources needed to produce the global public good of climate control. To be concrete, I denote this agency the International Bank for Emmisions Allowance Acquisition (IBEAA). Periodically, the COP to the FCCC would meet and determine the quantity of (dated)allowances to be purchased and retired. These purchases might be implemented in an active international market with lot of private traders....

In this institutional setup the amount of abatement that would be bought and the way that the cost of purchase would be shared would arise as a result of collective decision making:

> ... Just how the costs would be allocated among the participating countries would be determined in the negotiations that set up the GPGP system. The system, per se, is silent on the sharing arrangements. The analogy is sharing of costs of international peacekeeping. Cost shares may depend on per capita incomes or consumption levels, ..., benefits countries get from protection against climate change.

> ... the global total of emissions from participating countries would depend jointly on the evolving BAU_{tr} and the collective decision as to how much reductions to purchase from that level. These amounts could be specified in various ways.

Both the GPGP and CCCC approaches involve a separation of the finance of the purchase of the public good from its production, with the financing being done through contributions from the consumers of the public good. However, in the GPGP equilibrium of Guesnerie and Bradford, these contributions are exogenous and explained only to the extent that they satisfy budget balance and individual rationality conditions—the last two conditions in (4.2).

Using a partial general equilibrium model, Guesnerie shows that corresponding to any fixed profile of contributions there exists a GPGP equilibrium. The contributions in the CCCC approach adapted from Foley's work, on the other hand, are endogenously determined by the choice of the collective action, which takes the current prices p and t as given and chooses a budget proposal from $G(p,t)$; see (4.10). Thus a CCCC equilibrium refers to equilibrium levels of contributions and collective demand for abatement. Further Foley's model assumes a decentralized production sector where producers exhibit profit-maximizing behavior. The producing and consuming interests are separated in his model. This will not be the case for global public goods. In a GPGP approach national governments have a strong incentive to influence production decisions regarding the sale of abatement in the world markets in a way that maximizes the welfare of consumers of their countries. In this they could be influenced by strategic considerations, which may include incentives to free ride: taking the supply of abatement from all other countries as given, a country i chooses its level of abatement to maximize its consumer welfare subject to its technological constraint. This is an element of the noncooperative GPGP equilibrium of Guesnerie; see (4.1).

4.3 A Conjunction of Foley and Guesnerie

In this section I propose two models of GPGP that mix elements of the CCCC approach of Foley and the GPGP approach of Bradford and Guesnerie. In both the models the role of the collective action group is (as in Foley) to choose a budget proposal, given the current prices, through a political process that best represents the choice of all member countries. This choice also incorporates the individual rationality conditions to ensure participation of all countries. Model A involves nonstrategic behavior of countries, while model B incorporates the strategic element in the Nash-type supply behavior of the noncooperative GPGP equilibrium with fixed contributions. At the moment I will call these contestable GPGP equilibria.

Given a budget proposal $(z, (R_i)_i)$, in model A, country i solves

$$V_{Ai}(p, t, z, R_i) := \max_{x_i, y_i, a_i} U_i(x_i, z)$$

subject to

$$p \cdot x_i \leq p \cdot y_i + ta_i - R_i$$

$$f_i(y_i, a_i) \leq 0. \tag{4.12}$$

Suppose that the solution to this problem is, for all i,

$$(x_{Ai}, y_{iA}, a_{Ai}) = (\overset{*}{x}_{Ai}(p, t, z, R_i), \overset{*}{y}_{Ai}(p, t, z, R_i), \overset{*}{a}_{Ai}(p, t, z, R_i)). \tag{4.13}$$

In model B, given a budget proposal $(z, (R_i)_i)$, and abatement levels of all other countries $(a_j)_{j \neq i}$, country i's best response is obtained by solving

$$\hat{V}_{Bi}(p, t, z, R_i, (a_j)_{j \neq i}) := \max_{x_i, y_i, a_i} U_i\left(x_i, \sum_{j \neq i} a_j + a_i\right)$$

subject to

$$p \cdot x_i \leq p \cdot y_i + ta_i - R_i$$

$$f_i(y_i, a_i) \leq 0, \quad \sum_{j \neq i} a_j + a_i \leq z. \tag{4.14}$$

Note that the last constraint in (4.3) indicates that country i takes into account the decisions of collective action and all other countries regarding abatement levels, given that these decisions are common knowledge.

Suppose that the solution to this problem is, for all i,

$$(x_{Bi}, y_{Bi}, a_{Bi}) = (\hat{x}_{Bi}(p, t, z, R_i, (a_j)_{j \neq i}), \hat{y}_{Bi}(p, t, z, R_i, (a_j)_{j \neq i}),$$

$$\hat{a}_{Bi}(p, t, z, R_i, (a_j)_{j \neq i})). \tag{4.15}$$

Solving for a Nash equilibrium in $(a_i)_i$ obtains

$$(x_{Bi}, y_{Bi}, a_{Bi}) = (\overset{*}{x}_{Bi}(p, t, z, R_i), \overset{*}{y}_{Bi}(p, t, z, R_i), \overset{*}{a}_{Bi}(p, t, z, R_i)) \tag{4.16}$$

and the indirect utility function for country i for model B as

$$V_{Bi}(p, t, z, R_i) = \hat{V}_{Bi}(p, t, z, R_i, (\overset{*}{a}_{Bj}(p, t, z, R_j))_{j \neq i}). \tag{4.17}$$

Given a price system (p, t) and responses of the countries to its budget proposals, the collective action solves the following problem for models $l = A, B$:

$$\widehat{U}_{l1}(u_2, \ldots, u_I, p, t) := \max_{z, R_1, \ldots, R_I} V_{l1}(p, t, z, R_1)$$

subject to

$$V_{li}(p, t, z, R_i) \geq u_i \quad \forall i = 1,$$

$$tz = \sum_i R_i. \tag{4.18}$$

Suppose that the solution to this problem is given by the following collective demand and contribution functions:

$$z_l = \overset{*}{z}_l(u_2, \ldots, u_I, p, t),$$

$$R_{li} = \overset{*}{R}_{li}(u_2, \ldots, u_I, p, t) \quad \forall i. \tag{4.19}$$

The function $\widehat{U}_{l1}(\ )$ identifies a Pareto frontier for models A and B. It will correspond to the first-best frontier for $l = A$, while it will only be second best when $l = B$.

For a given p and t, the set of budget proposals that are incentive compatible and will not be unanimously rejected is defined for $l = A, B$ by[2]

$$G_l(p, t) := \{(z_l, R_{l1}, \ldots, R_{lI}) \mid z_l = \overset{*}{z}_l(u_2, \ldots, u_I, t, p),$$

$$R_{li} = \overset{*}{R}_{li}(u_2, \ldots, u_I, t, p) \; \forall i, u_i \geq \overset{*}{u}_i \; \forall i \neq 1, \text{ and}$$

$$\widehat{U}_{l1}(u_2, \ldots, u_I, p, t) > \overset{*}{u}_1\}. \tag{4.20}$$

Chapter 4 Design of Climate Change Policies

DEFINITION A contestable GPGP equilibrium for model $l = A, B$ is a set of levels of supply of abatement, contributions, private consumption, and production of individual countries $\bar{a}_i$, $\bar{R}_i$, $\bar{x}_i$, $\bar{y}_i$ for all i, an aggregate level of abatement $\bar{z}$, a market price of abatement $\bar{t}$, and a price vector for private goods $\bar{p}$, such that

$$(\bar{z}, \bar{R}_1, \ldots, \bar{R}_I) \in G_i(\bar{p}, \bar{t}), \tag{4.21}$$

for all i,

$$(\bar{x}_{li}, \bar{y}_{li}, \bar{a}_{li}) = (\overset{*}{x}_{li}(p,t,z,R_i), \overset{*}{y}_{li}(p,t,z,R_i), \overset{*}{a}_{li}(p,t,z,R_i)) \tag{4.22}$$

and

$$\sum_i \bar{x}_{li} = \sum_i \bar{y}_{li}. \tag{4.23}$$

A contestable GPGP equilibrium for model A can be called a nonstrategic contestable GPGP equilibrium, while in the case of model B, it can be called a strategic contestable GPGP equilibrium. Note also that at a contestable GPGP equilibrium, the demand for abatement from the collective action will be met by the supply of abatement from countries. In the strategic case, the market-clearing condition in the market for abatement can hold as a strict inequality. It remains to be seen (1) if contestable GPGP equilibria exist, (2) how contestable GPGP equilibria compare with the OFK arrangements defined in Guesnerie, and (3) the nature of participation under the contestable GPGP equilibria—especially, the role of LDCs in such an arrangement.

Notes

1. Other imperfect competition notions of equilibria, such as one compatible with a monopsony, could also be defined.

2. Recall, for all i, $\overset{*}{u}_i$ is the utility ensured in an initial noncooperative Nash equilibrium.

References

Foley, D. K. 1967. Resource allocation and the public sector. *Yale Economic Essays* 7: 43–98.

Foley, D. K. 1970. Lindahl solutions and the core of an economy with public goods. *Econometrica* 38: 66–72.

5

Untying the Climate-Development Gordian Knot: Economic Options in a Politically Constrained World

Jean-Charles Hourcade, P. R. Shukla, and Sandrine Mathy

Climate policies must deal with a contradiction generic to global environment policies. As was recognized as early as in 1972 at the UN conference on Human Environment at Stockholm, the participation of developing countries is essential. The current emissions of developing countries are, however, significant. If the trend continues, the future share of global emissions from developing countries will be even larger. Nevertheless, developing countries do not yet see the need to cooperate because they perceive environmental issues to be a form of Malthusianism. Thus, despite repeated calls for sustainable development at Rio (1992), the negotiations for framing a climate regime have remained disengaged from the debates on how to embark on sound development paths, in effect tying a Gordian knot through a succession of misunderstandings.

This unhappy turn in policy talks is all the more serious as the timing of the climate change issue is inopportune for developing countries. The increasing attention to the climate change phenomenon has coincided with the rapid economic growth being experienced today by many developing countries and even changing global power equations (of military might, of world markets, and of control over natural resources). No sword of a present-day Alexander can cut this knot tied by history. The aim of this chapter is to pick apart the threads that may untie the knot.

5.1 Intellectual Sources of North–South Misunderstandings, 1988 to 2005

It is fitting to recall the G7's haste, three years after the first prediction of global warming by three-dimensional climate models, to bring to the diplomatic agenda an affair for which, paradoxically, its members had borne a huge historical responsibility. This sudden change of mind cannot be explained without regarding a broad geopolitical reshaping of the petroleum playing field by a large block of developing countries that have emerged as major consumers in the twenty-first century (Schlessinger 1989).

Thus the impetus for a cap and trade architecture did not come from the deployment of an ex ante full-fledged vision but from a succession of diplomatic faits accomplis (Bodanski 2001). The critical moment came with the adoption at Berlin (1995) of a quantity-based approach to create an incentive system that would embrace all countries and sectors. The 1994 resolution rested on the economic conclusion that a single worldwide price of carbon is essential to minimize the cost of meeting a global target and to prevent distortion in international competition. Since 1992 the unpopularity of carbon taxes had left the cap and trade system as the sole contender for this purpose. Against protests from some quarters of the European Union that wanted caps with limits to trade, this system was advocated as best meeting the criteria of cost-effectiveness, environmental integrity, universal participation, and flexibility vis-à-vis national sovereignty.

On the North–South question of negotiating targets, the cap and trade system, as argued by Grubb (1989), has the advantage of organizing large enough North–South transfers to induce a significant commitment by the South in the short term. The commitment of the South could not adequately materialize, however, as the principle of common but differentiated responsibilities (article 3.1 of the UNFCCC) led to the cautious decision that only developed countries would adopt binding commitments in the first period. The silence on targets for the developing countries beyond 2012 had the perverse effect of treating them as pure spectators of the intra–annex-B debates (on target setting) and a supplementarity squabble, as if the South should be content with their absence of commitment or avoidance of damages, thanks to the Kyoto Protocol, and obtain a little financial and technological gain through jointly implemented abatement projects. The silence on the developing countries' future commitments and the disengagement of climate negotiations from other global governance issues of energy, trade, and technology encouraged skepticism if not outright distrust.

As became duly clear from subsequent events, the cap and trade option could not be a magic bullet for either side:

- The US Senate (Byrd-Hagel resolution 1997) called for developing countries to agree to "new specific scheduled commitments to limit or reduce greenhouse gas emissions." The large asymmetry of carbon constraints for developing economies was rejected by the no vote without a participation principle (Bodansky 2001).
- On the final day of COP3, the G7 and China concurred that emissions trading would not be possible until the question of emission rights and entitlements is addressed equitably (G7 and China 1997). For future quota allocations to be fair, emerging economies had to be politically engaged in the debate as well.

Why the cap and trade movement was not affected as a result is a complicated story. One explanation may be the great many communicated economic misinterpretations. Between COP3 and COP6, supporters of the Kyoto Protocol could not surmount the political divergences over the technical modalities of cap and trade enforcement to make credible offers to developing countries.

5.2 The Tabula Rasa Myth

The science of economics has evolved through uneasy twists of neglect, distrust, misuse, and abuse, the same as has any other science. Contributing to the confusion at present is the failure to recognize that optimal tools in a first-best world can be far from optimal if applied without discretion in a second-best world. In the real world there are no tabulae rasae; countries are full of imperfections, the hallmarks of which in conditions of underdevelopment are the existence of incomplete and fragmented markets with multiple discount rates and unequal marginal costs across sectors and regions, weak policy regimes, poor governance, underprotected property rights (land, technology, and IPRs), and dual economies in perpetual reformation. Carbon pricing for our second-best world has two main consequences.

First, the carbon price signal in countries experiencing multiple rapid transitions and technological shifts is swamped by noise from other signals. One possible perverse effect is that the carbon price will increase carbon emissions as the switch is made to carbon-intensive non-traded energy resources (Sagar 2005); another is that the rural sector's access to electricity will also be initially reduced as a result. For the developing countries carbon pricing is not just a matter of a relative price to service ratio. Their rural markets have informal lending mechanisms that provide credit to small enterprises and households, but with high interest rates in a context of low labor rates and no firm wage contracts. The benefits associated with energy resources from a formal market therefore do not exist. While benefits could be introduced by altering baseline parameters such as interest rates and wages, these mechanisms are beyond the reach of a climate centric policy.

Second, the benefits of carbon trading for development are uncertain. Not only are annex-B countries not able to accept allocation rules generating large external transfers,[1] they recognize that the revenues from carbon exports are not likely to materialize as higher income for the exporter. Indeed, to exploit the full potential of carbon export for a given world carbon price, the domestic energy price has to increase more, in relative value, in most developing countries than in the European Union and in the United States (respectively four and two times more in India). The negative income effect would be such that it could not be fully compensated by the inflow of carbon revenues especially given, in these countries, the market distortions, large inequalities, and political constraints to redistribution (Ghersi et al. 2003). To avoid the dampening of growth, governments would then export only that fraction of the export potential for which these adverse effects can be compensated by an efficient recycling of carbon revenues. So the availability of cheap carbon in developing countries would be significantly lower than that suggested by partial equilibrium analysis.

The lack of trust in carbon credits could be lessened by short-term benefits from the CDM. A change from the Kyoto joint implementation program to CDM would mean an inversion of priorities that would more credibly place sustainable development in the first rank and the facilitation of annex-B commitments in the third. But despite a recent rise in the number of projects, CDM participation is limited by the contradictory requirements of environmentalists seeking an environmental outcome additional to the baseline, on the one hand, and the funding agencies' reluctance to provide additional support from the fear of diversion of overseas development assistance funds for the environment, on the other. As a result both parties have put off funding (still untapped) and have excluded infrastructure projects from the CDM because the "avoided" emissions are difficult to ascertain.

Barring the pursuit of a Brazilian type of compliance fund or an extension of the share of proceeds on mechanisms that could provide funding complementary to carbon revenues, the developing countries have chosen to interpret the offer of annex-B countries as an empty pledge that would result in forfeiting real present welfare for unreal future gains.

5.2.1 Burden Sharing and Normative Equity Principles

In addition to the silence on equity issues relating to the climate debate, it is remarkable that economists have refrained from a theoretical framing of fairness on emission rights.[2] The interest has instead been expressed in studying the outcomes of competing ethical intuitions.

The more challenging of the equity issues is the equal per capita distribution of emissions rights (Agarwal and Narain 1991). Inarguably the allocation of distribution rights is inequitably charged. Individuals live in very different ambient climates, with different spatial constraints, and with different energy accesses (Godard 2000; Neumayer 2002). However, the rhetorical proximity between the notions of equity and equality has transformed the per capita principle into a political symbol for those who consider grandfathering to be environmental colonialism. This symbol was invoked strongly enough to be retained at Marrakech in 2001.

Grandfathering has stirred up intense debate. Grandfathering is widely considered to underlie the international agreements on, for example, multilateral fishing quotas (Sterner 2002), milk quotas in Europe (Burton 1985), and the SO_2 regime in the United States (Joskow et al. 1999). The real basis is the ethical legitimacy of any new environmental regulation as a renegotiation of a social contract, and it is all the more fair to account for interests vested under the existing contract[3] as previous generations were not informed of the consequences of their behaviors (Claussen et al. 1998). But distributing rights for the future use of the atmosphere based on the grandfathering principle would lead to inequitable future contracts that the developing countries cannot accept.

In the literature the search for consensus on such explicit principles has evolved mainly into two pragmatic positions, these being (1) starting from grandfathering to achieve a contraction and a convergence toward equal per capita emissions in the long run (Meyer 2000; Ghersi et al. 2003) or (2) to combine (at least indirectly) the ability to pay with other criteria (Jacoby et al. 1999). Also explored have been a triptych approach (Phylipsen et al. 1998; Groenenberg et al. 2000; Jansen et al. 2001) and proposals from Norway (1996), Australia (1997), and Iceland (1997) at the Ad hoc Group on the Berlin Mandate.

Because of conflicts of interest, no rule has yet reached consensus. The uncertainty about the links between the levels of the emissions caps and the carbon prices, and their macroeconomic consequences, is sure to generate fuzzy contracts. The present reasoning is that countries will have to accept one of these positions despite their divergent views. A thorough review by Lecocq and Crassous (2003) shows country preferences on allocation rules to be very unstable and conditional on baselines assumptions and time horizons. Both China and Europe, although starting from different emission levels, would reject rules with high weights to convergence toward 2030 but accept the rules if the weights are for only the immediate post-2012 period. It could be argued that setting targets for every five years creates a learning curve, but that notion addresses the issue only marginally. The possibility of drastic revision to the allocation rules may be further limited by the political cost of reversing diplomatic fait accomplis and by the risk of undermining the efficiency of the system, as countries could lower their abatement efforts in order to renegotiate lax targets for future periods (see Helioui 2002).

Added to the difficulties of burden sharing are two more obstacles. As governments face the task of trade-offs among various assessments criteria, maximizing the inflow funds from carbon trading and minimizing the costs for the low-income populations may not lead to the same vision of burden. Clearly, similar contradictions arise in cross-country comparisons, since poor citizens of a rich country contribute far less to global warming than do rich citizens of developing countries.

The problem with forms of trade-offs within a wide development agenda is what constitutes a fair burden. Political acumen could have made this obvious but for the vulgate of Kantian ethics used in international meetings that have eclipsed the Machiavellian idea that states are cold-hearted monsters designed to defend the selfish interests of their constituents. The only way out is to return to Pareto-improving policies; moving afar from the focus on "burden sharing" and "property rights entitlement."[4] Emissions quotas are then not treated as rights but as transitory allowances, and the relevant question becomes under what conditions cap and trade can be used as a tool for removing barriers to development. This re-framing does not eliminate debates about equity and responsibility; it puts them in the perspective of reshaping development instead of capping it. The challenge is to demonstrate that postulating the existence of win-win options is not another wishful way to reconcile contradictory interests.

5.2.2 Aligning Development Pathways and Long-term Climate Change Policies

An Intellectual Discipline: Starting from the Suboptimal and Real Baselines

The common practice of projecting secular growth baselines (often optimistic for reasons of political correctness) has hampered the detection of synergies between climate and development despite attempts through environment policies to alter this trend. It therefore is appropriate to start by delineating the baselines that incorporate barriers to the realization of a growth potential. In fact concern over disruptions of local environments opened the discussion at Stockholm (1972)[5] on the perverse effects of ongoing growth patterns, including environmental damage due to choice of equipment, structural unemployment, disregard of basic population needs, and erosion of soil.

The issues voiced at Stockholm in 1972 need to be revisited today as they relate to the evolving linkages between capital scarcity, infrastructure requirements, and social dualism. Cumulative energy investments between 2001 and 2030 are projected to reach 2.2 T$ in China, 2.1 T$ in the rest of Asia, and 1.3 T$ in Latin America (IEA 2003). Fay and Yepes (2003) estimate that from 2005 to 2010, 6.7 percent of the GDP in Asia and 5.5 percent in Africa will have to be invested in energy, transportation, water distribution, and sanitation infrastructures. Since 40 to 60 percent of savings will still have to be invested in buildings, funding infrastructure investments will remain critical for reason either of capital scarcity (Africa's saving rate is 8 percent of GDP) or of inefficient allocation in countries with high savings. The latter may in addition not sustain a capital-intensive growth when their savings rates decline such as in China where the inversion of age pyramid between 2020 and 2030 is expected to cause a decline of savings from the present 40 to 8 percent in 2050 (INGENUE 2005). An analogy may be the debt trap of Brazil in the late 1970s when 30 percent of the debt was from investments in the energy sector. Low savings can curtail infrastructure programs and enhance resource conflicts between populations with very different accesses to energy, water, and transportation.

Managing energy demand is becoming more critical as energy security is being ever more perceived as a prime development objective by nations competing for resources at a time of explosive energy prices. Recent trends only confirm the warnings of the World Energy Conference (WEC) in 1979, that the emergence of developing countries as major oil and gas consumers will be a source of major world tensions after 2000. Ten years later, in 1989, the WEC was reminded of this prediction in Montréal by James Schlessinger. Added to the controversies over the timing of peak oil production, the conventional oil reserves have remained increasingly concentrated in politically sensitive regions.

The question for a climate regime set on tackling energy-related obstacles is to what extent the value of carbon can help developing countries move upward on the production frontier. In principle, the sharing of a carbon benefit can produce a leverage effect by enhancing the profitability of foreign investments, on the one hand, and by compensating the transaction costs of Pareto-improving domestic policies, on the other. As we explain next, the mechanism to activate win-win options is not so remote; it indeed exists.

Development and Climate Synergies: Illustrations from India
The counterargument to the notion of no-regret policies that enhance both environment quality and economic income is that where a no-regret potential exists, it will be tapped anyway. The different examples of this section provide a perspective that is not restricted to the instruments and technologies of the 1990s (Jaffe and Stavins 1994; IPCC 1995); they show that climate policies can facilitate an upward shift in the development baselines with the new baselines presenting even more potential for decarbonization policies.

5.2.3 Conjoint Market for Local and Greenhouse Gas Emissions

Local air pollution reaches highest levels in large developing countries as they undergo transitition to urbanization and industrialization. Local air pollutants and greenhouse gas, which are often emitted conjointly, accumulate and create climate changes but also health problems for people, affecting their health and development. In India, for instance, 70 percent of coal consumption is by the electricity sector (CMIE 2003) with the emissions being mostly sulfur dioxide (SO_2). In fact, as of 2000, two-thirds of India's CO_2 and SO_2 emissions come from 500 large point sources, of which 82 are coal-based power plants and the rest mostly transport, steel, and cement manufacturing (Shukla et al. 2004).

Yet, although the electricity sector in India lacks efficient emissions controls, opportunities exist for creating conjoint emissions control mechanisms. Interestingly the relationship between sulfur and carbon control is asymmetric (Pandey and Shukla 2002; Garg et al. 2003). Cost-effective measures like better combustion efficiency and a fuel switch from coal to gas could reduce sulfur emissions to greater extent than carbon emissions, but the cost-effective sulfur control measures like the use of clean coal technologies or low-sulfur diesel fuel would have little impact on carbon emissions. The local pollution did draw the attention of national policy makers. However, the sequencing of SO_2 and CO_2 markets proved to be suboptimal because the single measures to control local pollution failed to net the co-benefits of concurrent SO_2 and CO_2 mitigation. There is nevertheless scope for a policy designed to align both markets and optimize co-benefits. At a low carbon price of $20/tCO_2$, the aggregate mitigation cost over the next 25 years in the conjoint system would be lower by $400 million compared to under the two separately operating markets. Further the conjoint system could deliver 520 Mt of additional CO_2 mitigation and thereby add $2.6 billion to the carbon benefits (Menon-Choudhary et al. 2004). Correcting such asymmetry and incongruent actions typically demands institutional financial arrangements at the national level that should benefit from a climate regime.

5.2.4 Synergy of Electricity Market Reforms and Revenues from Carbon Trading

After India's electricity sector reforms in early 1990s the sector became more dependent on domestic coal, as hydropower was confronted by high capital costs and the building of large dams besieged by protest movements and interstate water disputes. The barriers to hydro and the bottlenecks in coal supplies caused the electricity sector to shift to gas in a market where the combined cycle gas technology offered advantages of low investment, short gestation, and low local emissions. Despite this shift, the carbon content of electricity increased, and the hydro share of the market continued a secular decline (Shukla et al. 2004).

Globally energy reforms also remain centered on fossil fuel. However, in 2002 came the India Vision 2020 (Planning Commission 2002) plan, which proposed an alternative best-case scenario (BCS) that could bifurcate the sector's development. The Vision is directed toward an alternative pathway that includes modernizing existing plants, early adoption of advanced technologies, improved T&D efficiency, energy conservation, regional energy cooperation, and higher shares of hydro and renewable energy. Carbon emissions in 2020, under BCS, are projected to be 81 MtC below the BAU (822 MtC cumulated up to 2020).

In the case of India there could be envisaged a sector-based agreement by which the carbon abatements could be sold on the world carbon market.[6] The implementation of BCS policies (including carbon taxation and subsidies to renewable technologies) would have three conjoint effects: (1) lower profitability of coal plants and increased profitability of gas and renewable energy, (2) an inflow of foreign capital due to the relative profitability of investment in the sector and the lifting of barriers to foreign investors, which would replace part of the Indian investment in the power sector in the reference scenario, and (3) the ensuing domestic investment redirected toward other sectors. By this mechanism the carbon revenues would have a leveraging effect. For a given value of carbon, and regardless of the lower public expenditures to compensate local externalities, the main determinants would be as follows:

• The gap between the social marginal profitability of energy investment with or without BCS policies (and with taxation on coal).

• The gap between the social marginal profitability of energy and non-energy investment.

• The share of foreign investments as a function of a foreign investor's internal return rate.

With a value of 10$/tC and a linear increasing tax level reaching 30 percent of the coal price in 2035, the mechanism would generate an additional income of US$1.6 to US$7 for each dollar of carbon credit depending on assumptions of marginal productivity in the power sector and in the rest of the economy (see Mathy et al. 2001).

Climate Regime and the Co-benefits from Regional Cooperation
Regional cooperation, as promoted by the principle 9 of the Rio Declaration on the Environment, may be supported by compelling arguments. However, countries are so diverse in terms of institutional

capacities and political structure that deploying complementarities can become a very complicated business. In the South Asia region,[7] for example, there are diverse endowments in energy resources: coal in India, gas in Bangladesh, hydro potential in the Himalayan nations of Bhutan and Nepal, and strategic location of Pakistan for the transit routes linking South Asia with the vast gas and oil resources of Central Asia and the Middle East (Nair et al. 2003; Heller and Shukla 2004). Still there is little energy and electricity trade in the region.

Assuming a regional cooperation, Heller and Shukla (2004) show that the energy trade would yield direct economic benefits due to energy savings from improved fuel and technology choices and would lower investments in energy supply. The benefits are valued at US$319 billion from 2010 to 2030. The economic growth of the region would increase by 1 percent each year, benefiting an overwhelming number of the world's poor. Such cooperation would in addition deliver a cumulative carbon saving of 1.4 GT for the period 2010 to 2030, or 70 percent of the global mitigation by the 1997 Kyoto Protocol standard over the estimated baseline emissions (Chandler et al. 2002), including the original commitment by United States at Kyoto. The energy changes would also reduce loads of SO_2 in the region by nearly 30 percent. In addition the balanced hydro development would yield spillover benefits that are synergistic with adaptation needs, among which are enhanced water supply and flood control.

Infrastructure, Development, and Climate Vulnerability

Infrastructures designed to withstand and long endure the variabilities of current climate conditions can render them vulnerable to climate change. A good example of such long-lived assets is the recently constructed Konkan railway. Located on the coastal strip of land bounded by the Sahayadri hills and Arabian Sea on the western coast of India, the Konkan railway covers a distance of 760 km. It costed US$745 million to build and commenced operations in 1998. The railway passes through a mountainous region and has 179 major bridges, 1819 minor bridges, and 9 tunnels exceeding 2.2 km (KRCL 1999). Climate projections for the area (Mitra et al. 2002) show an increase in the mean and variability of the distribution of the key climate parameter of rainfall, the impact assessed in days with more than 200 mm of precipitation. Besides rainfall, the development pattern was considered as affecting the geology, soil structure, and vegetation cover in the region and also landslides.

In the monsoon season waterlogged tracks and landslides disrupt train schedules; 140 such incidents were reported in 2000 (Kapshe et al. 2003). The railway company spends 6 percent of its revenue on repair and maintenance, and 20 percent of this expenditure is for minimizing the severity of such climate-related incidents. Future disruptions are expected to be greater because of the compounded effect of climate change and an aging railway infrastructure. It would be prudent to protect the infrastructure but adaptation measures like improving climate predictions, reinforcing the vulnerable points, and making maintenance regimes more frequent require committing more public expenditure, which cannot be fully recovered by taxes in India. This case is representative of many such cases in the developing world that call for a climate regime that can deliver cumulative assistance in the form of development, mitigation, and adaptation co-benefits.

5.3 The Kyoto Architecture Reinterpreted, Amended, Completed

The complexity of climate and development nexus may be an argument for what Jacoby metaphorically calls a *favela* regime (Jacoby 2005), namely a self-organizing process instead of a hopeless pursuit of some grand architecture (Bodanski 2003). But *favelas* turned more often into self-reproducing pockets of violence, slavery, and poverty than into an innovative urban scheme. The transition to a common architecture may fare better if it is instead modest and flexible.

The main purpose for a common architecture is that the key sectors for the climate control are capital intensive with investments spanning over decades. Early and credible signals launched in this direction should therefore avoid lock-ins to carbon-intensive systems. This does not mean a full-fledged architecture by 2012 but the initiation of a process that can gain momentum and converge on some viable system. The flaws of Kyoto have been extensively pointed out (Victor 2001), but there are two reasons why Kyoto remains the framework that can support such a convergence. The first is political: it is diplomatically discomforting to write off a treaty ratified by all countries but one (since the ratification by Australia in 2007), as is now the geopolitical game. The second is the for lack of anything better argument, that no proposed substitute provides the same potential to untie the development-environment knot.

An internationally coordinated carbon tax (Cooper 1998; Nordhaus 1994) would confront the same equity issue as a cap and trade system (higher marginal welfare loss for a given carbon price in low-income populations) while not providing ways to compensate abatements through generous allocations of quotas (Chichilnisky and Heal 2000). In developing countries the middle class represents a small share of population compared with low-income classes, and this makes it difficult to operate emission compensation domestically through a reshaping of fiscal systems. Net foreign inflows may be helpful, however, and whether they come from overseas aid at a similar order of magnitude as the transfers from carbon trading is, to say the least, open to discussion.

An acceleration of R&D efforts disconnected from any economic signal may be a technological push, a model that works only for large scientific ventures (space exploration, conventional electronuclear fusion, etc.); a technological push is less effective when the innovation is to be deployed in hundreds of end-use services and under large controversies about the most promising technologies on the supply side. The six-country initiative, called the Asia–Pacific Partnership for Clean Development and Climate, proposes cooperative voluntary actions for the development, deployment, and transfer of technologies to meet these countries' own development objectives. The question remains nevertheless whether such cooperation can deliver its full potential in the absence of any economic signal.

In the absence of anything better, the question might be what reinterpretation and amendment of the Kyoto Protocol would alleviate the current flaws? There is no good answer other than to set the cap and trade system within a paradigm of climate negotiations.

5.3.1 Shifting the Bargaining Paradigm, Shifting the Status of the Climate Regime

In the reordering of the world since Kyoto, there have been far-reaching structural transitions in major developing countries that are sure to alter the dynamics of new rounds of any international negotiation. These countries' enrichments, of course, create ground for demanding from them acceptance of emissions limitations; their response will depend on whether they will consider that investing in backbone infrastructures will lock them into an unsound development pattern, and whether they will grasp the dire implications of climate change issues for energy security (Heller and Shukla 2003). The initial national communications to the UNFCCC of a number of developing countries suggest such an alignment of climate and development objectives, and their governments have made multiple declarations in that direction: Millennium Declaration at the UN Millennium Summit (2000), the Johannesburg Declaration at the World Summit on Sustainable Development (2002), and the Delhi Declaration on Sustainable Development and Climate Change at the Eighth Conference of Parties (2003).

The outcome will depend on an acceptance of so-called offers made by the North, but realistically, the timing is inconvenient; it should be recognized that climate action is more urgent today than in 1992 and 1997. Any significant transfers to developing countries in climate action have to be accepted as threatening jobs. This is why developed countries should be serious about why they want climate policies: Is it solely for environmental reasons? Or is it because of the political instability that climate change could create in areas of low adaptativity? Or is it part of the geopolitics of energy?

A recent declaration by the G8 suggests a broadening of the negotiation paradigm: "We will act with resolve and urgency now to meet our shared and multiple objectives of reducing greenhouse gas emissions, improving the global environment, enhancing energy security and cutting air pollution in conjunction with our vigorous efforts to reduce poverty" (G8 declaration, July 2005). Climate policies could be a unifying force in a transformation of economic globalization into a mutually benefiting process instead of a divisive one for all nations and thus avert threats to energy security, climate refugees, and local political stability. A big step forward would be to replace negative arguments by positive ones. The demand for infrastructures in developing countries offers a window of opportunity (G8 2005) and a type of Marshall Plan as mentioned in several world reports in the 1970s (Carter et al. 1976; Tinbergen 1976; Brandt 1980) could enable the developing world to rise to its full potential.

Such a broadening of the bargaining paradigm has two implications for a climate regime. First, this regime does not pretend to dictate the core decisions on decarbonization of an economy. It instead views all related policy processes as opportunities to internalize over the long term the social value of carbon abatements. Second, its architecture is "minimal," and not designed to dictate uniform solutions. It is flexible enough to initiate bottom-up regional or sector-based cooperation, and it is integrated enough to prevent a *favela* type of trap.

5.3.2 Basic Principles for a Minimal Architecture in a Dynamic World

The Kyoto Protocol provides some insight into how a minimal architecture could secure a flexible relationship between the environmental interests and development. It is in fact necessary to dispel the notion that any flexible mechanism represents a threat to the environmental integrity of any climate regime.

Using Carbon Prices as Inducement, not as a Unique Driver of Climate Policies
Decarbonization will depend on a wider range of signals than just the carbon price: interest rates, insurance premiums, certification of clean technologies, tax systems, regulation of the labor market, prices of land and housing, transport costs and regulations, and local environmental factors. This is not to minimize the role of carbon prices but to expand the argument, since price is a parameter against which there can be measured climate benefits from an initiative to change behavior.

In this view, carbon revenues do not ensure that additional carbon abatement accrue only from purely climate centric measures as would be the case if the world market could confront factories that specialize in a product with precisely measured GHG emissions. Indeed, because the Kyoto Protocol only creates a carbon market among countries,[8] it is up to governments to do the selecting, controlling, and redistributing of emission allowances among sectors, and to apply other policy tools that cannot be exchanged individually on a world carbon market.[9] Two examples are lowered emissions as a by-product of speed limits on vehicles that are intended for road safety or improved regional cooperation in the South Asia to save on capital investments. These are not tangibles that can be exchanged on a carbon market, but they affect carbon output in the world carbon market.

The concept of a world carbon market therefore provides countries with a lot of flexibility to overcome the unique obstacles that the carbon price may pose for them individually. Governments are not forced to increase all their domestic energy prices by the level of the international carbon price; they have the leverage to employ other policy parameters in delivering their domestic objectives or constraints.

Terms of the Negotiation: Diversified Pledges Aligned on Long-term Price Signals

A precondition for making cap and trade acceptable to developing countries is to abandon the Malthusianism notion that commitment binds a country to certain emission constraints. This can be done first through diversifying the menu of pledges. For the global commitments of annex-B countries (and countries reaching an agreed threshold level of a per capita income), the considerations most relevant to them might be as follows:

• Nonbinding global quotas (Philibert and Pershing 2002): Countries could export carbon if they meet their quotas and not be penalized in case of an overshoot.

• Sector-based targets: Countries could select sectors whose participation could bring development benefits.

• Clean development mechanisms: Countries and sectors not yet mature to pledge emission limits could receive supportive action through participation assistance programs.

Another argument for easing the commitment objectives might be that developing countries are generally not in a position to pledge to or negotiate even the lax quotas that annex-B countries may concede to in order to induce them to adopt a limit on carbon emissions. With the option of nonbinding commitments non–annex-B countries might be brought to a good faith dialogue; freed from the fear of economic burden, they may also consider regulating the system to prevent the carbon value and the economic gains they secure by entering the regime from being deflated.

Another way around the cap on development may be to base targets on performance criteria rather than on absolute caps for non–annex-B countries. The advantage here is that this limits the risk of volatility in economic growth rates,[10] so whether fast-growing countries experience an 8 percent growth rate or only 6 percent makes a significant 9.8 percent difference over a five-year period.[11]

Linking Incentives to Comply with Assistance to Developing Countries

The efficacy of a climate regime should not depend on the credibility of a government's commitment, particularly since many countries (and not just the United States) will not accept a system limiting the sovereignty of their legislative institutions. The European experience of sovereignty transfers is a very specific case, and international affairs will likely remain a matter of pledges and review. The question is how to secure compliance in this regard.

For countries with binding commitments the difficulty is that given uncertainty about compliance costs, good faith governments will commit themselves only to lax targets. To facilitate an accord on ambitious targets, a price cap was proposed in 1997 (Kopp, Morgenstern, and Pizer 2000); operating as a safety valve (the worst-polluting countries would pay an agreed-on price), it would hedge against bad surprises. Although the optimists (i.e., the proponents of ambitious climate action) may not be happy with an arrangement where abatement costs are low, a safety valve is secured. A price cap can result in a hybrid system à la Roberts and Spence (1976) where a floor price provides disclosure about the price–quantity relationship and thus guides long-term expectations and facilitates a tightening of the system.

The critics against this idea point out that the parties may not be sanctioned for not fulfilling their "legally binding" objectives. This is symptomatic of a misperception of the notion of "legally binding." Military actions set aside, the only effective sanction in international affairs can ultimately come from economic and political reprisals, but these reprisals depend on other tools of international coordination than climate conventions and will never be activated regardless of any related issues. Without linkages the compliance provisions cannot proceed, as in the Marrakech Accord, beyond allowing de facto an accumulation of environmental debts.

A price cap could add economic "teeth" to the system because missed abatements get paid, and the collected funds are used to restore the environment. Funds earmarked for the environment could provide an incentive for developing countries to participate, as is the case of the Compliance Fund proposed by Brazil (UNFCCC 1997) whereby in the non–annex-B countries the selecting of abatement projects would be through auctions.

Special Treatment of Energy-Intensive Industries Exposed to Global Competition
When passing from a general declaration to its implementation, policy makers may face an implicit type of protest from companies in energy and carbon-intensive industries in the form of an argument about asymmetrical constraints in international competition. The risks are

often overstated as regards the products markets: the likely impact of carbon prices on production costs are one order of magnitude lower than the large oscillations in exchange rates experienced over the last three decades (Quirion and Hourcade 2004) and the increase in transportation cost operates as a countervailing factor. The risk is far higher in terms of equity in allocations of quotas; though in a closed economy, this can be eliminated by allocating a minor share of quotas for free (Goulder 2004); in an open economy, the risks can be covered by the harmonization of quota allocation rules, especially the share between free allocated and auctioned quotas.

Although it could be argued that there are many other sources of differences in competitive conditions (including wages), the political economics of negotiation amply show that no government is in a position to resist the pressure to protect jobs. The way out is to take stock of the fact that ultimately governments will operate internally a differentiation of targets and carbon prices to households and industry (as they do for energy prices) and can, in this way accept price equalization as an inescapable condition for exposed energy-intensive industry. The potential for conflicts at the WTO is large in this domain, and this is why it may be important to rely on international sector-based agreements in the few concerned industries on the quota allocation rules. Since national policies would be charged with endorsing these agreements, most of the concerns about international competition could be addressed.

5.3.3 Reconciling Long-term Goals and Immediate Pressures from Shifting Context of ODA

Because the benefits that accrue from decarbonization (carbon revenues, avoided damages) are long term, the political agents may feel blocked by the absence of tangible short-term gains. For countries under urgent pressure to switch to more carbon-intensive technologies (introducing motorized transport for bicycling or animal traction, deforestation to increase food production) short-term action could help prevent future carbon-intensive lock-ins.

Inevitably there will be need for annex-B countries to fashion overseas aid and multilateral financing as climate policy tools to accompany such reforms in non–annex-B countries. The debate framed so far is in terms of avoiding a crowding-out of public aid by transfers submitted based on environmental conditions. This prospect could be avoided by the provision of additional resources to the carbon-trading regime (price caps, shares of proceeds from carbon trading, taxes on bunkers and international aviation). Indeed how to raise money is far less difficult than to guarantee their efficient use in conjunction with other funding mechanisms. This has become all the more complex as developing countries are evolving quickly in very different directions.

Rapidly emerging countries are now main recipients of international private investments. But the quantity and quality responses to the flow of funds are not meeting the corresponding investment requirements on decarbonization objectives. Public funds could be used to provide support for critical project financing and, more important, for public–private technical and institutional partnerships (structural reforms, multilateral agencies, credit exportation agencies, third-party financing). Developing countries whose reduction potentials are only limited volumes of emissions would in this way receive assistance from ODA for the construction of infrastructures and for enlarging basic structural needs.

Climate policies offer the opportunity of adding a quantifiable dimension (the emissions levels) for evaluating ODA efficiency, for disciplining the investor and the host country, and for ensuring quality results for money expended. The monitoring capacity can also be used indirectly by development finance institutions to provide risk mitigation instruments. The PLANTAR project (sustainable fuelwood under the PCF) in Minas Gerais (Brazil) could not obtain any currency risk insurance beyond two years; with carbon finance revenues (US$ or € denominated) placed in an offshore escrow account, an OECD commercial bank agreed to a five-year loan and the amortization is structured to match expected payments for the CERs. Emission reduction purchase agreements (ERPA) can serve as a type of insurance package that can be dedicated to funding project completions. ERPA funds can generate the necessary cash and prevent currency risk, because the lenders must provide the cash up front.

5.4 Conclusion

Climate policies cannot be an isolated item on the international agenda. Objectives of environmental integrity present opportunities for engaging developing countries in pro-active climate strategies. The problem with envisaging the cap and trade system as an architecture encompassing all gases, sectors, economic actors, and governments is that the focus remains on securing a traded ton of carbon that is precisely measured, whereas the core challenge is to curtail future emissions from quickly expanding infrastructures where the counterfactual baselines are impractical to measure. The cap and trade system misses opportunities for accruing mutual benefits of climate and development actions that relate primarily to these new infrastructures. The system will merely polarize the debate on sharing the burden of carbon abatement rather than on mainstreaming the climate actions with development agenda. There should be concern about the constraint on development that a cap and trade system would bring.

There are mechanisms through which climate policies can exert a leverage effect on development that are absent in the cap and trade scheme. A mix of price signals, capital inflows, and technology transfers could be generated by carbon trading systems, but most important is institutional design by which revenues from carbon trading can be directed to removing obstacles to development and integrated into development policies. The configurations for such integration might vary, but in any case require a bottom-up design for facilitating a coordination of diverse initiatives and cost-effective and welfare-maximizing actions for mutual benefits vis-à-vis the development and carbon abatement objectives.

The nature of the problem makes it neither politically feasible nor economically prudent to start from a full-fledged "grand architecture" nor to rely only on self-organizing processes such as the Madisonian approach (Victor et al. 2005) that set ambitious but nonbinding goals and do not generate the credible and stable policy signals such as could secure carbon saving investment in developing infrastructures.

The way out is not to dismiss the Kyoto Protocol, though it was ratified by overwhelming majority countries, but to re-interpret it by inverting the climate centric view that has prevailed so far. The technical instruments of inversion would involve the diversification of pledges, nonbinding commitments, safety valves, voluntary agreements in some key sectors of the world industry, re-design of the CDM in the direction of infrastructures programs. All these tools can be aligned to assemble a set of initiatives that will not result in a fragmentation of effort. But the challenge will be the many linkages of energy security, local environment issues, debt traps or social dualism, and reshaping of international funding and overseas aid that need to be recognized. The climate regime should be part and parcel of Pareto-improving policies that seek beneficial exchange in the globalization of economies and to narrow the North–South divide (Stiglitz 1998).

Notes

1. This reluctance was explicit in the US position; the supplementarity condition insisting on domestic abatements, advocated by the European Union, is an indirect form, although motivated by the political virtue of a demonstration effect.

2. For the few exceptions, see Chichilisky and Heal (2000).

3. The Brazilian attempt in translating the principle of historical responsibilities (den Elzen et al. 1999) confronts directly this difficulty.

4. Note that little progress has been made so far on rights over the global commons and that negotiations on forests (and related biodiversity) have reinforced national sovereignty over natural resources.

5. Myrdal's "Asian Drama," Sen's early contributions, R. Dumont's "Afrique Noire est mal partie," and the UNCTAD group (R. Prebisch) were used in questions raised about the trickling down of Western economic growth to developing countries (Rostow) and about the replication in these countries of the socialist primitive capital accumulation.

6. For a detailed description of the methodology, refer to Mathy et al. (2001).

7. Nearly a quarter of the world population resides in this region, which comprises Bangladesh, Bhutan, India, Maldives, Nepal, Pakistan, and Sri Lanka.

8. To minimize market powers and the strategic use of carbon trading by governments, one important addition to the Marrakech accord would be to have all imports and exports between governments take place through transparent auctions run by a state-regulated clearinghouse.

9. In this regard the European Carbon Trading System is not a small-scale model of an international trading system; it is a specific modality adopted by some governments to meet their targets.

10. A floor price of carbon could provide an additional hedge against deflated carbon prices.

11. The Chinese per capita growth rate was 2 percent in 1990 and 13 percent in 1992; Argentina experienced −8 percent growth rate in 1989 and 9 percent in 1991.

References

Agarwal, A., and S. Narain. 1991. Global warming in an unequal world, a case of environmental colonialism. Center for Science and Environment, Delhi.

Bodansky, D. 2001. Bon voyage. Kyoto's uncertain revival. *National Interest* (Fall): 45–55.

Bodansky, D. 2003. Climate commitments, assessing the options. In *Beyond Kyoto, Advancing the International Effort against Climate Change*. Washinton, DC: Pew Center on Global Climate Change, pp. 37–59.

Brandt, W. 1980. *Nord–Sud: Un Programme de survie*. Rapport de la commission indépendante sur les problèmes de développement international. Paris: OECD.

Burton, M. 1985. Implementation of the EC milk quota. *European Review of Agricultural Economics* 12(4): 461–71.

Carter, A. P., W. Leontieff, and P. Petri. 1976. *The Future of the World Economy*. New York: United Nations.

Chandler, W., R. Schaffer, Z. Dadi, P. R. Shukla, F. Tudela, O. Davidson, and S. Alpan-Atamar. 2002. Washington, DC: *Climate Change Mitigation in Developing Countries*. Report. Pew Center on Global Climate Change.

Chichilnisky, G., and G. Heal. 2000. *Environmental Markets: Equity and Efficiency*. New York: Columbia University Press.

Claussen, E., and L. Mc Neilly. 1998. *Equity and Global Climate Change: The Complex Elements of Global Fairness*. Washington, DC: Pew Center on Global Climate Change.

CMIE (Center for Monitoring Indian Economy). 2003. *Energy*. Mumbai: Center for Monitoring Indian Economy.

Cooper, R. 1998. Toward a real global warming treaty. *Foreign Affairs* 77(2): 66–79.

Den Elzen, M., M. Berk, M. Schaeffer, J. Olivier, C. Hendricks, and B. Metz. 1999. The Brazilian proposal and other options for international burden sharing: An evaluation of methodological and policy aspects using the FAIR model. RIVM report N°728001011. Bilthoven.

Fay, M., Yepes, F. 2003. Investing in infrastructure: What is needed from 2000 to 2010. World Bank Research working paper 3102.

Garg, A., Shukla, P. R., D. Ghosh, M. Kapshe, and R. Nair. 2003. Future GHG and local emissions for India: Policy links and disjoints. *Mitigation and Adaptation Strategies for Global Change* 8(1): 71–92.

G7 and China. 1997. Position paper of the group G7 and China on the mechanism of the Kyoto Protocol for the second meeting of the contact group on mechanism. 2p.

Garg, A., P. R. Shukla, D. Ghosh, M. Kapshe, and R. Nair. 2003. Future GHG and local emissions for India: Policy links and disjoints. *Mitigation and Adaptation Strategies for Global Change* 8(1): 71–92.

Ghersi, F., J. C. Hourcade, and P. Criqui. 2003. Viable responses to the equity-responsibility dilemna: a consequentialist view. *Climate Policy* 3(S1): S115–33.

Godard, O. 2000. Sur l'éthique, l'environnement et l'économie: La justification en question. *Cahier du Laboratoire d'Econométrie de l'Ecole Polytechnique*: 513.

Goulder, L. (with A. Lans Bovenberg and D. J. Gurney). 2004. Efficiency costs of meeting industry-distributional constraints under environmental permits and taxes. *Rand Journal of Economics*, forthcoming,

Groenenberg, H., D. Phylipsen, and K. Blok. 2000. Differenciating commitments world wide: Global differenciation of GHG emissions reductions based on the Triptych approach—A preliminary assessment. *Energy Policy* 29(12): 1007–30.

Grubb, M. 1989. *The Greenhouse Effect: Negotiating Targets*. London: Royal Institute for International Affairs.

Heller, T. C., and P. R. Shukla. 2003. Development and climate—Engaging developing countries. In B*eyond Kyoto: Advancing the International Effort against Climate Change*. Washington, DC: Pew Center on Global Climate Change, pp. 111–40.

Heller, T. C., and P. R. Shukla. 2004. Financing the climate-friendly development pathway: With illustrative case studies from India. Workshop on "Development and Climate" organized by RIVM, The Netherlands and Indian Institute of Management, Ahmedabad (IIMA), New Delhi, September.

Helioui, K. 2002. On the dynamic efficiency and environmental integrity of GHG tradable quotas. In J Albrecht, ed., *Instruments for Climate Policy: Limited versus Unlimited Flexibility*. Aldershot: Edward Elgar, ch. 11.

INGENUE Team. 2005 Scenarios for global ageing: An investigation with the INGENUE 2 world model. Working paper. 42p.

International Energy Agency (IEA). 2000, 2002, 2003. *World Energy Outlook*. OECD, Paris.

IPCC. 1995. *Climate Change 1995: Mitigation*. Contribution of Working Group III to the Second Assessment Report. Cambridge: Cambridge University Press

Jacoby, H. D., R. Schlamensee, and I. S. Wing. 1999. Toward a useful architecture for climate change negotiations. Report 49. Joint Program on the Science and Policy of Global Change, Massachusetts Institute of Technology.

Jacoby, J. 2005. Climate *favela*: Regime building with no architect. *Post 2012 Climate Policy: Architectures and Participation Scenarios*. Venice: FEEM.

Jaffe, A. B., and R. N. Stavins. 1994. The energy paradox and the diffusion of conservation technology. *Resource and Energy Economics* 16: 91–122.

Jansen, J. C., J. J. Battjes, J. P. M. Sijm, C. H. Volkers, and J. R. Ybema. 2001. The multi-sector convergence approach: A flexible framework for negotiating global rules for national greenhouse gas emissions mitigation targets. CICERO Working paper 2001:4, ECN-C—01-007.

Joskow, P., R. Schmalensee, and E. M. Bailey. 1999. The market for sulfur dioxide emissions. *American Economic Review* 88(4): 669–85.

Kapshe, M., P. R. Shukla, and A. Garg. 2003. Climate change impacts on infrastructure and energy systems. In P. R. Shukla, S. Subodh, N. H. Ravindranath, A. Garg, and S. Bhattacharya, eds., *Climate Change and India: Vulnerability Assessment and Adaptation*. Hyderabad, India: Universities Press Pvt, Ltd.

Kopp, R., R. D. Morgenstern, and W. A. Pizer. 2000. Limiting cost, assuring effort, and encouraging ratification: Compliance under the Kyoto Protocol. CIRED/RFF Workshop on Compliance and Supplemental Framework ⟨http://www.weathervane.rff.org/features/parsconf0721/summary.html⟩.

KRCL. 1999. *Treatise on Konkan Railway*. Mumbai: Konkan Railway Corporation, Ltd.

Lecocq, F., and R. Crassous. 2003. International climate regime beyond 2012. Are quota allocation rules robust to uncertainty? World Bank Policy Research working paper 3000. World Bank, Washington, DC.

Mathy, S., J. C. Hourcade, and C. de Gouvello. 2001. Clean development mechanism: Leverage for development? *Climate Policy* 1(2): 251–68.

Menon-Choudhary, D., P. R. Shukla, T. Nag, and D. Biswas. 2004. Electricity reforms, firm level responses and environmental implications. In J. Ruet, and P. Kalra, eds., *Studies on Electricity Reforms in India*. New Delhi: Manohar Publishers, forthcoming.

Meyer, A. 2000. Contraction and convergence. The global solution to climate change. Schumacher briefing 5. The Schumacher Society, Bristol.

Mitra, A. P., D. Kumar, K. Rupa Kumar, Y. P. Abrol, K. Naveen, M. Velayuthan, and S. W. A. Naqvi. 2002. Global change and biogeochemical cycles: The South Asia region. In P. D. Tyson, R. Fuchs, C. Fu, L. Lebel, A. P. Mitra, E. Odada, J. Perry, W. Steffen, and H. Virji, eds., *The Earth System: Global Regional Linkages*. Berlin: Springer, pp. 75–108.

Nair, R., P. R. Shukla, M. Kapshe, A. Garg, and A. Rana. 2003. Analysis of long-term energy and carbon emission: Scenarios for India. *Mitigation and Adaptation Strategies for Global Change* 8: 53–69.

Neumayer, E. 2002. Can natural factors explain any cross-country differences in carbon dioxide emissions? *Energy Policy* 30: 7–12.

Nordhaus, W. D. 1994. *Managing Commons*. Cambridge: MIT Press.

Pandey, R., and P. R. Shukla. 2003. Methodology for exploring co-benefits of CO_2 and SO_2 mitigation policies in India using AIM/end use model. In M. Kainuma, Y. Matsuoka, and T. Morita, eds., *Climate Policy Assessment: Asia–Pacific Integrated Modeling*. Berlin: Springer, pp. 113–22.

Philibert, C., and J. Pershing. 2002. *Beyond Kyoto, Energy Dynamics and Climate Stabilisation*. Paris: OECD/AIE.

Planning Commission. 2002. *Report of the Committee on India Vision 2020*. New Delhi: Government of India.

Phylipsen, G. J. M., J. W. Bode, and K. Blok. 1998. A triptych approach to burden differentiation: GHG emissions in the European bubble. *Energy Policy* 26(12): 929–43.

Quirion, P., and J.-C. Hourcade. 2004. Does the CO_2 emission trading directive threaten the competitiveness of European industry? Quantification and comparison to exchange rates fluctuations. European Association of Environmental and Resource Economists Annual Conference, Budapest, June.

Roberts, M. I., and M. Spence. 1976. Effluent charges and licences under uncertainty. *Journal of Public Economics* 5: 193–208.

Sagar, A. D. 2005. Alleviating energy poverty for the world's poor. *Energy Policy* 33: 1367–72.

Schlessinger, J. R. 1989. Energy and geopolitics in the 21st century. World Energy Conference 14th Congress Montreal, Quebec.

Shukla, P. R., T. Heller, D. Victor, D. Biswas, T. Nag, and A. Yajnik. 2004. *Electricity Reforms in India: Firm Choices and Emerging Generation Markets*. New Delhi: Tata McGraw-Hill.

Sterner, T. 2002. Policy instruments for environmental and natural resource management. RFF, World Bank, SIDA, Washington, DC.

Stiglitz, J. E. 1998. The private uses of public interests: Incentives and institutions. *Journal of Economic Perspectives* 12(2): 3–22.

Tinbergen, J. 1976. *Reshaping the International Order*. Report to the Club of Rome. New York: Duttow.

UNFCCC. 1997. Proposed elements of a protocol to the United Nations framework. Convention on Climate Change. Presented by Brazil in response to the Berlin Mandate. AGBM/1997/misc.1/add.3.

Victor, D. G., J. House, and S. Joy. 2005. A Madisonian approach to climate policy. *Science* 309 (September).

II Stability of Outcomes

6

Transfer Schemes and Institutional Changes for Sustainable Global Climate Treaties

Johan Eyckmans and Michael Finus

Voluntary provision of public goods is a well-known problem in economics (for references, see Cornes and Sandler 1996). Nonexcludability from positive externalities of public goods leads to underprovision of public goods by private entities. In the national context, governments can mitigate this problem. They can provide the appropriate levels of public goods with financial resources from taxation. However, in the international context, this is more difficult because no "world government" exists to take on this role. Consequently international treaties have to rely on voluntary participation and must be designed in a self-enforcing way. In the presence of free-rider incentives, this frequently means that not all countries participate in an international environmental treaty and/or the agreed level of public provision only marginally exceeds noncooperative levels (e.g., Böhringer and Vogt 2004; Finus and Tjøtta 2003; Murdoch and Sandler 1997a, b). The mitigation of global warming exemplifies this problem. In 1997 the Kyoto Protocol was signed by the United States (but never officially ratified) after more than ten years of difficult negotiations. This treaty aimed at reducing global greenhouse gas emissions by 5.2 percent compared to the 1990 levels by 2008 through 2012—which is far below what would be advisable according to cost-benefit analyses.[1] Even worse, the United States declared in 2001 that they were withdrawing from the Protocol In the aftermath of this decision, other signatories started demanding reductions in their previously accepted moderate abatement targets. Only eight years after the conference in Kyoto, the Kyoto Protocol finally entered into force after official ratification by Russia.

From the bumpy road toward an international climate agreement, it is evident that there are some fundamental characteristics associated with climate change that makes this environmental problem more difficult to solve than other transboundary pollution problems. This chapter has two purposes: first, to shed light on some fundamental forces that hamper successful treaty-making in the context of greenhouse gas mitigation and, second, to consider measures to improve the success of self-enforcing climate treaties. For this purpose we combine two modules in our analysis: one is an integrated assessment model that captures the feedback between the economy and the environmental damages to the climate system, and the other is a noncooperative game-theoretic model that determines stable coalitions in the presence of free-riding incentives.

The first measure that we consider to enhance the success of global climate treaties is transfers aimed at balancing strong asymmetries among the actors involved in climate change. We consider twelve transfer schemes. Many of these transfer schemes are related to the literature that pays much attention to the philosophical and moral motivation of various schemes (Rose and Stevens 1998; Rose et al. 1998; Stevens and Rose 2002). Most of these papers, however, focus on the welfare effects in terms of abatement costs, giving an incomplete picture of the incentives to form self-enforcing agreements. The game-theoretical literature, on the other hand, focuses on the incentive aspects but often has to make strong simplifying assumptions in order to keep the analysis tractable. Most papers assume symmetric players, rendering the analysis of transfers uninteresting (e.g., Carraro and Siniscalco 1993; Barrett 1994). Some papers consider asymmetric players but impose a very particular form of asymmetry (e.g., two types of countries) or base their analysis on a simplified climate model that captures in an incomplete way the stock externality aspect of climate change (e.g., Barrett 1997, 2001; Botteon and Carraro 1997; Chander and Tulkens 1997; Hoel and Schneider 1997; Weikard et al. 2006). Notable exceptions are Eyckmans and Tulkens (2003), Germain and van Steenberghe (2003), and Germain et al. (2003) who use a more sophisticated dynamic stock externality model of greenhouse gas emissions and concentrations. Moreover most papers consider only a small portfolio of transfer rules and these rules are mainly related to stylized solutions of bargaining theory (e.g., Barrett 1997; Botteon and Carraro 1997; Germain et al. 1998).[2] Therefore it is one of our main objectives in this chapter to analyze a large variety of transfer schemes and to study in a systematic way their effects on coalition formation and stability based on a full-fledged integrated assessment model with very asymmetric players.

Our second objective is to study what change of institutional rules would make it more difficult to upset the stability of a treaty. We contrast open membership, as is typical for public goods, with exclusive membership, as is typical for club goods. For this purpose we consider a modification of the concept of internal and external stability by d'Aspremont et al. (1983) as has frequently been applied in the noncooperative game-theoretic analyses of international environmental agreements (e.g., Barrett 1994; Carraro and Siniscalco 1993; Hoel 1992). We were inspired by the recent literature in this field, finding that exclusive membership may be conducive to the stability of treaties (e.g., Carraro 2000; Finus 2003; Finus and Rundshagen 2003). However, so far no sound evidence can be presented on the impact of "more stability" on welfare and the environment; this would require simulations and a departure from the assumption of symmetric players as well as a static payoff structure.

6.1 Integrated Assessment Model

The CLIMNEG world simulation model (in the sequel referred to as CWS model) is an integrated assessment, economy-climate model that resembles closely the seminal RICE family of models; see Nordhaus and Yang (1996) and Nordhaus and Boyer (2000).[3] The CWS model captures the endogenous feedback of climate change damages on production and consumption possibilities. The economic part of the CWS model consists of a long-term dynamic perfect foresight Ramsey type of optimal growth model with endogenous investment and carbon emission reduction decisions. The environmental part consists of a carbon cycle and temperature change module.

In the CWS model the world is divided into six regions: USA, JPN (Japan), EU (European Union), CHN (China), FSU (former Soviet Union), and ROW (rest of the world). In every region i, and at every time period t, the following budget equation describes how "potential GDP," $Y_{i,t}$, can be allocated to consumption, $Z_{i,t}$, investment, $I_{i,t}$, emission abatement costs, $Y_{i,t}C_i(\mu_{i,t})$, and climate change damages, $Y_{i,t}D_i(\Delta T_t)$:

$$Y_{i,t} = Z_{i,t} + I_{i,t} + Y_{i,t}C_i(\mu_{i,t}) + Y_{i,t}D_i(\Delta T_t). \tag{6.1}$$

Output $Y_{i,t}$ is produced with capital and labor. Capital is built up through investment, and it depreciates at some fixed rate. Labor supply is assumed to be inelastic. Therefore investment $I_{i,t}$ is the only endogenous production input, and it constitutes the first choice variable in the model.

Abatement costs $Y_{i,t}C_i(\mu_{i,t})$ are expressed as "loss of potential GDP": C_i is the share of "potential GDP" devoted to abatement, which is a function of $\mu_{i,t} \in [0,1]$, measuring the relative emission reduction compared to the business-as-usual scenario without any abatement policy. $\mu_{i,t}$ is the second choice variable in the model. Damages $Y_{i,t}D_i(\Delta T_t)$ are also expressed as "loss of potential GDP": D_i is the share of "potential GDP" destroyed by climate change damages and is a function of temperature change ΔT_t. Temperature change depends on the stock of greenhouse gases, which in turn depends on emissions that accumulate in the atmosphere. Finally, emissions depend on emissions released through production minus abatement. Hence the second choice variable in this model is the abatement level $\mu_{i,t}$.

Both choice variables (investment and abatement) affect output, abatement costs, damage costs, and therefore also consumption—not only domestically but also abroad. For abatement this is immediately evident as remaining emissions (after abatement) increase the stock of greenhouse gases and effect environmental damage in every country. However, this is also true for investment, since capital is an input in the production process and atmospheric emissions are proportional to production. Technological progress is captured by the CWS model but only exogenous where the time path is taken from RICE. New technology that increases production possibilities and energy efficiency thus would decrease the emission output ratio over time.

We measure welfare of a region i as total lifetime discounted consumption:

$$W_i(s) = \sum_{t=0}^{\Omega} \frac{Z_{i,t}}{[1+\rho_i]^t}, \tag{6.2}$$

where ρ_i stands for the discount rate of region i, Ω denotes the time horizon, and s is a strategy vector. Vector $s = \{I_{i,t}, \mu_{i,t}\}_{i \in N; t=0,\ldots,\Omega}$ consists of a very long time path of 35 decades[4] for emission abatement and investment for all six regions and hence is of length $2 \times 35 \times 6 = 420$.

The version of the CWS model used in this chapter is an updated version of the model used in Eyckmans and Tulkens (2003). The new version of the CWS model uses the more sophisticated carbon cycle model of RICE99 described in Nordhaus and Boyer (2000). In addition to a better representation of the climate system, the economic database and parameters of the CWS model have been updated and the reference year is now 2000 instead of 1990. As is evident from the input data displayed in the appendix, we assume a relatively low discount rate of 1.5 percent, except for CHN and ROW where we assume the rate to be 3 percent so as to account for the generally accepted view that policy makers in rapidly developing countries are more "impatient" compared to their counterparts in developed countries. These discount rates seem roughly in line with Weitzman (2001), who suggests that in the context of global warming a constant discount rate of 2 percent (or less) is appropriate to capture long-term effects.

In the coalition formation analysis below we relate, where possible, our results to the basic cost and damage characteristics of the different regions. To do so, we first draw a clear picture of the cost–benefit ratios of the different regions in the CWS model. We compute these ratios for the reference year 2000 for every region with a corresponding marginal cost for a 10 percent reduction in greenhouse gas emissions and a marginal climate change damage estimate for a 5 degree Celcius temperature increase. These marginal abatement cost and damage estimates are reported in table 6.2 of the appendix (as marginal cost and marginal damage indexes). The parameters of the model imply that JPN and EU face steep abatement costs while and CHN and ROW get by with flat marginal abatement costs. USA and ROW are characterized by intermediate marginal cost estimates. The regional differences in abatement costs mainly reflect differences in energy efficiency. That is, energy-efficient regions face higher costs when cutting back emissions than regions characterized by low energy efficiency. Finally, damage functions are particularly steep in EU, USA, and ROW, to a lesser extent in JPN, but are relatively flat in FSU and CHN. The high damage estimate (as a percentage of potential GDP) for ROW is because the climate change is believed to affect developing countries more radically than industrialized countries as their agriculture, fishery, and forestry economies tend to depend more on climate-related production processes (IPCC 2001). The low damage estimate for FSU is due to some expected benefits from moderate temperature increase such as an expansion of arable land.

6.2 Implications of Coalition Formation

In this section we discuss some fundamental features of coalition formation, excluding stability as this is dealt with in a later section.

6.2.1 No Cooperation and Full Cooperation

We consider two benchmarks for our argument: no cooperation and full cooperation. *No cooperation* means that each region maximizes only its lifetime consumption and only with respect to its particular strategy vector regardless of the strategies of all other regions. The result is a noncooperative or Nash equilibrium strategy vector s^N. In terms of the concept of coalition formation, this situation can be treated as each region forming a singleton of a coalition structure $c^N = \{\{1\}, \{2\}, \ldots, \{n\}\}$. In contrast, under *full cooperation*, regions choose abatement and investment levels in order to maximize the sum of lifetime consumption of all countries in the world. This way every region's emissions externalities are fully internalized; that is to say, the optimal abatement and investment strategies take into account spillover effects to all other regions. The result is a fully cooperative or socially optimal strategy vector s^F. From the standpoint of coalition formation, this situation can be treated as all regions forming a coalition, or a grand coalition of structure $c^F = \{N\}$ with N being the set of all players.

Because emissions constitute a negative externality,[5] strategy vector s^F differs from strategy vector s^N (Müller-Fürstenberger and Stephan 1997). Global emission abatement is higher in the social optimum than in Nash equilibrium. However, also in the Nash equilibrium some abatement is undertaken unilaterally since—compared to the business as usual scenario (BAU)—at least national damages are taken in consideration when a region chooses a noncooperative strategy vector s^N.

The difference between the Nash equilibrium and the social optimum shows up in very different global emissions over time. Whereas Nash equilibrium emissions grow steadily with economic growth, socially optimal emissions rise only until 2100, level off, and then tend to decrease afterward. For instance, in the year 2200, global emissions in the social optimum (16 GtC; gigatons of carbon) are only 43 percent of the emissions in Nash equilibrium (37 GtC). By the year 2300, global accumulated emissions from the year 2000 onward amount to 10,346 GtC in Nash equilibrium but only to 5,775 GtC in the social optimum, implying a reduction by 45 percent. This difference also shows up in the development of carbon concentration. Starting from an atmospheric carbon concentration of 810 GtC in 1990, the Nash equilibrium concentration rises steadily and reaches 3,708 GtC in 2300. By contrast, the socially optimal concentration grows at a much slower rate, reaching a value of 1,967 GtC in 2300. The socially optimal atmospheric carbon concentrations amount to only 53 percent of the Nash equilibrium concentration level. We can conclude, in general, that ecological indicators like emissions and concentration levels are halved when countries shift from no cooperation to full cooperation.

However, in terms of global welfare (i.e., discounted lifetime world consumption; see table 6.1 for details) the difference between full cooperation (3,859.0 trillion US$) and no cooperation (3,846.7 trillion US$) is not so pronounced. Although the difference is not small in absolute magnitude (approximately 12.3 trillion US$), it appears not so big in relative terms (0.32 percent). As was argued already by Eyckmans and Tulkens (2003), the reason for this small relative difference is that the substantial abatement efforts required initially in the social optimum are only matched by lower damages in later periods, given the stock pollutant nature of greenhouse gases. Further, in economic terms, the effect of full cooperation compared to no cooperation is large. However, because the negative externalities of carbon emissions occur mainly in the future and receive less weight due to discounting, the effect turns out to be relatively small when global welfare is measured as net present value.

Finally, it is not because global welfare under full cooperation is higher that this scenario consititues a Pareto improvement over no cooperation in the absence of transfers. In our model we observe that the major winners from full cooperation are USA (+8.4 trillion US$), EU (+5.8 trillion US$) and JPN (+1.4 trillion US$). Also FSU (+0.5 trillion US$) gains but ROW (−2.7 trillion US$) and CHN (−1.1 trillion US$) loose substantially. The reason is USA, JPN, and EU have to contribute below average to socially optimal abatement because of steep marginal abatement cost curves, but the benefit is above average in each case because of the steep marginal damage cost curves. By contrast, CHN and ROW face just the opposite incentive structure, and therefore in the social optimum, ROW and CHN have to contribute well above average to joint abatement and the benefit to them is relatively small because of the flat marginal abatement costs and flat marginal damage cost curves.

6.2.2 Partial Cooperation

In this subsection we consider intermediate steps between no cooperation and full cooperation. *Partial cooperation* means that a subgroup of regions—at least two regions but not all regions—forms a coalition.

Members of such a subgroup are assumed to maximize the sum of their members' lifetime consumption while taking as given the strategies of the nonmember regions. That is, coalition members coordinate their choice of abatement and investment strategies by taking into account the spillover effect on fellow members but ignoring the effect on outsiders. Outsiders are assumed to act as singletons, as described in the noncooperative equilibrium scheme. Hence the partially cooperative equilibrium strategy vector s^P can be interpreted as partial Nash equilibrium between the coalition and outsiders (see Chander and Tulkens 1995, 1997). The associated coalition structure is $c^P = \{S, \{k\}, \ldots, \{n\}\}$, with S being a nontrivial coalition (i.e., a coalition of at least two regions), and all regions that do not belong to S are singletons.[6]

Table 6.1 displays a selection of coalition structures, the payoffs of the different countries (without transfers among coalition members), and the associated welfare and ecological implications. Overall, among the six players the total number of coalition structures is 58: one coalition structure with no cooperation c^N, one coalition structure with full cooperation c^F, and 56 coalition structures with partial cooperation c^P. There are 15 coalition structures with a coalition of 2 members, 20 with a coalition of 3 members, 15 with a coalition of 4 members, and 6 with a coalition of 5 members. The coalition structures listed in table 6.1 are sorted in descending order of global welfare. Welfare, concentration, and emissions are expressed as a closing the gap index, that is, by how much they close the gap between the social optimum and Nash equilibrium in relative terms (see the legend of table 6.1 for details).

From table 6.1 it is evident that all coalition structures, which exclude the singletons, generate higher global welfare (and lower global emissions and concentration). This is due to two important properties of coalition formation that apply to our model. The first property is called superadditivity. *Superadditivity* means that if a region joins a singleton or a coalition, the *aggregate* welfare of all regions involved in the merger increases. In other words, there is a coalitional gain from cooperation. Superadditivity stems from the fact that coalition members can always implement the strategies they would have chosen before the merger. Hence cooperation cannot but increase the joint payoffs. The second property is called positive externality. *Positive externality* means that if a region joins a singleton or a coalition to form a bigger coalition, all outsiders that are not involved in this merger benefit from the merger. The positive externality effect can be decomposed into two effects. First, global abatement will increase after the merger because the new coalition will produce more abatement. Although outsiders to the merger will respond to the coalition's increase of abatement by reducing their own abatement effort, this response is less than proportional. The resulting net increase of global abatement is beneficial to all players because it reduces climate change damages. Second, outsiders to the merger incur lower abatement costs because they reduce their abatement efforts in response to the higher coalitional effort.

It also appears from table 6.1 that the identity of members in partial cooperation matters more than the number of participants for the success of cooperation. Put differently, the sometimes held view that a high participation indicates success of an IEA proves to be wrong. For instance, the coalition including the five members USA, JPN, EU, CHN, and FSU ranks much lower (position 32 to be precise) than many coalitions of only three or four members and even lower than some coalitions with only two members: {JPN, ROW} at position 30 or {FSU, ROW} at position 31.

From a quick glance at the 15 highest ranked coalition structures, it is evident that participation of CHN and ROW is crucial for achieving a high score on the welfare indicator. ROW's and CHN's important role stems from the fact that they can provide cheap abatement. However, there is also an additional dimension related to environmental damages. In a given coalition the higher the marginal damages of a coalition member, the higher are the joint abatement efforts, all else being equal. This explains the relative importance of EU and USA for effective (i.e., welfare-improving) cooperation. Together these two arguments clearly show that effective climate policy requires the cooperation of both industrialized and developing countries. This explains why the coalition comprising USA, JPN, EU, and FSU, which we labeled "old Kyoto," ranks relatively low (position 45), as the two key low-cost players—CHN and ROW—are outsiders. A similar conclusion can be made about the coalition we labeled "new Kyoto" after the withdrawal of USA (position 52). This decision led to a dramatic drop in welfare and ecological variables, to almost noncooperative levels. Thus our model provides evidence that supports the efforts of many governments and nongovernmental organizations to induce the United States to rejoin the Kyoto Protocol. However, it also supports the US concern, and that of many other countries, that an effective climate policy should include developing countries.

Table 6.1
Welfare and ecological implications of different coalition structures

Rank	Composition	Size	USA	JPN	EU	CHN	FSU	ROW	WORLD	CGI-W	CGI-E	CGI-M
1	Grand coalition (full cooperation)	6	1,487.37	307.22	1,087.48	359.66	97.74	519.55	3,859.02	100.00	100.00	100.00
2	USA, EU, CHN, FSU, ROW	5	1,486.98	307.23	1,087.19	359.85	97.72	519.90	3,858.87	98.70	94.90	94.70
3	USA, JPN, EU, CHN, ROW	5	1,487.08	307.17	1,087.27	359.73	97.86	519.69	3,858.80	98.20	96.70	96.80
4	USA, EU, CHN, ROW	4	1,486.68	307.18	1,086.98	359.92	97.83	520.05	3,858.63	96.80	91.50	91.70
5	USA, JPN, CHN, FSU, ROW	5	1,485.69	306.93	1,086.39	360.43	97.64	521.08	3,858.16	93.00	77.30	77.60
6	USA, JPN, CHN, ROW	4	1,485.39	306.88	1,086.17	360.48	97.71	521.22	3,857.84	90.40	73.60	75.00
7	USA, CHN, FSU, ROW	4	1,485.23	306.90	1,086.03	360.57	97.62	521.39	3,857.73	89.50	71.50	72.40
8	JPN, EU, CHN, FSU, ROW	5	1,484.98	306.78	1,085.61	360.71	97.59	521.68	3,857.34	86.30	65.90	67.20
9	USA, CHN, ROW	3	1,484.92	306.84	1,085.80	360.61	97.67	521.50	3,857.33	86.30	68.70	68.90
10	JPN, EU, CHN, ROW	4	1,484.64	306.73	1,085.39	360.74	97.63	521.78	3,856.91	82.80	63.40	63.70
11	EU, CHN, FSU, ROW	4	1,484.43	306.73	1,085.26	360.82	97.56	521.92	3,856.71	81.20	59.60	61.20
12	EU, CHN, ROW	3	1,484.09	306.68	1,085.03	360.85	97.59	522.00	3,856.23	77.30	56.90	57.60
13	USA, JPN, EU, FSU, ROW	5	1,484.72	306.79	1,085.64	361.76	97.56	519.41	3,855.88	74.50	74.40	71.80
14	USA, EU, FSU, ROW	4	1,484.44	306.80	1,085.43	361.69	97.55	519.79	3,855.70	73.00	70.50	68.00
15	USA, JPN, EU, ROW	4	1,484.49	306.75	1,085.48	361.71	97.66	519.56	3,855.65	72.60	71.80	69.30
...												
30	JPN, ROW	2	1,481.07	306.19	1,083.05	361.13	97.36	522.43	3,851.23	36.70	23.40	24.00
31	FSU, ROW	2	1,480.83	306.15	1,082.88	361.09	97.33	522.46	3,850.76	32.80	19.80	20.90
32	USA, JPN, EU, CHN, FSU	5	1,481.33	306.24	1,083.33	359.27	97.29	522.93	3,850.40	29.90	24.80	22.90
...												

	Membership	Size										
45	USA, JPN, EU, FSU ("old" Kyoto)	4	1,481.18	306.21	1,083.13	361.07	97.36	522.55	3,851.49	5.90	2.90	4.70
...												
52	JPN, EU, FSU ("new" Kyoto)	3	1,483.46	306.59	1,084.74	361.41	97.53	521.48	3,855.21	2.30	0.60	1.10
...												
57	JPN, FSU	2	1,479.04	305.87	1,081.67	360.80	97.21	522.21	3,846.78	0.50	0.10	0.20
58	Only singletons (no cooperation)	1	1,479.01	305.86	1,081.65	360.79	97.21	522.20	3,846.71	0.00	0.00	0.00

Note: Size: enrollment size of coalition S in coalition structure $c = \{S, \{k\}, \ldots, \{n\}\}$. Membership: only members in coalition S are listed. Welfare: global welfare expressed in relative terms: $(\sum_{i=1}^{N}(W_i(c^P) - W_i(c^N)))/(\sum_{i=1}^{N}(W_i(c^F) - W_i(c^N)))$, where W_i is discounted lifetime consumption integrated over 1990 to 2300, global welfare with full cooperation is $\sum_{i=1}^{N} w_i(c^F) = 3,846.71$ trill US\$$_{1990}$ (trillion US dollars expressed in 1990 levels), global welfare with no cooperation is $\sum_{i=1}^{N} w_i(c^N) = 3,859.02$ trill US\$$_{1990}$ and global welfare with partial cooperation is denoted by $\sum_{i=1}^{N} w_i(c^P)$. Concentration: atmospheric carbon concentration M at time $t = 2300$ expressed in relative terms: $(M(c^N) - M(c^P))/(M(c^N) - M(c^F))$, where concentration with full cooperation is $M(c^F) = 1967.025$ GtC (gigatonnes carbon), concentration with no cooperation is $M(c^N) = 3707.538$ GtC, and concentration with partial cooperation is denoted by $M(c^P)$. Emissions: total accumulated emissions over 1990 to 2300 at time $t = 2300$ expressed in relative terms: $(\sum_{i=1}^{N}(E_i(c^N) - E_i(c^P)))/(\sum_{i=1}^{N}(E_i(c^N) - E_i(c^F)))$, where emissions with full cooperation are $\sum_{i=1}^{N} E_i(c^F) = 5,775$ GtC, emissions with no cooperation are $\sum_{i=1}^{N} E_i(c^N) = 10,346$ GtC, and emissions with partial cooperation are denoted by $\sum_{i=1}^{N} E_i(c^P)$.

At the level of individual regions, all single regions k are better off in every partially cooperative coalition structure $c^P = \{S, \{k\}, \ldots, \{n\}\}$ than in the singleton coalition structure $c^N = \{\{i\}, \ldots, \{j\}, \{k\}, \ldots, \{n\}\}$. This is due to the positive externality property mentioned above, which also implies that the most favorable condition for a singleton is if all other regions form a coalition. It is this property that makes free-riding attractive and that makes it difficult to form large stable coalitions as we will demonstrate in section 6.3.

For regions that are members of a coalition, things are different. Although coalitions can reap substantial gains from cooperation due to superadditivity, individual members may be worse off than in the singleton coalition structure c^N. This was earlier illustrated for the full cooperation scheme in subsection 6.2.1, and the same appears to be true for partial cooperation. In fact only 10 out of 56 coalition structures that constitute partial cooperation are *individually profitable*, that is, imply a gain from cooperation to *all* members. None of the top 15 ranked coalition structures in table 6.1 are individually profitable. No coalition with five members and only one with four members is individually profitable. This finding suggests that without transfers, even moderate partial cooperation will prove very difficult.

6.3 Stability of Coalition Formation

6.3.1 *A First Approach*

In this subsection we have a first look at the stability of coalition structures. A necessary condition for stability is individual profitability: a region can always remain a singleton, which gives it at least as much than in the singleton coalition structure because of the positive externality property. However, even if individual profitability holds, a coalition member may have an incentive to leave its coalition. This is not the case provided the following condition holds:[7] internal stability:

$$W_i(\{S, \{k\}, \ldots, \{n\}\}) \geq W_i(\{S\setminus\{i\}, \{i\}, \{k\}, \ldots, \{n\}\}) \quad \forall i \in S. \quad (6.3)$$

If a coalition member i leaves coalition S to become a singleton, it saves abatement costs. However, not only the deviator will reduce its abatement effort but also the remaining regions in coalition $S\setminus\{i\}$ will abate less, leading to an increase of damages in every region. The relative importance of both welfare effects determines the incentive for a region i to remain in or to leave coalition S. As a tendency, given a coalition S, the more regions join S, the higher the incentive of current members to leave their coalition will be. The reason is that more members mean higher abatement effort and hence higher abatement costs and lower damages. Hence the incentive to leave a coalition increases gradually due to the convexity of abatement cost functions and the concavity of benefits from reduced damages. However, there is also a second dimension of stability: external stability:

$$W_j(\{S, \{j\}, \{k\}, \ldots, \{n\}\}) \geq W_j(\{S \cup \{j\}, \{k\}, \ldots, \{n\}\}) \quad \forall j \notin S. \quad (6.4)$$

External stability is the mirror image of internal stability: no singleton should have an incentive to join coalition S. The advantage of joining is that damages decrease: global abatement increases and, in particular, one's own efforts are matched by those of other members. However, higher abatement means also higher abatement costs. Again, the relative importance of both welfare effects determines the incentive of joining coalition S or remaining an outsider. For the same reason as mentioned above, as a tendency, the more regions already joined in coalition S, the less attractive it becomes to follow suit.

Testing for stability, it turns out that only one of the 10 individually profitable coalition structures that qualify as potential candidates is stable. Although 7 coalitions are internally stable, only the couple {USA, EU} is also externally stable. However, this combination comes only at position 50 according to the welfare indicator. Hence partial cooperation—at least without transfers and/or change of institutional rules—leads to very few stable coalitions and only small welfare gains compared to no cooperation.

It is interesting to observe that both USA and EU have very similar marginal cost–benefit ratios (see the marginal cost and marginal damage indexes in table 6.2). Hence the only stable coalition without transfers is a couple of regions that are very similar in terms of global warming cost and benefit parameters. Although it is dangerous to make general claims on the basis of only one observation, we believe this illustrates nicely the difficulty in the formation of effective climate agreements. On the one hand, we saw in section 6.2.2 that one needs a diversity of regions combining low-cost abatement producers with high climate change damage (i.e., high benefits from emission abatement) regions in order to achieve substantial welfare improvements. On the other hand, such diverse interests create instability in the absence of transfers. Therefore combinations of regions with very different cost–benefit ratios are very unlikely in the absence of transfers.

Table 6.2 Background information of scenarios

		USA	JPN	EU	CHN	FSU	ROW	Total
Base data	GDP_i	7,564	3,388	8,447	969	558	6,633	27,559
	POP_i	282	127	377	1,263	288	3,716	6,052
	E_i	1.574	0.330	0.888	0.947	0.626	2.192	6.556
Ratios	GDP_i/POP_i	26,823	26,677	22,406	767	1,938	1,785	4,554
	E_i/POP_i	5,582	2,598	2,355	750	2,174	590	1,083
	E_i/GDP_i	208	97	105	977	1,122	330	238
Marginal cost index	MC_i	0.18	0.27	0.25	0.08	0.07	0.16	1.00
Marginal damage index	MD_i	0.27	0.13	0.33	0.02	0.02	0.23	1.00
Scenario 0: No transfers	λ_i	—	—	—	—	—	—	—
	t_i	—	—	—	—	—	—	—
Scenario 1: Egalitarian $\tilde{\lambda}_i = POP_i$	λ_i	0.05	0.02	0.06	0.21	0.05	0.61	1.00
	t_i	−7.79	−1.10	−5.07	+3.70	+0.05	+10.21	+13.96
Scenario 2: Right to pollute $\tilde{\lambda}_i = [E_i/POP_i]^{-1}$	λ_i	0.04	0.09	0.09	0.30	0.10	0.38	1.00
	t_i	−7.87	−0.30	−4.66	+4.79	−0.73	+7.31	+12.10
Scenario 3: Ability to pay ($\eta = 1$) $\tilde{\lambda}_i = [GDP_i/POP_i]^{-\eta}$	λ_i	0.01	0.01	0.02	0.52	0.21	0.22	1.00
	t_i	−8.18	−1.17	−5.61	+7.55	+2.00	+5.41	+14.96
Scenario 4: Ability to pay ($\eta = 10$) $\tilde{\lambda}_i = [GDP_i/POP_i]^{-\eta}$	λ_i	0.00	0.00	0.00	1.00	0.00	0.00	1.00
	t_i	−8.36	−1.36	−5.83	+13.43	−0.54	+2.65	+16.08
Scenario 5: Ecological subsidy $\tilde{\lambda}_i = E_i/GDP_i$	λ_i	0.06	0.03	0.03	0.46	0.32	0.09	1.00
	t_i	−5.24	−2.69	−8.67	+11.05	+3.35	+2.19	+16.59

Scenario 6: Status quo	$\tilde{\lambda}_i = E_i/POP_i$		0.36	0.16	0.16	0.04	0.25	0.04	1.00
	t_i		+0.05	−0.36	−6.44	+3.54	+2.04	+1.17	+6.80
Scenario 7 MC × MD	$\tilde{\lambda}_i = MC_i \times MD_i$		0.24	0.17	0.40	0.01	0.01	0.18	1.00
	t_i		−5.45	+0.77	−0.93	+1.24	−0.47	+4.83	+6.85
Scenario 8 inverse MC × MD	$\tilde{\lambda}_i = [MC_i \times MD_i]^{-1}$		0.01	0.02	0.01	0.34	0.60	0.02	1.00
	t_i		−8.19	−1.12	−5.73	+5.31	+6.86	+2.88	+15.05
Scenario 9 MC/MD	$\tilde{\lambda}_i = MC_i/MD_i$		0.05	0.17	0.06	0.28	0.37	0.06	1.00
	t_i		−7.70	+0.77	−5.04	+4.55	+4.06	+3.37	+12.75
Scenario 10 inverse MC/MD	$\tilde{\lambda}_i = MD_i/MC_i$		0.29	0.09	0.25	0.06	0.04	0.27	1.00
	t_i		−4.76	−0.23	−2.80	+1.83	−0.02	+5.98	+7.81
Scenario 11: Chander-Tulkens	$\tilde{\lambda}_i = MD_i^F$		0.37	0.07	0.27	0.10	0.02	0.17	1.00
	t_i		−3.81	−0.51	−2.56	+2.42	−0.27	+4.74	+7.16

Note: Indicators are calculated using base data in the appendix with $GDP_i = Y_i^0$ (gross domestic product), $POP_i = L_i$ (population), and $E_i = E_i^0$ (emissions). Emissions per capita (E_i/POP_i) are measured in kilogram carbon per capita, GDP per capita (GDP_i/POP_i) in US$ per capita and emission intensity of GDP (E_i/GDP_i) in gram carbon per US$. In order to compute the marginal cost and damage indexes, we computed first $\widetilde{MC}_i = [Y_i^0/\alpha_{i,2000}]^{b_{i,1}} b_{i,2} [0.10]^{b_{i,2}-1}$ and $\widetilde{MD}_i = [Y_i^0/p_i]\theta_{i,1}\theta_{i,2}[2]^{\theta_{i,2}-1}$ and normalized these estimates such that $MC_i = \widetilde{MC}_i/\sum_{j \in N} MC_j$ and $MD_i = \widetilde{MD}_i/\sum_{j \in N} MD_j$. λ_i is the weight of the transfer scheme (see section 6.3.2), t_i is the transfer (trillion US$), "total" for transfers, in the last column, refers to the sum of all positive transfers. Finally, the weights MD_i^F used for the Chander-Tulkens scenario are based on the discounted stream of marginal climate change damages from 2000 to 2300 for the full cooperation scenario of the CWS model.

6.3.2 A Second Approach: Transfers

Preliminaries

In the previous discussion it became clear that individual profitability is a necessary condition for stability but that this requirement is frequently violated because of large asymmetries between regions. Chander and Tulkens (1995, 1997) suggested a way out of this dilemma by using a transfer scheme that guarantees every country at least its no-cooperation payoff. For instance, consider a coalition structure $c = \{S, \{k\}, \ldots, \{n\}\}$ and assume that every coalition member $i \in S$ receives additionally to its welfare without transfers $W_i(c)$ a transfer $t_i(c)$ of the following type:

$$t_i(c) = [W_i(c^N) - W_i(c)] + \lambda_i \sum_{j \in S}[W_j(c) - W_j(c^N)], \qquad (6.5)$$

where $W_i(c^N)$ is welfare in the singleton coalition structure and $W_i(c)$ is welfare without transfers in a coalition structure c that may be partial cooperation ($c = c^P$; $2 < \#S < 6$) or full cooperation ($c = c^F$; $\#S = 6$). The first term on the right-hand side in (6.5) puts everyone back to its no-cooperation payoff, the second allocates the aggregate gain to the coalition from cooperation to its members where λ_i is a weight, $1 \geq \lambda_i \geq 0$, $\sum_{i \in S} \lambda_i = 1$. Substituting (6.5) into $\hat{W}_i(c) = W_i(c) + t_i(c)$ gives welfare with transfers:[8]

$$\hat{W}_i(c) = W_i(c^N) + \lambda_i \sum_{j \in S}[W_j(c) - W_j(c^N)]. \qquad (6.6)$$

Because $\sum_{j \in S}[W_j(c) - W_j(c^N)] > 0$ is true due to superadditivity, individual profitability holds for every member in S. Note that (6.6) can be interpreted as the outcome of a general Nash bargaining solution with different weights. Typically in the game-theoretic literature, these weights are interpreted as bargaining power. In contrast, in the environmental literature, weights are usually derived from various moral concepts and therefore have a normative flavor. Below, we consider several alternative formulas for the weights in (6.6).

Motivation and Fundamental Features

As mentioned at the start of this chapter, some of the transfer schemes considered here are closely related to the literature that has given much attention to their moral and philosophical motivation (e.g., Rose and Stevens 1998 Rose et al. 1998; Stevens and Rose 2002). In the following, first discussed are the input data used to compute weights of the various transfer schemes. Then our schemes are introduced, and briefly mentioned is their motivation.

In order to save space, some indicators that help explain the results of our stability analysis (reported in the next subsection) are illustrated in table 6.2. The first three rows in table 6.2 show some commonly used indicators of economic and ecological performance of different regions. They are computed using the input data gross domestic product (GDP_i), population (POP_i), and emissions (E_i) for the base year 2000 as it enters our model (see table 6.A4 in appendix). Emissions per capita (E_i/POP_i) indicate that USA citizens are the largest and ROW citizens the smallest emitters per head. GDP per capita (GDP_i/POP_i) indicates that CHN is the poorest and USA the richest region in our model. Emissions per unit of GDP (E_i/GDP_i) is a commonly used indicator to measure emission intensity of production. It is evident that JPN is the most and FSU the least emission efficient region in our model.

The subsequent rows in table 6.2 display different scenarios. *Scenario 0* is the benchmark case of no transfers, and scenarios 1 to 11 represent different transfer schemes, resulting from different weights. The scenarios 1 to 5 refer to what we would like to call "equity motivated" surplus-sharing rules. The scenarios 6 to 10 are more pragmatic schemes. The final scenario 11 can be interpreted as combining both equity and incentive arguments and is therefore considered separately.

The first column lists the numbers and names that we attach to each scenario and the second column provides the formula for computing gross weights $\tilde{\lambda}_i$. Subsequent columns display normalized weights ($\lambda_i = \tilde{\lambda}_i / \sum_{j \in N} \tilde{\lambda}_j$) and transfers ($t_i$) under the assumption of full cooperation. Although the values will differ for other partially cooperative coalition structures, they give at least a rough indication of the welfare implications of different scenarios.

Scenario 1 assumes that weights are proportional to population. The normative idea behind this rule may be summarized as "one person, one vote," and we therefore call scenario 1 "egalitarian." For the grand coalition, egalitarian means a larger share of the gains from cooperation for CHN and ROW compared to "no transfers." Transfers flow from the USA, JPN, and EU toward CHN and ROW, and to a lesser extend to FSU.

Scenario 2 relates weights to the inverse of the emissions per capita ratio: the higher the emissions per capita ratio of a region, the lower is its share in the surplus of cooperation. Scenario 2's motivation is that every citizen should be entitled to the same "right to pollute," and it can also be interpreted as a reflection of historical responsibility for the current stock of greenhouse gases. Since USA has the highest emissions per capita ratio (and also the highest share in cumulative historic emissions), it receives the lowest weight under scenario 2. This is reversed for CHN and ROW: both regions receive the highest weights under scenario 2.

Scenarios 3 and *4* allocate the gains from cooperation to the "poor" and thus use environmental policy as a vehicle to transfer money from the "rich" to the "poor." The parameter η is usually referred to as the "degree of inequality aversion" where $\eta \to +\infty$ would correspond to the "Rawlsian maximin rule."[9] In our model, already $\eta = 10$ approximates this rule since all weights are zero, except China's weight that is equal to 1. However, even a value of "only" $\eta = 1$ implies a substantial reshuffling of the gains from cooperation from the "rich" to the "poor." Following the literature we call scenarios 3 and 4 "ability to pay" scenarios.

Scenario 5 calls for an ecological subsidy that relates weights directly to emissions per dollar of output. "Ecological subsidy" means that regions with low emission efficiency, measured by the ratio of carbon emissions over GDP, will receive higher gains from cooperation. The subsidy should motivate these regions to invest in less emission intensive production capital. In line with the mainstream literature, in our model, emission efficiency is highest (i.e., low ratio E_i/GDP_i in table 6.2) in JPN and EU and very low in CHN and FSU. ROW's low emissions are associated with its low stage of economic development, so its estimated emission efficiency is relatively high.

We come to the so-called pragmatic sharing rules in *scenario 6*, which is the mirror image of scenario 2. In this scheme the coalition members share the surplus of cooperation in proportion to each country's emissions per capita. Scenario 6 is about regions preserving their status quo; regions acknowledge the political reality that substantial monetary transfers are politically not feasible. Of all scenarios considered in table 6.2, maintaining the status quo rule results in the lowest level of resources to be transferred (as noted in the "total" column in the table).

Scenarios 7 through 10 relate the surplus sharing weights to two fundamental characteristics of the regions: their marginal emission abatement costs and marginal climate change damages. Reviewed are the four possible combinations of these characteristics: the product (plus its inverse) and the ratio (plus its inverse) of marginal cost and marginal damage indexes. In *scenario 7*, the surplus of the cooperation is distributed proportional to the product of the marginal cost and marginal damage indexes, and in *scenario 8*, it is the inverse of this ratio. As is intuitively clear, the countries with low marginal emission abatement costs and high marginal climate change damages are free-riders in an IEA. Although upon joining the coalition they may be able to make a lot of reductions (since they are cheap producers of abatement), they nevertheless do not place much value on lowering environmental damage. Typically the cost of joining an IEA exceeds the benefit for such countries, so they prefer to remain outsiders. CHN and ROW are two examples because under full cooperation their losses are highest. Scenario 8 corrects for free-riding due to the incentive of low cost–low damage as it allocates the surplus of cooperation inversely in proportion to the product of marginal costs and marginal damages indexes. So scenario 7 can be expected to perform badly and scenario 8 to perform better in terms of coalitional stability.

Barrett (1994) investigated participation in voluntary IEA using a stylized theoretical model with symmetric countries. He compared the slopes of the marginal abatement cost function and marginal benefit (i.e., from avoiding environmental damage) function. He showed that if marginal abatement costs are relatively steep compared to marginal damages the number of signatories of a stable coalition will be small, and vice versa. Barrett's work led us to consider the ratio of the marginal cost over marginal damage indexes in *scenario 9* and the inverse of this ratio in *scenario 10* to determine the surplus-sharing weights. It was especially difficult to formulate hypotheses for the expected performances of these scenarios because Barrett's (1994) original analysis did not account for asymmetries or for transfers between regions.

In *scenario 11* we refer to a transfer scheme introduced by Chander and Tulkens (1995, 1997). The weights in this scenario are proportional to the marginal climate change damages in a full cooperation scheme. A normative interpretation of this transfer rule can be supported by noting that countries that experience relatively strong climate change damages receive a larger share of the surplus. However, Chander and Tulkens (1997) also provide a strategic motivation for their solution. They show, for their particular choice of weights, that the resulting allocation belongs to the core of the global emission game: no individual country, nor any coalition of countries, can do better by breaking away from the grand coalition and forming an agreement on their own. Because both a normative and a strategic interpretation of the Chander-Tulkens transfer rule are possible, we consider this rule separately from the equity-motivated and pragmatic schemes.

As table 6.2 shows, except for the pragmatic schemes of scenarios 6, 7, and 10, the total transfers (last column) can be much larger than the total gains from cooperation, amounting to around 12.3 trillion US$.[10] As is also clear from the table, most transfer schemes entail an extreme redistribution of the surplus. In some schemes, the original asymmetry due to the diverging costs and benefits in the no-transfer scenario is replaced by an asymmetry resulting from the transfer schemes. So apparently transfer schemes cannot be depended on to foster stable cooperation.

Results

Table 6.3 lists internally and externally stable coalition structures that are stable under at least one scenario. With exception of scenarios 1, 4, and 10, only one or two coalitions are stable with transfers. Only one coalition with four members is stable. Three stable coalitions consist of three players and most stable coalitions comprise only two members. This relatively poor improvement in stability compared to the no-transfer case (scenario 0) shows that individual profitability is a necessary but by no means a sufficient condition for stability in the sense of d'Aspremont et al. (1983). Also with transfers, the free-rider incentive increases with the size of the coalition. A small coalition may make a difference by closing the gap between no cooperation and full cooperation by 50 or more percent. The most prominent case is the three-player coalition {USA, CHN, ROW}, which is stable under the Chander-Tulkens transfer scheme (scenario 11), as it achieves a welfare score of 86.3 percent. No other coalition, with or without transfers, performs better in terms of welfare.

Scenario 4 (ability to pay, with $\eta = 10$) and scenario 10 (inverse of MC–MD ratio) show substantial progress in the number of stable coalitions compared to the no-transfer case. Both scenarios stabilize five two-player coalitions each. However, they stabilize rather different coalitions, and {CHN, ROW} is the only coalition that is stable under both scenarios. In general, the stable coalitions under the scenario 10 outperform in welfare terms the stable coalitions under scenario 4, the ability to pay scenario. Maximal global welfare attainable under sce-

Table 6.3
Stable coalition structures

Rank	Size	Membership	Welfare	No transfers 0	Egalitarian 1	Ability to pollute 2	Ability to pay ($\eta=1$) 3	Ability to pay ($\eta=10$) 4	Ecological subsidy 5	Status quo 6	$MC \times MD$ 7	Inverse $MC \times MD$ 8	MC/MD 9	Inverse MC/MD 10	Chander-Tulkens 11
8	3	USA, CHN, ROW	86.3												o e
18	3	USA, JPN, ROW	65.1							e					
19	3	USA, FSU, ROW	64.4							e					
21	2	USA, ROW	61.6				e	e	e	e		e	e		
25	2	EU, ROW	54.7				e	e	e			e	e		
27	3	CHN, FSU, ROW	49.3	e		o e			o e					o e	
28	2	CHN, ROW	42.1		o e			o e	e			e	e	o e	
29	3	JPN, FSU, ROW	41.4	e	o e										
30	2	JPN, ROW	36.7	e	o e		e	e	e		o e	e	e	o e	
31	2	FSU, ROW	32.8		o e		e	e				o e		o e	
36	4	USA, JPN, CHN, FSU	24.5							o e					
37	3	USA, JPN, CHN	22.4							e					
40	2	USA, CHN	19.8				o e	o e				e			
43	2	EU, CHN	16.0				o e	o e				e			
46	2	JPN, CHN	5.8				o e	o e				e			
49	2	CHN, FSU	3.5				o e	o e				o e			
50	2	USA, EU	3.0	o e				e							
56	2	JPN, EU	0.9					e							
Open membership, internal and external				1	3	1	1	5	1	2	1	2	1	5	1
Exclusive membership, internal and external				4	3	1	5	11	5	6	1	9	5	5	1

Note: The coalition structures listed are internally and externally stable in at least one scenario and sorted in descending order of global welfare; welfare is measured in relative terms as described in the legend of table 6.1. "o" stands for internally and externally stable open membership, and "e" for internally and externally stable exclusive membership unanimity voting.

nario 10 amounts to 61.6 percent compared with 42.1 percent under scenario 4.

In terms of membership, CHN is a member of every stable coalition under the ability to pay (scenario 4). This is not so surprising, since CHN is the sole beneficiary of transfers under the extreme ability to pay scenario, receiving close to 100 percent of the cooperation surplus. In terms of internal stability this implies that CHN is unlikely to defect from any coalition it belongs to. At the same time, other countries are reluctant to join CHN because they know with certainty that they will receive almost nothing of the cooperation surplus. Hence coalitions including CHN are often externally stable. But the mirror image of this external stability is that CHN can hardly find partners to form an internally stable coalition. In particular, we observe that CHN finds at most one partner to form a stable coalition and that most of such couples do not perform well in terms of global welfare. CHN's and ROW's performance with respect to closing the welfare gap between no cooperation and full cooperation ranges between 3.5 and 42 percent.

Another striking result is the importance of the ROW group of countries, which always belongs to the stable coalitions under the scenarios 1 and 10, as a result of ROWs high share in the cooperative surplus under both scenarios. This implies that ROW is not likely to defect from the coalitions to which it belongs. Regarding external stability, the argument is the same as that for CHN under scenario 4. Few countries are willing to join ROW because then they become net contributors instead of beneficiaries under transfer scenarios 1 and 10. Hence, while the high surplus share for ROW promotes external stability, the mirror image is internal stability of coalitions, including ROW, and that becomes problematic. Coalitions comprising ROW under scenarios 1 and 10 find at most one partner to form a stable partnership. Because of ROW's high benefits or rather avoidance of climate change damages, these partnerships imply, in general, higher abatement targets than any partnership with CHN under scenario 4. Therefore coalitions including ROW perform better in terms of global welfare, closing the gap with welfare indexes ranging between 32.8 and 61.6 percent.

In sum, transfers can make a difference, but they fail to stabilize the grand coalition in a full cooperation scenario. Membership in stable coalitions largely depends on the design of the transfer rule. It appears that for a climate change policy to be successful, the transfer scheme should strike a balance between motivating low damage cost (i.e., typically developing) countries to join a prospective coalition and not scare away high damage cost (i.e., typically industrialized) countries. None of the schemes we considered in this chapter meet this objective. Even the best-performing scenarios 1, 4, and 10 (allowing open membership) fall short of this objective because the cooperation remains limited to only two players Because of the rather extreme redistribution entailed by these transfer scenarios, the main beneficiaries of the transfers find at most one partner. Some of these partnerships may manage to close the welfare gap by 50 percent or more, but it is evident that more could be achieved by coalitions of three or more players. The Chander-Tulkens transfer rule performs well in this respect as it is the only scheme capable of stabilizing a strongly welfare-improving three-player coalition that combines two cheap abatement producers (CHN and ROW) with a relatively high benefit region (USA). Although the Chander-Tulkens rule stabilizes only one coalition, it closes the welfare gap between no cooperation and full cooperation by 86.3 percent.

6.3.2 A Third Approach: Changing Membership Rules

Motivation and Fundamental Features
As noted in the discussion above, the stability problem has two dimensions: internal and external stability. To ensure internal stability, regions opting to leave the coalition could be harshly punished. Further it is important that punishment be credible. In our model punishment is implicit in that the coalitions and single players play a partial Nash equilibrium in economic strategies (see section 6.2). Hence it is assumed that if a region i leaves coalition S, the remaining regions $S\backslash\{i\}$ reoptimize their economic strategies. That is to say, the remaining regions revise their abatement targets downward, which has the negative impact on the deviating country through higher climate change damages. While such punishment may not the harshest possible, it is a credible punishment. After a deviation, regions $S\backslash\{i\}$ revise their equilibrium economic strategies as a best reply to the changed condition.

It may be argued that the problem of internal instability can be simply mitigated by not allowing members to leave. However, such a change of institutional rule would be counter to the notion of voluntary participation. Alternatively, an argument could be made that stability would be ensured by barring outsiders from joining a particular coalition. However, while at least in theory such a modification appears possible, to have, literally, external stability means to have an *open membership rule*. That is, every outsider can join the coalition without bidding for acceptance as is the opposite of an institutional scheme that has an *exclusive membership rule*. In an exclusive membership coalition, if an outsider wants to accede to an agreement, the current members of the coalition vote either by majority or unanimity on the accession of the new member. While such voting procedures are usually not part of international environmental agreements, there are many examples among other international treaties and organizations such as the World Trade Organization (WTO), the European Union (EU), and the Security Council of the United Nations Organization (UNO) where membership is exclusive and tightly controlled. At least a priori there seems to be no reason why exclusive membership in environmental agreements should not be considered as an alternative rule. A close example is the UNO, whose mission is to foster security and political stability worldwide through peace-keeping, and which also provides a global public good.

Under an exclusive membership rule, an outsider that wants to join a coalition must first obtain approval from the original members, who can vote either "accept" or "not accept." We assume that they will vote "accept" if they gain from accession by the newcomer. Under the unanimity voting rule, all coalition members must vote "accept" before the outsider can join. Under the majority voting rule, any majority in favor of accession is sufficient for an outsider to join. Because unanimity voting is a more demanding criterion, it makes joining a coalition more difficult for outsiders. Hence it can yield more (external) stability.

By so technical an institutional change, internally stable coalitions that are externally unstable under open membership can become externally stable under exclusive membership. In our example such will be the case if an outsider has an incentive to join a coalition S, and then its application is turned down because at least one member (unanimity voting) or a majority of members (majority voting) would fear a loss from the accession. Thus the question that cannot be answered by theoretical reasoning, as mentioned at the start of this chapter, is whether "more stability" translates into higher global welfare.

Before turning to the results in our model, we should note that a change of membership is not a mere "technical" trick. For instance, it could be argued that current members never turn down an application because more members means more contributors to cooperation. However, in our model a coalition $S \cup \{i\}$ increases abatement efforts compared to coalition S, which may be regarded as "too ambitious" by at least some of the current members. Consequently, one could argue that all current members of S could just allow region i to join, asking it to increase its abatement effort, and at the same time that all members of S do not change their economic strategies. Alternatively, one could argue that region i must not join coalition S but may just increase its abatement effort, which current members neither can avoid nor have an interest to do so. However, none of these alternatives can be in the interest of the "potential accessor" i. The reasoning is simple: under a coalition structure $c = \{S, \{i\}, \{k\}, \ldots, \{n\}\}$, $s = (s_i, s_{-i})$ is the equilibrium economic strategy vector that implies that s_i is a best reply to s_{-i} (and vice versa). Consequently, if under coalition structure $c' = \{S \cup \{i\}, \{k\}, \ldots, \{n\}\}$ s_{-i} is the same as under $c = \{S, \{i\}, \{k\}, \ldots, \{n\}\}$, it cannot be an improvement for region i to change its strategy to s'_i when joining coalition S. A similar argument applies if region i only changes its strategy without joining coalition S.

Results

In table 6 3 the stable coalition structures under exclusive membership are restricted to unanimity voting, that is, to the most stringent exclusive membership arrangement. In the scenarios we see that exclusive membership more than doubles the number of stable coalitions (from 24 to 56). However, hidden behind this general observation are large differences in the individual transfer schemes. In the case of no transfers (scenario 0), exclusive memberships raises the number of stable coalitions to four compared to only one under open membership. The most successful stable coalition achieves a welfare level of 49.3 percent without transfers.

Also for many of the transfer schemes the maximum welfare attainable by a stable coalition is raised substantially. The largest differences can be observed for scenario 4 (ability to pay $\eta = 10$) and scenario 8 (inverse MC $\times$ MD) where exclusive membership raises the number of stable coalitions from 5 to 11 and from 2 to 9 respectively. Interestingly the stability results for the Chander-Tulkens transfer scheme are not affected by the move to exclusive membership rule. Also under exclusive membership only the coalition {USA, CHN, ROW} is stable for the Chander-Tulkens transfer scheme. However, as we observed earlier, it is an extremely successful coalition in terms of global welfare.

Taken together, these results suggest that institutional rules are as important as transfers in ensuring stability of international environmental agreements. In particular, if transfers are not available, difficult to implement, or politically not feasible, then a change of institutional rules may be an alternative means to successful cooperation. What is more, such institutional changes can basically be effected at no cost.

6.4 Summary and Conclusion

We have analyzed coalition formation in the context of global warming. Our approach combines a game-theoretic analysis with numerical simulations based on a dynamic integrated assessment model that captures the feedback to the economy of the climate system and environmental damages. The model comprises six world regions: the United States, Japan, European Union, China, former Soviet Union, and "rest of the world." Stability of coalitions was tested with the concept of internal and external stability. From our simulations, we discuss six key results.

First, in the context of global warming, the difference between full and no cooperation is large in ecological terms. Emissions and concentrations of greenhouse gases are cut by half approximately. However, the difference is not so large in welfare terms due to discounting and to the benefits from cooperation occurring mainly in the future.

Second, partial cooperation can also be an important step to mitigate global warming. However, the identity may be more important than the number of members for partial cooperation to succeed. Our simulations indicated that success viewed only in terms of high participation without measuring the effectiveness of an agreement can be misleading. Moreover we found that coalitions that do not comprise key players with low marginal abatement costs (e.g., CHN and ROW) and/or high marginal damages (e.g., EU, USA, and ROW) will not achieve much at the global level. Thus it is critical for developing countries to be included in future climate treaties.

Third, open membership rule without transfers provides for only one stable coalition in our model. In this simulation cooperation on global warming proves difficult because of the large asymmetries in costs and damages between regions.

Fourth, the success of cooperation seems to improve in the later transfer schemes reported in this chapter. At most four out of six world regions participate in stable cooperation, but in terms of welfare, only a handful of coalitions can close the gap between no cooperation and full cooperation by more than 50 percent. Hence we argued that making agreements individually profitable to all participants through transfers may be a necessary but by no means sufficient condition to establish successful self-enforcing treaties. Free-rider incentives remain strong, and a big obstacle to high participation and effectiveness.

Fifth, we find it difficult to observe systematic differences in performance between morally motivated and pragmatic transfer schemes. While we do observe that extremely redistributive transfer schemes can mitigate the asymmetric distribution of the gains from cooperation before transfers, it is at the expenses of introducing a new asymmetry. This does not allow for effective stable coalitions with participation of key industrialized countries that are so important for the success of joint climate policy.

Sixth, changing the institutional design from open to exclusive membership can make a big difference, and institutional redesign may be more important than transfers. It therefore is apparent that in future environmental treaties open membership should not be taken for granted even though the context is concern over a public good.

Clearly, models are limited in their applicability. We mention three important missing aspects. First, our models divide the world into six regions, whereas in reality we can roughly count 200 countries. Because the ROW's aggregation is so large, we risk the prospect of an overestimation in the findings of stable cooperation. Future analysis should disaggregate this region into its key players like India and Brazil. Second, our integrated CWS model, as well as most CGE-models analyzing climate change, assumes damages to be known with certainty. On the one hand, this can lead to an overestimation of the incentives for cooperation. On the other hand, provided some agents attach some probability to uncertain catastrophic events caused by global warming, our assumptions imply an underestimation. Third, our long time horizon runs the risk of overestimating the foresight of politicans. A much shorter period would, however, overlook the climate problem and provide no incentive for cooperation. Moreover discounting would adjust foresight much too much downward.

For future consideration we would like to mention two of many possible research options. First, it would be interesting to have more theoretical insight into the design of transfer schemes that would mitigate free-rider incentives in an "optimal way." For this we would need more understanding of the relation between individual region's characteristics (e.g., cost–benefit ratios) and their free-riding incentives. Although we made a first attempt in this chapter in that direction, we believe much remains to be done in this respect. Second, in the absence of transfers, it would be interesting to learn more about how abatement duties could be allocated to improve upon the success of self-enforcing treaties. In the presence of strong free-rider incentives, this may well imply less ambitious abatement targets and/or a departure from an efficient abatement allocation within a coalition, as we assumed, if this is compensated by larger participation.

Appendix

The CLIMNEG world simulation model (version 2) consists of three blocks of equations: the economy, the carbon cycle, and a radiative forcing and temperature module. The full listing of the equations is given below:

Economy module
$$\begin{cases} Y_{i,t} = Z_{i,t} + I_{i,t} + Y_{i,t}C_i(\mu_{i,t}) + Y_{i,t}D_i(\Delta T_t) \\ Y_{i,t} = a_{i,t}K_{i,t}^{\gamma}L_{i,t}^{1-\gamma} \\ C_i(\mu_{i,t}) = b_{i,1}\mu_{i,t}^{b_{i,2}} \\ D_i(\Delta T_t) = \theta_{i,1}\Delta T^{\theta_{i,2}} \\ K_{i,t+1} = [1 - \delta_K]K_{i,t} + I_{i,t} \quad\quad K_{i,0} \text{ given} \\ E_{i,t} = \alpha_{i,t}[1 - \mu_{i,t}]Y_{i,t} \end{cases}$$

Carbon cycle module
$$\begin{cases} M_{t+1}^{at} = M_t^{at} + \tau_{11}M_t^{at} + \tau_{21}M_t^{uo} + \sum_{j \in N} E_{j,t} \quad M_0^{at} \text{ given} \\ M_{t+1}^{uo} = M_t^{uo} + \tau_{12}M_t^{at} + \tau_{22}M_t^{uo} + \tau_{32}M_t^{lo} \quad\quad M_0^{uo} \text{ given} \\ M_{t+1}^{lo} = M_t^{lo} \quad\quad\quad\quad\quad\quad + \tau_{23}M_t^{uo} + \tau_{33}M_t^{lo} \quad\quad M_0^{lo} \text{ given} \end{cases}$$

Radiative forcing and temperature module
$$\begin{cases} F_t = \dfrac{4.1 \ln(M_t^{at}/M_0^{at})}{\ln(2)} + F_t^x \\ T_t^o = T_{t-1}^o + \tau_3[T_{t-1}^{at} - T_{t-1}^o] \\ T_t^{at} = T_{t-1}^{at} + \tau_1[F_t - \lambda T_{t-1}^{at}] - \tau_2[T_{t-1}^{at} - T_{t-1}^o] \\ \Delta T_t = \dfrac{T_t^{at}}{2.50} \end{cases}$$

Chapter 6 Transfer Schemes and Institutional Changes

Table 6.A1
Variables

$Y_{i,t}$	Production (billion 1990 US$)
$Z_{i,t}$	Consumption (billion 1990 US$)
$I_{i,t}$	Investment (billion 1990 US$)
$K_{i,t}$	Capital stock (billion 1990 US$)
$C_{i,t}$	Cost of abatement (fraction production)
$D_{i,t}$	Damage from climate change (fraction of production)
$E_{i,t}$	Carbon emissions (gigatonnes = billion tonnes of C)
$\mu_{i,t}$	Emission abatement (fraction)
M_t^{at}	Atmospheric carbon concentration (gigatonnes = billion tonnes of C)
M_t^{uo}	Carbon concentration upper ocean (gigatonnes = billion tonnes of C)
M_t^{lo}	Carbon concentration lower ocean (gigatonnes = billion tonnes of C)
F_t	Radiative forcing (watts per m^2)
F_t^x	Exogenous radiative forcing (watts per m^2)
T_t^{at}	Temperature increase in the atmosphere (°C)
T_t^o	Temperature increase in the deep ocean (°C)
ΔT_t	Change of temperature increase in the atmosphere (°C)

Table 6.A2
Global parameter values

$a_{i,t}$	Productivity	RICE
$L_{i,t}$	Population	RICE
$\alpha_{i,t}$	Emission-output rate	RICE
δ_K	Capital depreciation rate	0.10
γ	Capital productivity parameter	0.25
F_t^x	Exogenous radiative forcing (watts per m^2)	RICE
τ_{11}	Parameter carbon cycle module	−0.033384
τ_{12}	Parameter carbon cycle module	+0.033384
τ_{21}	Parameter carbon cycle module	+0.027607
τ_{22}	Parameter carbon cycle module	−0.039103
τ_{23}	Parameter carbon cycle module	+0.011496
τ_{32}	Parameter carbon cycle module	+0.000422
τ_{33}	Parameter carbon cycle module	−0.000422
τ_1	Parameter temperature module	0.226
τ_2	Parameter temperature module	0.44
τ_3	Parameter temperature module	0.02

Table 6.A2
(continued)

λ	Parameter temperature module	1.41
M_0^{at}	Initial carbon concentration atmosphere 1990	783
M_0^{uo}	Initial carbon concentration upper strata ocean 1990	807
M_0^{lo}	Initial carbon concentration lower strata ocean 1990	19,238

Table 6.A3
Regional parameter values

	$\theta_{i,1}$	$\theta_{i,2}$	$b_{i,1}$	$b_{i,2}$	ρ_i
USA	0.01102	2.0	0.07	2.887	0.015
JPN	0.01174	2.0	0.05	2.887	0.015
EU	0.01174	2.0	0.05	2.887	0.015
CHN	0.01523	2.0	0.15	2.887	0.030
FSU	0.00857	2.0	0.15	2.887	0.015
ROW	0.02093	2.0	0.10	2.887	0.030

Table 6.A4
Variables in 2000 (reference year)

	Y_i^0	(%)	K_i^0	(%)	L_i^0	(%)	E_i^0	(%)
USA	7,564	0.274	19,741	0.280	282	0.047	1.574	0.240
JPN	3,388	0.123	9,754	0.138	127	0.021	0.330	0.050
EU	8,447	0.307	22,804	0.323	377	0.062	0.888	0.135
CHN	969	0.035	2,686	0.038	1,263	0.209	0.947	0.144
FSU	558	0.020	1,490	0.021	288	0.048	0.626	0.095
ROW	6,633	0.241	14,105	0.200	3,716	0.614	2.192	0.334
World	27,559	1.000	70,580	1.000	6,052	1.000	6.556	1.000

Note: Y_i^0 and K_i^0 billion US$, L_i^0 million people, and E_i^0 gigatonnes carbon equivalent.

Notes

This chapter has been written while M. Finus was a visiting scholar at the Katholieke Universiteit Leuven, Centrum voor Economische Studiën (Leuven, Belgium) and at Fondazione Eni Enrico Mattei, FEEM (Venice, Italy). He acknowledges the financial support by the CLIMNEG 2 project funded by the Belgian Federal Science Policy Office and the kind hospitality of FEEM. Both authors acknowledge research assistance by Carmen Dunsche and François Gérard. The chapter has benefited from comments by Sylvie Thoron and other participants to the David Bradford Memorial Workshop in Venice in July 2005. Comments by two anonymous referees and by the editors are gratefully acknowledged.

1. For instance, see Nordhaus and Boyer (2003) and Kolstad and Toman (2005) for an overview of cost–benefit studies on climate change.

2. A notable exception is, for instance, Bosello et al. (2003).

3. An overview of the equations and parameters of the model is provided in the appendix. The model was initially introduced by Eyckmans and Tulkens (2003), where a detailed exposition can be found.

4. We choose a sufficiently long time period to avoid endpoint bias. However, due to discounting, only a shorter period is strategically relevant for players.

5. The mirror image is that abatement constitutes a positive externality; see subsection 6.3.2.

6. This assumption is widely made in coalition theory; see Bloch (1997) and Yi (1997).

7. The concept of internal and external stability that we apply is due to d'Aspremont et al. (1983). It is Nash equilibrium where no player has an incentive to revise a decision about membership. For an overview of other concepts applied in the environmental context, see Finus (2003).

8. This procedure is valid because we assume a lump-sum transfer of discounted payoffs for simplicity. Because payoffs after transfers are just an affine transformation of the original payoffs, the transfers have no effect on equilibrium economic strategy vectors in our framework. The game is therefore a TU-game as assumed almost throughout the literature on coalition formation.

9. See Eyckmans, Proost, and Schokkaert (1993) for details and for an application of this approach to the problem of sharing the burden of global greenhouse gas emission reduction.

10. The total transfers in the last column of table 6.3 are the sum of all positive transfers (= sum of all negative transfers) but not the sum of all transfers that is zero by definition. Total transfers are an indicator of the amount of financial resources redistributed by the transfer scheme. It is important to note that because transfers are computed according to (6.5), it can happen that total transfers exceed the total gain from cooperation. For instance, suppose $W_i(c^N) - W_i(c) > 0$ and $\lambda_i = 1$, then $t_i(c) > \sum_{j \in S}[W_j(c) - W_j(c^N)]$.

References

Barrett, S. 1994. Self-enforcing international environmental agreements. *Oxford Economic Papers* 46: 804–78.

Barrett, S. 1997. Heterogeneous international environmental agreements. In C. Carraro, ed., *International Environmental Negotiations: Strategic Policy Issue*. Cheltenham: Edward Elgar, pp. 9–25.

Barrett, S. 2001. International cooperation for sale. *European Economic Review* 45: 1835–50.

Bloch, F. 1997. Non-cooperative models of coalition formation in games with spillovers. In C. Carraro and D. Siniscalco, eds., *New Directions in the Economic Theory of the Environment*. Cambridge: Cambridge University Press, pp. 311–52.

Böhringer, C., and C. Vogt. 2004. The dismantling of a breakthrough: The Kyoto Protocol as symbolic policy. *European Journal of Political Economy* 20: 597–618.

Bosello, F., B. Buchner, and C. Carraro. 2004. Equity, development, and climate change control. *Journal of the European Economic Association* 1: 601–11.

Botteon, M., and C. Carraro. 1997. Burden-sharing and coalition stability in environmental negotiations with asymmetric countries. In C. Carraro, ed., *International Environmental Negotiations: Strategic Policy Issue*. Cheltenham: Edward Elgar, pp. 26–55.

Carraro, C. 2002. Roads towards international environmental agreements. In H. Siebert, ed., *The Economics of International Environmental Problems*. Tübingen: Mohr Siebeck, pp. 169–202.

Carraro, C., and D. Siniscalco. 1993. Strategies for the international protection of the environment. *Journal of Public Economics* 52: 309–28.

Chander, P., and H. Tulkens. 1995. A core-theoretic solution for the design of cooperative agreements on transfrontier pollution. *International Tax and Public Finance* 2: 279–93.

Chander, P., and H. Tulkens. 1997. The core of an economy with multilateral environmental externalities. *International Journal of Game Theory* 26: 379–401.

Cornes, R., and T. Sandler. 1996. *The Theory of Externalities, Public Goods and Club Goods*, 2nd ed. Cambridge: Cambridge University Press.

d'Aspremont, C., A. Jacquemin, J. J. Gabszewicz, and J. A. Weymark. 1983. On the stability of collusive price leadership. *Canadian Journal of Economics* 16: 17–25.

Eyckmans, J., S. Proost, and E. Schokkaert. 1993. Efficiency and distribution in greenhouse negotiations. *Kyklos* 46: 363–97.

Eyckmans, J., and H. Tulkens. 2003. Simulating coalitionally stable burden sharing agreements for the climate change problem. *Resource and Energy Economics* 25: 299–327.

Finus, M. 2003. Stability and design of international environmental agreements: The case of transboundary pollution. In H. Folmer and T. Tietenberg, eds., *International Yearbook of Environmental and Resource Economics, 2003/4*. Cheltenham: Edward Elgar, pp. 82–158.

Finus, M., and B. Rundshagen. 2003. Endogenous coalition formation in global pollution control: A partition function approach. In Carlo Carraro, ed., *Endogenous Formation of Economic Coalitions*. Cheltenham: Edward Elgar, pp. 199–241.

Finus, M., and S. Tjøtta. 2003. The Oslo Protocol on sulfur reduction: The great leap forward? *Journal of Public Economics* 87: 2031–48.

Germain, M., P. L. Toint, H. Tulkens, and A. de Zeeuw. 2003. Transfers to sustain dynamic core-theoretic cooperation in international stock pollutant control. *Journal of Economic Dynamics and Control* 28: 79–99.

Germain, M., P. L. Toint, and H. Tulkens. 1998. Financial transfers to ensure cooperative international optimality in stock pollutant abatement. In S. Faucheux, J. Gowdy, and I. Nicolaï, eds., *Sustainability and Firms: Technological Change and the Changing Regulatory Environment*. Cheltenham: Edward Elgar, pp. 205–19.

Germain, M., and V. van Steenberghe. 2003. Constraining equitable allocations of tradable CO_2 emission quotas by acceptability. *Environmental and Resource Economics* 26: 469–92.

Hoel, M. 1992. International environment conventions: The case of uniform reductions of emissions. *Environmental and Resource Economics* 2: 141–59.

Hoel, M., and K. Schneider. 1997. Incentives to participate in an international environmental agreement. *Environmental and Resource Economics* 9: 153–70.

Intergovernmental Panel on Climate Change IPCC. 2001. *Climate Change 2001: Mitigation. Contribution of Working Group III to the Third Assessment Report of the Intergovernmental Panel on Climate Change*. Cambridge: Cambridge University Press.

Kolstad, C. D., and M. Toman. 2005. The economics of climate policy. In Karl-Göran Mäler and Jeffrey R. Vincent, eds., *Handbook of Environmental Economics*, vol. 3. New York: Elsevier, pp. 1561–618.

Müller-Fürstenberger, G., and G. Stephan. 1997. Environmental policy and cooperation beyond the nation state: An introduction and overview. *Structural Change and Economic Dynamics* 8: 99–114.

Murdoch, J. C., and T. Sandler. 1997a. Voluntary cutbacks and pretreaty behavior: The Helsinki Protocol and sulfur emissions. *Public Finance Review* 25: 139–62.

Murdoch, J. C., and T. Sandler. 1997b. The voluntary provision of a pure public good: The case of reduced CFC emissions and the Montreal Protocol. *Journal of Public Economics* 63: 331–49.

Nordhaus, W. D., and Z. Yang. 1996. A regional dynamic general-equilibrium model of alternative climate-change strategies. *American Economic Review* 86: 741–65.

Nordhaus, W. D., and J. Boyer. 2000. *Warming the World: Economic Models of Global Warming*. Cambridge: MIT Press.

Rose, A., and B. Stevens. 1998. A dynamic analysis of fairness in global warming Policies: Kyoto, Buenos Aires, and beyond. *Journal of Applied Economics* 1: 329–62.

Rose, A., B. Stevens, J. Edmonds, and M. Wise. 1998. International equity and differentiation in global warming policy. *Environmental and Resource Economics* 12: 25–51.

Rose, A., and B. Stevens. 2002. A dynamic analysis of the marketable permits approach to global warming policy: A comparison of spatial and temporal flexibility. *Journal of Environmental Economics and Management* 44: 45–69.

Weikard, H.-P., M. Finus, and J.-C. Altamirano-Cabrera. 2006. The impact of surplus sharing on the stability of international climate agreements. *Oxford Economic Papers* 58: 209–32.

Weitzman, M. L. 2001. Gamma discounting. *American Economic Review* 91: 260–71.

Yi, S.-S. 1997. Stable coalition structures with externalities. *Games and Economic Behavior* 20: 201–37.

7 Parallel Climate Blocs: Incentives to Cooperation in International Climate Negotiations

Barbara Buchner and Carlo Carraro

While awareness of the potential scope of climate change seems to be growing across countries, consensus is missing on the design approach and implementation of climate change control. Recent developments in international climate policy stress the importance of taking individual countries' incentives and specificities into account. After the top-down process of the Kyoto negotiations, where overall binding emissions targets were assigned without regard to country differences, there has been an increasing tendency for countries to focus more on domestic and/or bilateral climate-friendly activities than on global emissions targets. Small groups of countries have also been observed to cooperate in taking this initiative.

In July 2005, the Asia-Pacific Partnership on Clean Development and Climate stressed the sharing of technology in effecting climate policy. The agreement signed by the United States, Australia, Japan, China, India, and South Korea sanctioned a voluntary, technology-based initiative to reduce greenhouse gas emissions without legally binding emissions targets. The basic strategy is for these countries to cooperate in developing new technologies that can be deployed in developing countries. Other such bilateral agreements on technology and climate change have followed, and they attest to the attractiveness of this strategy. For example, in September 2005, the European Union and China agreed to strengthen cooperation and dialogue on climate change and energy issues, with a special focus on clear coal technology (see Buchner and Carraro 2005a for a description of the main technological agreements).

The cooperation that has emerged among small groups of countries can be also observed in recent developments in emissions trading markets. In Europe, an emissions trading scheme was officially launched in January 2005, with roughly 15,000 installations in 25 countries and 6 major industrial sectors. In August 2005, Canada announced the establishment of the Canadian Offset System. The Canadian government released its plan to encourage the creation of greenhouse gas emission reduction (GhG ER) domestic offsets, including the introduction of domestic emissions trading. In September 2005, after months of internal government debates, Japan's Ministry of Environment announced its decision to set up a voluntary domestic emissions trading scheme, beginning in April 2006 with an initial coverage of 34 manufacturing companies.

Similarly a wide variety of initiatives are in preparation at state and local levels in the United States, despite the opposition of the federal government to commit to binding international climate policy goals. In particular, several northeastern states are attempting to set up an emissions trading scheme comprising a regional greenhouse gas initiative (RGGI); many other states are implementing serious climate change strategies as well, with California taking the lead in its call for a return to 1990 emissions levels by 2020. California is also the leader in the negotiations on a GHG trading scheme among three Pacific Coast states. In addition New Mexico has become the first US state in September 2005 to sign up for voluntary emissions trading at the Chicago Climate Exchange, pledging to reduce its state's GHG emissions by 4 percent by 2006.

From these policy measures to combat climate change the movement is poised to evolve toward multiple initiatives/agreements. Yet, to our knowledge, no sound analysis has been performed to substantiate whether such events would correspond to some basic underlying economic incentives that make it convenient for countries to commit to unilateral or small group policy measures or whether these tendencies simply reflect a new round of political noise on a global climate change control.

Therefore the main objective of this chapter is to appropriately assess the main countries' incentives to cooperate on GHG emission control, namely to participate in a climate coalition. This will enable us to understand whether the equilibrium of the policy process is actually characterized by a global climate coalition or rather by a set of fragmented climate regimes in which several small coalitions emerge. To this aim, this chapter uses the FEEM-RICE model, a well-known integrated assessment climate-economy model, and some tools of noncooperative coalition theory to identify the equilibrium coalition structure that could emerge out of climate negotiations. In order to be able to unravel the basic economic incentives, we keep the policy framework as simple as possible, without capturing all the details of the recent policy developments. For example, in our setting, climate negotiations only center around the stringency of the environmental target.

The main assumption in our analysis, and its major novel feature with respect to previous empirical analyses of climate policy, is that a global agreement is only one of the possible outcomes of climate negotiations. According to their own economic interests, countries are also free to form regional or subglobal agreements. In particular, a given country can (1) join one of the existing climate coalitions, (2) propose a new coalition, or (3) simply decide to free-ride on the other countries' cooperative abatement efforts. Each climate coalition allows the implementation of the flexible mechanisms (emission trading) within the coalition in order to guarantee an efficient implementation of the environmental targets adopted within the coalition. This framework is meant to mimic the recent developments in climate policy described above, where we can observe the emergence of various carbon markets around the world, with countries participating in one of these markets (see Victor 2007 for a discussion of this policy framework).

Let us finally stress that the focus of our research is on economic incentives faced by countries. There are several other political, cultural, and environmental factors that could influence a country's decision to join a given climate coalition, which will not be addressed in this chapter. However, the economic dimension of climate negotiations has evolved as one of the key aspects in the international climate debates (and has often been considered as the most important one in the United States). As a consequence this chapter can provide a relevant, albeit partial, contribution to the analysis of the future evolution of international climate policy.

7.1 Regional and Subglobal Climate Blocs: Lessons from Coalition Theory

The strategic choice of players who decide whether or not to form a coalition with other players and, if they do, with which specific players to cooperate has been the subject of recent research in game theory.[1] Many of these recent studies are based on a noncooperative approach for which binding commitments are excluded. This approach is particularly suitable for analyzing the likely outcomes of future negotiations on climate change control because no supranational authority exists that can force countries to adopt policy measures to reduce their GHG emissions. Let us therefore examine the indications that the noncooperative theory of coalition formation provides for the analysis of climate negotiations.

The study of coalition formation poses three basic questions (see Bloch 1996): Which coalitions will be formed? How will the coalitional payoffs be divided among members? How does the presence of other coalitions affect the incentives to cooperate? The traditional cooperative game theory (see Aumann and Drèze 1974) focuses on the second question—the division of payoffs among coalition members. The first question can be assumed away in most cooperative game theory, and the third can be simply ignored, since a coalitional function cannot take into account externalities among coalitions.

These limitations have led to a new branch of the game-theoretic literature on the formation of coalitions as a noncooperative, voluntary, process. In the noncooperative approach, a player's decision to join a coalition is often modeled as a two-stage game. In the first stage, a player independently decides whether or not to join, by anticipating the consequence of his/her decision on the economic variables under control. In the second stage, he/she sets the value of these variables, given the coalition structure formed in the first stage. Under the simplifying assumption that the second-stage equilibrium is unique for any coalition structure, the first-stage game can be reduced to a partition function, which assigns a value to each coalition in a coalition structure as a function of the entire coalition structure. This enables us to capture the important effects of externalities across coalitions.

The theoretical literature on the noncooperative coalition formation has shown that some countries will form a coalition even without a commitment to cooperate and even in the presence of positive spillovers (i.e., in the case where the formation of a coalition by some players increases the payoff of the players outside the coalition, as in public good provision).

The equilibrium coalition structure depends on several key assumptions: identical agents, a membership rule, the order of moves, the players' reactions, and the slopes of their reaction functions (see Carraro and Marchiori 2003). Nevertheless, some conclusions seem to be robust with respect to these assumptions (except for the first one) and the related equilibrium concepts. For example, when a treaty is signed by many countries (i.e., a large coalition is formed), the amount of public good provided by the coalition (e.g., the amount of GHG abatement) is very close to the noncooperative business-as-usual activity (Barrett 2002). For the purpose of our analysis, the most important conclusion is as follows: if countries are free to decide not only whether or not to sign a treaty but also which treaty (i.e., which coalition to join), there is generally more than one coalition at equilibrium. For example, in the case of trade negotiations, there may be several trade blocs, and in the case of environmental negotiations, several regional or subglobal climate agreements.

This conclusion agrees, for example, with those of Bloch (1995, 1996), Ray and Vohra (1997, 1999), Yi (1997, 2003), and Yi and Shin (1995), who analyzed the formation of multiple coalitions by adopting different notions of stability. Bloch (1995, 1996) considered an infinite-horizon "coalition unanimity" game, in which a coalition forms if and only if all potential members agree to form the coalition. Ray and Vohra (1997) used the "equilibrium binding agreement" rule, under which coalitions are allowed to break up into smaller sub-coalitions only, and Yi and Shin (1995) the "open membership" game, in which nonmembers can join an existing coalition even without the consensus of the existing members. The different membership rules do lead to different predictions about stable coalition structures (see Carraro and Marchiori 2003). For example, the "open membership" rule is unlikely to support the grand coalition as an equilibrium outcome. The equilibrium coalition structure is generally very fragmented. By contrast, the "coalition unanimity" rule and the "equilibrium binding agreements" rule support more concentrated coalition structures at the equilibrium, but quite often not the grand coalition (see Finus and Rundshagen 2003).

These results were used by Carraro (1998, 1999) to argue that the Kyoto Protocol was not likely to be signed by all the relevant countries and that the emergence of parallel climate blocs was likely. However, economists are not alone in suggesting that climate negotiations may lead to multiple fragmented climate agreements. Some indications that multiple climate blocs could be the outcome of future climate negotiations can also be found in the political science and legal literature (e.g., see Egenhofer, Hager and Legge 2001; Stewart and Wiener 2003; Reinstein 2004). As Victor (2007, p. 134) says: "evidence that governments are taking the climate challenge seriously will come in the form of fragmented and variegated markets rather than integrated systems."

However, no economic analysis has yet quantified the incentives for negotiating countries to form multiple climate coalitions and identified which countries would belong to which coalition. This study is a first attempt to fill this gap.

7.2 The Model

7.2.1 *The Theoretical Framework*

Coalition formation is modeled as a two-stage game. There are n players $N = \{1, \ldots, n\}$ consisting of countries or world regions in our empirical model. In the first stage, countries choose their membership: a country can either join coalition S_i and become a signatory or remain a singleton and nonsignatory. These decisions lead to a coalition structure $S = (S_1, S_2, \ldots, S_k, l_k)$ if k coalitions form and the remaining players are singleton.

In the second stage, countries choose their economic strategies. At this stage it suffices to denote the vector of economic strategies by $\omega(S) = (\omega_1(S), \ldots, \omega_n(S))$, given that k coalitions have formed in the first stage. In the second stage, countries receive individual payoffs $\pi_i(\omega(S))$ that depend on the economic strategies of all countries.[2]

The subgame-perfect equilibria of this two-stage game can be computed by backward induction. To do so, it is sufficient for strategies to constitute a Nash equilibrium at every stage. For the second stage, this entails that economic strategies form a Nash equilibrium between coalitions S_i, $i = 1, \ldots, k$, and nonsignatories.[3] That is,

$$\sum_{i \in S} \pi_i(\omega_S^*(S), \omega_{-S}^*(S)) \geq \sum_{i \in S} \pi_i(\omega_S(S), \omega_{-S}^*(S)) \quad \forall \omega_S(S), \tag{7.1}$$

$$\forall i \notin S: \pi_i(\omega_S^*(S), \omega_i^*, \omega_{-i}^*(S)) \geq \pi_i(\omega_S^*(S), \omega_i(S), \omega_{-i}^*(S)) \quad \forall \omega_i(S),$$

where $\omega_S(S)$ is the economic strategy vector of coalitions $S_1 \ldots S_k$, $\omega_{-S}(S)$ the vector of all other countries not belonging to $S_1 \ldots S_k$, $\omega_i(S)$ the strategy of nonsignatory i, and $\omega_{-i}(S)$ the strategy vector of all other nonsignatories except i under coalition structure S. An asterisk denotes equilibrium strategies.

Computationally, this implies that nonsignatories will choose their economic strategies so as to maximize their individual payoff $\pi_i(\omega)$, whereas all signatories $j \in S_h$ jointly maximize the aggregate payoff of their coalition S_h. Strategically, this means that the behavior of nonsignatories toward all other countries is selfish and noncooperative; signatories behave cooperatively toward their fellow members (otherwise cooperation would not be worthwhile analyzing) but noncooperatively toward outsiders. Economically, this means strategies are group (but not globally) efficient within coalition S.

Given that the second stage of the game has been solved, let us define $v_i(S) = \pi_i(\omega^*(S))$ as the valuation of country i if the coalition structure S forms. This definition succinctly summarizes all information relevant to the second stage.

For the first stage, a Nash equilibrium in terms of participation can be computed.[4] The following two conditions must be met:

Profitability
A multiple coalition structure S is profitable if, when a coalition $S_i \in S$ forms, all players $j \in S_i$, $\forall i$, receive a payoff larger than when no coalition forms, meaning $v_j(S_i; S) \geq v_j(1; S^1)$, where S^1 is the coalition structure where all players are singleton.

Stability
A multiple coalition structure S is stable if each coalition $S_i \in S$ is internally stable, externally stable, and intracoalition stable. It is *internally stable* if no cooperating player would be better off by leaving the coalition to form a singleton. Formally, $v_i(S_i; S) \geq v_i(1; S')$ for all players in the coalition S_i and all coalitions in S, where $S' = S\backslash\{S_i\} \cup \{S_i - 1, 1\}$. It is *externally stable* if no singleton would be better off by joining any coalition belonging to the coalition structure S. Formally, $v_i(1; S) \geq v_i(S_i; S'')$ for all players who do not belong to S_i or to any other nontrivial coalition in S, where $S'' = S\backslash\{S_i, 1\} \cup \{S_i + 1\}$. It is *intracoalition stable* if no player belonging to S_i would be better off by leaving S_i to join any other coalition $S_j \in S$. Formally, $v_i(S_i; S) \geq v_i(S_j + 1; S°)$ for all players in the coalition S_i and all coalitions in S, where $S° = S\backslash\{S_i, S_j\} \cup \{S_i - 1, S_j + 1\}$.

That is, at the equilibrium, no signatory belonging to coalition $S_i \in S$, $\forall i$, has an incentive to leave its coalition in order to become a nonsignatory, given the participation decisions of all other countries. By the same token, no nonsignatory has an incentive to join coalition S_i, given the decisions of all other countries. And no player wants to move from one coalition to another one.

As shown below the set of stable coalitions may be empty. That is to say, the asymmetries between players and/or their different incentives to free-ride and/or to join a specific coalition may be such that no coalition structure is both profitable and stable.

7.2.2 The Empirical Model

The analysis of the possible outcomes of the dynamic process that defines the incentives to participate in a climate agreement has been carried out by using a modified version of Nordhaus's RICE model (see Nordhaus and Yang 1996) in which endogenous and induced technical change are modeled. In our version of the model, called FEEM-RICE (see Buonanno et al. 2002), technical change produces a twofold action: on the one hand, via increasing returns to scale, it obtains endogenous growth; on the other hand, by affecting the emission/output ratio, it accounts for the adoption of cleaner and energy-saving technologies.[5]

In the model, six countries/regions—US, EU, Japan (JPN), former Soviet Union (FSU), China (CHN), and the rest of the world (ROW)—optimally set the intertemporal values of four strategic variables: investment, R&D expenditure, abatement effort, and net demand for emission permits. Countries play the two-stage game described in section 7.3.1. Given the interdependency of countries' decisions, and the dynamic nature of the RICE model, the equilibrium value of the control variables is the solution of a dynamic Nash game. More precisely, we adopt the PANE-equilibrium concept introduced by Eyckmans and Tulkens (2003) and used in several other papers (see Eyckmans and Finus 2007; Carraro, Eyckmans, and Finus 2006).

We can provide here only a short overview of the FEEM-RICE model, a more detailed description can be found in the appendix (see also Eyckmans and Tulkens 2003 for a description of a modified version of the RICE model). The economic module of FEEM-RICE consists of a long-term dynamic, perfect foresight Ramsey-type optimal growth model with endogenous technical change. The decision variables are investment, R&D expenditure, and carbon emission abatement. The carbon cycle and temperature change module are the same as in RICE. In each world region and at every time period t, a regional budget equation describes how gross production, $Y_{i,t}$, can be allocated to consumption, $Z_{i,t}$, investment, $I_{i,t}$, emission abatement costs, $Y_{i,t}C_i(\mu_{i,t})$, and climate change damages, $Y_{i,t}D_i(\Delta T_t)$.

Gross production can be interpreted as "potential GDP," that is, what could be produced in the absence of the climate change problem. Abatement costs are an increasing and convex function of emission abatement effort. Abatement effort measures the relative emission reduction compared to the business-as-usual scenario (BAU) without any abatement policy. Climate change damages are an increasing and convex function of temperature change ΔT_t. Abatement costs and climate change damages are treated as proportions of "potential production." Hence total costs and damages are the product of costs and damages with potential production $Y_{i,t}$, respectively.

Every region is characterized by a production function that maps combinations of capital stock $K_{i,t}$ and labour input $L_{i,t}$ into output. The production technology is assumed to satisfy constant returns to scale of the Cobb-Douglas type. Labor supply is assumed to be inelastic. Capital accumulation is described in the standard way. In our version of the RICE model, investment is divided into physical and R&D investments. The former adds up to the existing stock of capital. The second modifies the productivity parameter in the production function and also affects the emission–output ratio (see the appendix for a detailed presentation of the equations).

Production gives rise to emissions of greenhouse gases. In FEEM-RICE, emissions are proportional to "potential" output. Emissions accumulate in the atmosphere according to a standard linear stock externality accumulation process. Carbon concentration is translated into temperature change according to an increasing function. Welfare of each country is measured by its aggregate lifetime discounted consumption.

In addition to the model structure, two assumptions qualify our results.[6] First, all countries/regions that adhere to the Kyoto/Bonn agreement are assumed to meet their Kyoto target from 2010 onward.[7] We therefore adopt the so-called Kyoto forever hypothesis (Manne and Richels 1999). Our reference to the Kyoto/Bonn agreement is partly imprecise since, for the sake of brevity, we will at times call the Kyoto Protocol or Kyoto/Bonn agreement a Kyoto forever scenario.

Second, cooperating countries are assumed to adopt cost-effective environmental policies. In particular, cost-effective market mechanisms (e.g., emission trading) are chosen over command-and-control measures in order to guarantee an efficient implementation of the environmental targets adopted within the coalition. Note that annex-B countries that belong to a coalition and therefore engage in emissions trading face their Kyoto targets, whereas China is assumed to agree to a 10 percent reduction of emissions with respect to the BAU scenario over the whole time horizon if it accepts to participate in a coalition (and in emissions trading). If various subglobal coalitions form, then they are assumed to behave independently, without a link between them (i.e., there is no trade between all regional blocs on a common market). This latter assumption mimics the present configuration of permit markets (see Victor 2007).

Using the FEEM-RICE model, we will analyze the incentives to move away from the present situation where the EU, Japan and Russia are committed to complying with their Kyoto targets and where the other countries/regions are free to determine their climate policy unilaterally. Therefore our benchmark case, or business-as-usual scenario, to which we compare different potential climate regimes, is the coalition formed by the annex-B$_{-US}$ countries.

Our focus is on post-2012 scenarios. We assume that a global agreement is only one of the possible outcomes of climate negotiations. Countries are also free to form regional or subglobal agreements. Therefore we consider situations where countries that now belong to the Kyoto coalition may decide, according to their own economic interests, to leave the Kyoto coalition and cooperate on GHG emission control with other countries/regions.[8] The time horizon over which climate policy is optimized is 2010 to 2100. When analyzing the decision to leave or join a coalition, we adopt the "open membership" rule, which implies that nonmembers can join and leave an existing coalition even without the consensus of the other members.

7.3 Results

In this section we plan to identify the stable coalition structures of the game. For this purpose we compare different regimes both with respect to the benchmark climate coalition formed by the annex-B$_{-US}$ countries, and with respect to all other different regimes. Comparisons will be performed in terms of each region's payoff (total domestic welfare). In order to limit the number of coalition structures to be compared, we simplify the policy framework as follows: First, given that the inclusion of the least-developed countries is very unlikely in the next stages of climate negotiations, the rest of the world (ROW) is exempted from potential short- to medium-term emission reduction commitments. The player ROW is thus a free rider in all our policy scenarios. Symmetrically, given its strong commitments to emission reductions, Europe cannot be a free rider, meaning it always belongs to a climate coalition.

Given these constraints and the FEEM RICE model, our first conclusion is that no coalition structure involving the five negotiating players

(ROW is always a free rider) is internally and externally stable. This result is consistent with previous findings using the RICE model (e.g., see Bosello et al. 2003 where only one coalition structure is found to be stable, and Carraro, Eyckmans, and Finus, 2006 where no stable coalition can be found in the absence of transfers).[9]

This result suggests that stability analysis does not seem to be a useful exercise in analyzing the future prospects of climate policy unless one wants to conclude that only unilateral measures are likely to be implemented in the future. This conclusion obviously depends on the model used for the empirical analysis and on the abatement targets assumed for the next negotiation stages (in Buchner and Carraro 2005b, we analyzed whether different future emission targets could increase the prospect for a stable coalition, but this was not the case). It also depends on the absence of transfers (in Carraro, Eyckmans, and Finus 2006, the introduction of transfers is shown to yield some stable coalition structures) and on the initial allocation of emission rights (Bosello et al. 2003 show that more equitable allocations slightly increase the number of stable coalition structures).

In this chapter we adopt a different approach. Rather than looking for additional instruments (transfers, allocations of emission rights, linkages with other economic issues[10]) that may yield a stable coalition structure, we use the payoffs associated to all feasible coalition structures as indicators of each region/country's preferences for different alternative coalition structures. This way we can identify the most preferred coalition structures for each country/region of the world and then analyze whether the preferences can help identify a coalition that is likely to be chosen by a sizable set of countries. In doing so, we can check whether a given coalition structure is at least profitable to all countries/regions belonging to that coalition.

Our results are summarized in tables 7.1a, 7.1b, and 7.2. Tables 7.1a and 7.1b show the ranking of climate coalition structures according to domestic welfare. Table 7.2 shows the ranking of climate regimes according to global welfare and global emissions. These tables enable us to identify the most likely behavior of countries in future climate negotiations (at least to the extent that economic incentives affect climate negotiations).

Let us first focus on the United States. The two most preferred coalition structures are those where the United States is not involved in any climate coalition (see table 7.1a). Note that for the United States the most preferred regime is the present annex-B$_{US}$ coalition. However,

Table 7.1a
Ranking of climate regimes according to domestic total welfare: US, JPN, and EU

USA	JPN	EU
(JPN, EU, FSU)	(EU, FSU)	(JPN, EU, CHN, FSU)
(JPN, CHN) and (EU, FSU)	(JPN, CHN) and (EU, FSU)	(USA, JPN, EU, CHN, FSU)
(JPN, EU)	(JPN, EU, CHN, FSU)	(JPN, CHN) and (EU, FSU)
(EU, FSU)	(USA, JPN, EU, CHN, FSU)	(EU, FSU)
(JPN, EU, CHN, FSU)	(JPN, EU, FSU) and (USA, CHN)	(JPN, EU, FSU) and (USA, CHN)
(JPN, EU, FSU) and (USA, CHN)	(JPN, EU, FSU)	(JPN, EU, FSU)
(USA, JPN, EU, CHN, FSU)	(USA, JPN, EU, FSU)	(USA, JPN, EU, FSU)
(JPN, EU) and (USA, FSU)	(JPN, EU) and (USA, FSU)	(JPN, EU) and (USA, FSU)
(USA, JPN, EU, FSU)	(JPN, EU)	(JPN, EU)

Table 7.1b
Ranking of climate regimes according to domestic total welfare: CHN and FSU

CHN	FSU
(JPN, EU) and (USA, FSU)	(JPN, EU)
(USA, JPN, EU, FSU)	(USA, JPN, EU, FSU)
(JPN, EU, FSU)	(JPN, EU, FSU)
(JPN, EU)	(JPN, EU, FSU) and (USA, CHN)
(EU, FSU)	(JPN, EU, CHN, FSU)
(JPN, EU, FSU) and (USA, CHN)	(EU, FSU)
(USA, JPN, EU, CHN, FSU)	(JPN, CHN) and (EU, FSU)
(JPN, EU, CHN, FSU)	(JPN, EU) and (USA, FSU)
(JPN, CHN) and (EU, FSU)	(USA, JPN, EU, CHN, FSU)

the most preferred regime by the United States when it participates in a climate coalition is that in which it cooperates (and trades) with China. A second cooperative bloc is formed by the annex-B$_{US}$ coalition; that is to say, US prefers the coalition structure [(JPN, EU, FSU), (USA, CHN)].

The ranking of climate coalition structures for the other two main industrialized countries—European Union and Japan—shows some similarities. Both the European Union and Japan rank the importance of the present Kyoto coalition very low, thus suggesting that a post-

Table 7.2
Ranking of climate regimes according to global welfare and global GHG emissions

Global GHG emissions	Global welfare
(USA, JPN, EU, CHN, FSU)	(JPN, EU, CHN, FSU)
(JPN, EU, FSU) and (USA, CHN)	(JPN, CHN) and (EU, FSU)
(JPN, EU) and (USA, FSU)	(EU, FSU)
(USA, JPN, EU, FSU)	(JPN, EU, FSU)
(JPN, CHN) and (EU, FSU)	(USA, JPN, EU, CHN, FSU)
(JPN, EU, FSU)	(JPN, EU, FSU) and (USA, CHN)
(JPN, EU, CHN, FSU)	(JPN, EU)
(EU, FSU)	(USA, JPN, EU, FSU)
(JPN, EU)	(JPN, EU) and (USA, FSU)

2012 change is likely to change this attitude. And both rank cooperation with China high. Indeed the coalition (JPN, EU, CHN, FSU) is the European Union's most preferred regime, whereas Japan ranks first the coalition structure consisting of two blocs, the Asian bloc (JPN and CHN) and the European bloc (EU plus FSU). In both cases the European Union and Japan can profit from either a large emissions permit market or at least the presence of an important permit supplier, which implies a low permit price and thus low abatement costs. The worst regime for the European Union (and for Japan) is one where they form a coalition without having any large permit supplier at their disposal.

Note that the coalition structure [(JPN, EU, FSU), (USA, CHN)], which is the climate regime where the US and CHN cooperate within one bloc while EU, FSU and JPN cooperate within a second bloc, is ranked fifth both by the European Union and Japan. However, there are also some differences in the preferences of the European Union and Japan. In particular, large coalitions are more preferred by the European Union than by Japan.

Let us analyze the preferences of less developed countries—CHN and FSU. China acts as a rational free rider. Its preferred regime is the two-bloc regime in which EU cooperates with JPN and US with FSU, and its second-best option is also a regime in which CHN free rides. CHN's most preferred regime when it participates in a climate coalition is one where it cooperates (and trades) with US, while a second cooperative bloc is formed by the annex-B$_{US}$ coalition. The possibility of an Asian bloc appears to restrict its potential advantage with respect to gains from the emission market.

Finally, FSU is penalized by CHN's participation in a climate regime because CHN has lower marginal abatement costs and therefore replaces FSU as the large supplier of emissions permits. Clearly, FSU should avoid coalitions in which CHN also participates. FSU's most preferred regime is the annex-B$_{US}$ coalition, where CHN is not involved and FSU represents the only permit seller.

What are the policy lessons that can be derived from tables 7.1a and 7.1b? As seen above, US and CHN have a strong incentive to free ride, namely to set their environmental policy unilaterally and profit from the abatement levels set for the Kyoto coalition countries. In particular, the annex-B$_{US}$ coalition is US's most preferred regime and is ranked third by CHN. The annex-B$_{US}$ coalition is also good for FSU, for which this is the second-best outcome when it decides not to free-ride.

EU and JPN have a strong incentive to maintain cooperation with a large permit seller, at least with FSU. Indeed the worst coalition structures for the EU and for JPN are where they both form a coalition without either CHN or FSU.

In short, the climate coalition structure where only EU, JPN, and FSU cooperate is fairly robust in terms of economic incentives (though not stable); it is highly ineffective from an environmental viewpoint, as is demonstrated by table 7.2. Paradoxically, this regime is not particularly welcome by the EU and Japan but is the most preferred regime of other countries.

What other possible climate regime has some economic incentives for the participating countries/regions? It is clear that FSU does not like to cooperate with CHN, because of the losses that it would suffer in the permit market. CHN would like to free ride, but if it cooperates, its preferred coalition structure is [(JPN, EU, FSU), (USA, CHN)], and the same goes for US. The EU would prefer a large coalition, whereas JPN likes a regional two-bloc coalition (when it does not free ride). Therefore, if US and CHN should decide to cooperate to control their GHG emissions, they would sign a bilateral agreement rather than join a large global coalition. The conclusion is that the coalition structure [(JPN, EU, FSU), (USA, CHN)] has an increased likelihood of replacing the present fairly robust coalition structure [(JPN, EU, FSU), USA, CHN].

Let us look at the coalition structure [(JPN, EU, FSU), (USA, CHN)] more closely. Is it profitable to all countries? Is it environmentally effective? Are there elements in the real world policy process that suggest this coalition structure to be feasible? Let us first look at this latter aspect. China's decision to ratify the Kyoto Protocol demonstrates that the country expects benefits from ratification to be high because China is the largest permit seller. Chinese officials have already claimed that the government will voluntarily try to restrict the growth of CO_2 emissions, though it is strictly opposing binding GHG reduction targets (*The Japan Times*, January 26, 2002). Overall, the CHN strategy appears to be strongly linked to the moves of US, and together these two countries could accomplish a breakthrough in international climate cooperation.

Without binding commitments or with very mild abatement targets, and given the consequent high amount of permits that could be supplied, CHN is a very attractive partner in climate change control activities. This is why US should choose to cooperate with CHN under a joint climate pact. This way US could achieve two goals: (1) satisfy domestic political requirements by involving developing countries in their climate strategy, and (2) reap high benefits from a large joint emissions market (today US and CHN together account for more than one-third of the world CO_2 emissions and this share is growing). In particular, US could drastically decrease its abatement costs through emission trading, and CHN could profit from selling a large amount of permits.

What are the main economic and environmental implications of this coalition structure? Figure 7.1 shows that both US and CHN lose if either chooses to free ride. Therefore their coalition structure is not profitable However, the loss for US is small, and it can be largely compensated by ancillary benefits from GHG emission abatement that are not taken into account in our model.

The loss for CHN is also small given that the ancillary benefits, both on the environmental and economic side, can be large. As noted above, this coalitions structure is the most favorable to both US and CHN provided that they do not free ride and attempt at some form of cooperative emissions abatement.

The coalitions structure [(JPN, EU, FSU), (USA, CHN)] is a bit more beneficial for the Kyoto climate bloc consisting of EU, JPN, and FSU because of the enhanced environmental effectiveness of this two-bloc regime. Indeed GHG emissions are almost 20 percent lower than in the benchmark case (see figure 7.1). This two-bloc climate regime is also characterized by a large expansion of CHN's R&D investments. CHN overinvests in R&D to increase its share of sales in the bilateral emissions trading market. The segmentation of the trading market

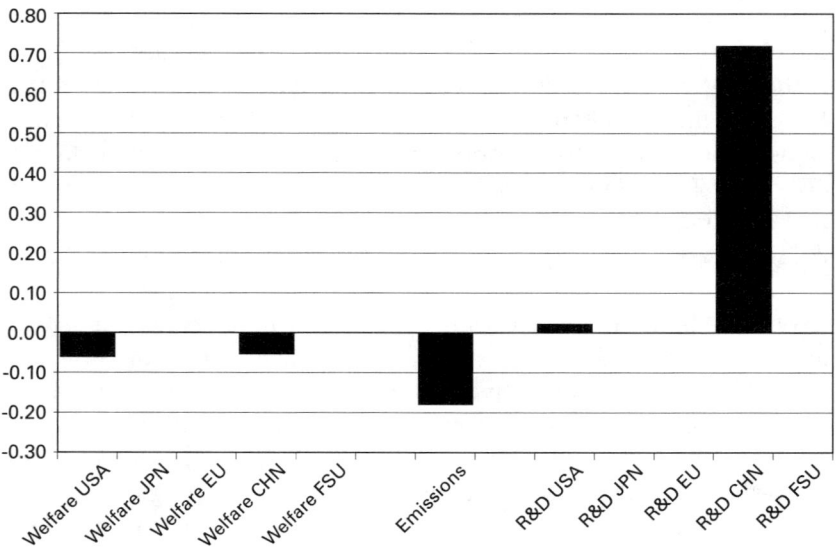

Figure 7.1
A third climate regime with two blocs: (1) US and CHN; (2) EU, FSU, and JPN

explains why R&D investments within the benchmark annex-B$_{US}$ coalition do not change. However, if the comparison is made with the coalition (annex B$_{US}$ + China), then it can be seen that R&D investments in this climate regime are higher for all annex-B$_{US}$ countries. The reason is again the larger marginal abatement costs when CHN is not a seller in the permit market. This induces higher investments in R&D in EU and JPN and also strategic R&D investments in FSU, which will find it optimal to increase its supply of permits.

Summing up, the coalition structure [(JPN, EU, FSU), (USA, CHN)] is neither stable nor profitable according to the definitions of section 7.1, but the coalition (JPN, EU, FSU) is profitable and stable. CHN and US suffer a small loss when cooperating inside this coalition structure, but the loss is the smallest one among all possible coalition structures in which US and CHN cooperate. The conclusions above are based on a decentralized analysis of each country's incentives to join a climate coalition. However, it is important to assess what a central planner would do when faced with the goal of maximizing global welfare. The answer is provided by table 7.2, where in the second column global welfare, which includes the welfare of free riders, is maximized by the coalition structure [(JPN, EU, CHN, FSU), USA].[11] Next is the coalition structure [(JPN, CHN), (EU, FSU), USA], where the climate regime that allows the US to cooperate with CHN and the annex B$_{-US}$ forms as a second bloc. The [(JPN, EU, FSU), (USA, CHN)] bloc ranks only sixth. Notice that global welfare seems to be maximized when US does not belong to a climate coalition, and this suggests that unconstrained growth of US economy is beneficial to the world economy despite the damage to the environment. Next notice that the two-bloc climate regime [(JPN, EU, FSU), (USA, CHN)] can provide the second largest possible benefit for the environment (see table 7.2, first column).

Summarizing, if for some environmental, economic, or political reason US and CHN decide to cooperate to control their GHG emissions, then the probability is quite high that they will sign a bilateral agreement rather than join a large global coalition. This coalition structure would correspond to [(JPN, EU, FSU), (USA, CHN)], which slightly increases welfare in EU, JPN, and FSU, at least with respect to the present Kyoto coalition (see figure 7.1). Therefore the annex-B$_{-US}$ countries could accept a two-bloc regime, where US and CHN cooperate on emission abatement and trade permits in a bilateral permit market. The economic loss for US and CHN would be small with respect to the situation where they free ride, but the global environmental benefits would be large.

7.4 Conclusions

The conclusions emerging from the chapter analysis can be summarized as follows: A move from the present climate regime in which the United States and China do not cooperate to reduce their GHG emissions is not likely, at least in the near future. The United States is more likely to adopt unilateral policies than to join a coalition to control GHG emissions. However, it is unlikely that, at least in the medium term, the United States will continue to reject any form of cooperation on climate change control. If the United States decides to cooperate, the climate regime that is least opposed (in terms of net economic benefits) by the negotiating countries is the one in which China and the United States cooperate bilaterally and the annex-B$_{-US}$ countries form a parallel coalition.

Of course, this argument must be taken cautiously. First, the analysis is based only on economic incentives, whereas political decisions would constitute another incentive base. Second, we did not account for the link between climate negotiations and other international negotiation processes (on crime, trade, terrorism, technology, etc.). Third, the FEEM-RICE model used in this study is a simplified representation of the world economic system, even though it captures the main economic mechanisms and the related incentive schemes. Additional research using other models would be beneficial.

Still the results derived from the empirical analysis are consistent with suggestions and results proposed in the game-theory literature (where country asymmetries are usually neglected). So the relevant economic incentives are likely to be captured in the analysis of this chapter, which may therefore serve to provide indications on the prospective future evolution of negotiations on climate change control.

Appendix: The FEEM-RICE Model

The FEEM-RICE model is an extension of Nordhaus and Yang's (1996) regional RICE model of integrated assessment, which is one of the most popular and manageable integrated assessment tools for the study of climate change (e.g., see Eyckmans and Tulkens 2003). It is basically a single-sector optimal growth model that has been extended to incorporate the interaction between economic activities and climate. One such model has been developed for each macro region into which the world is divided (the United States, Japan, Europe, China, former Soviet Union, and rest of the world).

Within each region a central planner chooses the optimal paths of fixed investment and emission abatement that maximise the present value of per capita consumption. Output (net of climate change) is used for investment and consumption and is produced according to constant returns Cobb-Douglas technology, which combines the inputs from capital and labor with the level of technology. Population (taken to be equal to full employment) and technology levels grow over time in an exogenous fashion, whereas capital accumulation is governed by the optimal rate of investment. There is a wedge between output gross and net of climate change effects, the size of which is dependent on the amount of abatement (rate of emission reduction) as well as on the change in global temperature. The model is completed by three equations representing emissions (which are related to output and abatement), carbon cycle (which relates concentrations to emissions), and climate module (which relates the change in temperature relative to 1990 levels to carbon concentrations) respectively.

In our extension of the model, technical change is no longer exogenous. The issue of endogenous technical change is tackled instead by following the ideas contained in both Nordhaus (1999) and Goulder and Mathai (2000), and accordingly modifying Nordhaus and Yang's (1996) RICE model. Doing so requires the input of a number of additional parameters, some of which have been estimated using information provided by Coe and Helpman (1995), while the remaining parameters are calibrated so as to reproduce the business-as-usual scenario generated by the RICE model with exogenous technical change.

In particular, the following factors are included: first, endogenous technical change affecting factor productivity is introduced. This is done by adding the stock of knowledge in each production function and by relating the stock of knowledge to R&D investments. Second, induced technical change is introduced, by allowing the stock of knowledge to affect the emission-output ratio as well. Finally, international technological spillovers are accounted for in the model.

To determine the optimal value of all control variables, including their own GHG abatement strategy, countries play a noncooperative Nash game in a dynamic setting, which yields an open loop Nash equilibrium (see Eyckmans and Tulkens 2003 for an explicit derivation of first-order conditions of the optimum problem). This is a situation where, in each region, the planner maximizes social welfare subject to the individual resource and capital constraints and the climate module, given the emission and investment strategies (in the base case) and the R&D expenditure strategy (in the endogenous technological change case) of all other players.

The Standard Model without Induced Technical Change

As previously mentioned, it is assumed for the purpose of this model that innovation is brought about by R&D spending that contributes to the accumulation of the stock of existing knowledge. Following an approach pioneered by Griliches (1979, 1984), it is assumed that the stock of knowledge is a factor of production, which therefore enhances the rate of productivity (see also the discussion in Weyant 1997; Weyant and Olavson 1999). In this formulation, R&D efforts prompt nonenvironmental technical progress, but with different modes and elasticities. More precisely, the RICE production function output is modified as follows:

$$Q(n,t) = A(n,t)K_R(n,t)^{\beta_n}[L(n,t)^{\gamma}K_F(n,t)^{1-\gamma}], \tag{7A.1}$$

where Q is output (gross of climate change effects), A the exogenously given level of technology, and K_R, L, and K_F are respectively the inputs from knowledge capital, labor, and physical capital.

In (7A.1) the stock of knowledge has a region-specific output elasticity equal to β_n ($n = 1, \ldots, 6$). It should be noted that as long as this coefficient is positive, the output production process is characterized by increasing returns to scale, in line with current theories of endogenous growth. This implicitly assumes the existence of cross-sectoral technological spillovers within each country (Romer 1990). In addition it should be noted that while allowing for R&D-driven technological progress, we maintain the possibility that technical improvements can also be determined exogenously (the path of A is the same as that specified in the original RICE model). The stock accumulates in the usual fashion:

$$K_R(n, t+1) = R\&D(n,t) + (1 - \delta_R)K_R(n,t), \tag{7A.2}$$

where $R\&D$ is the expenditure in Research and Development and δ_R is the rate of knowledge depreciation. Finally, it is recognized that some resources are absorbed by R&D spending. That is,

$$Y(n,t) = C(n,t) + I(n,t) + R\&D(n,t), \tag{7A.3}$$

where Y is net output (net of climate change effects as specified in the RICE model), C is consumption, and I gross fixed capital formation.

At this stage the model maintains the same emissions function as Nordhaus's RICE model, which will be modified in the next section:

$$E(n,t) = \sigma(n,t)[1 - \mu(n,t)]Q(n,t), \tag{7A.4}$$

where σ can be loosely defined as the emissions–output ratio, E stands for emissions, and μ for the rate of abatement effort. The policy variables included in the model are rates of fixed investment and of emission abatement. For the other variables the model specifies a time path of exogenously given values. Interestingly this is also the case for technology level A and of the emissions–output ratio σ. Thus the model presented so far assumes no induced technical change, namely no exogenous environmental technical change, and a formulation of productivity that evolves both exogenously and endogenously. In the model, investment fosters economic growth (thereby driving up emissions) while abatement is the only policy variable used for reducing emissions.

Induced Technical Change

In the second step of our model formulation, endogenous *environmental* technical change is accounted for. It is assumed that the stock of knowledge—which in the previous formulation was only a factor of production—also serves the purpose of reducing, ceteris paribus, the level of carbon emissions. Thus, in the second formulation, R&D efforts prompt both environmental and nonenvironmental technical progress, although with different modes and elasticities.[12] More precisely, the RICE emission–output relationship is modified as follows:

$$E(n,t) = [\sigma_n + \chi_n \exp(-\alpha_n K_R(n,t))][1 - \mu(n,t)]Q(n,t). \qquad (7A.4')$$

In (7A.4') knowledge reduces the emissions–output ratio with an elasticity of α_n, which is also region-specific; the parameter χ_n is a scaling coefficient, whereas σ_n is the value to which the emission–output ratio tends asymptotically as the stock of knowledge increases without limit. In this formulation, R&D contributes to output productivity, on the one hand, and affects the emission–output ratio—and therefore the overall level of pollution emissions—on the other.

Knowledge Spillovers

Previous formulations do not include the effect of potential spillovers produced by knowledge, and therefore ignore the fact that both technologies and organizational structures disseminate internationally. Modern economies are linked by vast and continually expanding flows of trade, investment, people, and ideas. The technologies and choices of one region are and will inevitably be affected by developments in other regions.

Following the work of Weyant and Olavson (1999), who suggest that the definition of spillovers in an induced technical change context be kept plain and simple (in the light of a currently incomplete understanding of the problem), disembodied, or knowledge, spillovers are modeled (see Romer 1990). The spillovers of knowledge refer to the R&D carried out and paid for by one party that produces benefits to other parties, which then have better or more inputs than before or can somehow benefit from R&D carried out elsewhere. Therefore, in order to capture international spillovers of knowledge, the stock of world knowledge is introduced in the third version of the FEEM-RICE model, both in the production function and in the emission–output ratio equation. Equations (7A.1) and (7A.4′) are then revised as follows:

$$Q(n,t) = A(n,t) K_R(n,t)^{\beta_n} WK_R(n,t)^{\varepsilon_n} [L(n,t)^{\gamma} K_F(n,t)^{1-\gamma}] \tag{7A.1′}$$

and

$$E(n,t) = [\sigma_n + \chi_n \exp(-\alpha_n K_R(n,t) - \theta_n WK_R(n,t))][1 - \mu(n,t)] Q(n,t), \tag{7A.4″}$$

where the stock of world knowledge

$$WK_R(j,t) = \sum_{j \neq i} K_R(i,t) \tag{7A.5}$$

is defined in such a way as not to include a country's own stock.

Emission Trading

As mentioned above, throughout the analysis we assume the adoption of efficient policies. As a consequence the model includes the possibility of emissions trading. When running the model in the presence of emissions trading, two additional equations are considered:

$$Y(n,t) = C(n,t) + I(n,t) + R\&D(n,t) + p(t) NIP(n,t), \tag{7A.3′}$$

which replaces equation (7A.3) and

$$E(n,t) = Kyoto(n) + NIP(n,t), \tag{7A.6}$$

where $NIP(n,t)$ is the net demand for permits and $Kyoto(n)$ are the emission targets set in the Kyoto Protocol for the signatory countries and the BAU levels for the nonsignatory ones. According to (7A.3′), resources produced by the economy must be devoted, in addition to consumption, investment, and research and development, to net purchases of emissions permits. Equation (7A.6) states that a region's emissions may exceed the limit set in Kyoto if permits are bought, and vice versa in the case of sales of permits. Note that $p(t)$ is the price of a unit of tradable emission permits expressed in terms of the *numéraire* output price. Moreover there is an additional policy variable to be considered in this case, which is net demand for permits *NIP*.

In terms of the possibility of emission trading, the sequence whereby Nash equilibrium is reached can be described as follows: each region maximizes its utility subject to its individual resource and capital constraints, now including the Kyoto constraint and the climate module for a given emission (i.e., abatement) strategy of all the other players and for a given price of permits $p(0)$ (in the first round this is set at an arbitrary level). When all regions have made their optimal choices, the overall net demand for permits is computed at the given price. If the sum of net demands in each period is approximately zero, Nash equilibrium is obtained; otherwise, the price is revised as a function of the market disequilibrium and each region's decision process starts again.

Notes

This chapter is part of the research work being carried out by the Climate Change Modelling and Policy Unit at Fondazione Eni Enrico Mattei. The authors are grateful to Christian Egenhofer, Frank Convery, Johan Eyckmans, Henry Tulkens, two anonymous referees, and the participants at the FEEM-Stanford Conference on Post 2012 Climate Policy: Architectures and Participation Scenarios in Venice, 20–21 June 2005 for helpful suggestions and remarks. The usual disclaimer applies.

1. Most papers have been presented at the annual workshops of the Coalition Theory Network (see www.feem.it/ctn). Some of them are published in Carraro (2003) and in Demange et al. (2005).

2. This simple theoretical framework has often been adopted in the literature on international environmental agreements where the assumption of a coalition structure with a single coalition is the most obvious and realistic and where the game is characterized by positive externalities. A more general framework is sometimes used in coalition theory (Bloch 2003) but would not be useful for the purpose of this chapter.

3. This has been called a partial agreement Nash equilibrium by Chander and Tulkens (1997). Our analysis is in line with the mainstream literature on coalition theory. For an overview, see Bloch (2003) and Yi (2003).

4. This definition of coalitional stability is due to d'Aspremont et al. (1983) and has been frequently applied in the literature on international environmental agreements, for instance, by Barrett (1994), Carraro and Siniscalco (1993), Hoel (1992), and followed by many scholars.

5. The FEEM-RICE model has already been used in Bosello et al. (2003), Bosetti et al. (2005), Buchner and Carraro (2005a, b, 2006), Buchner et al. (2005), and Buonanno et al. (2002).

6. Note also that our analysis focuses only on CO_2. There are other greenhouse gases caused by humans and the Kyoto Protocol takes some of them into account. Moreover both the Bonn agreement and the subsequent Marrakech deal emphasize the role of sinks in meeting the Kyoto targets. As shown by several recent analyses (e.g., Manne and Richels 2001; Jensen and Thelle 2001), the inclusion of the other greenhouse gases and of sinks would further reduce mitigation costs.

7. The use of the "Kyoto forever" hypothesis may be seen as a strong assumption. However, the CO_2 concentration levels implicit in this assumption (if FEEM-RICE is a good description of the world) coincide with those in the A1B scenario (IPCC 2001), which can be considered the "median" scenario among those currently proposed. We thus use the "Kyoto forever" hypothesis not because it represents a realistic scenario but as a benchmark with respect to which policy alternatives can be compared.

8. Notice that the rest of the world (ROW) has been exempted from possible future climate commitments because the policy indications are that these countries is very unlikely to be included in the next stage of climate negotiations.

9. The result in this chapter is slightly stronger than in Bosello et al. (2003) because of the additional constraints imposed on some players of the game (e.g., ROW is always is a free rider) and because of the more demanding target assumed for China.

10. The role of issue linkage in explored in Buchner et al. (2005).

11. The fact that the grand coalition does not appear first in the ranking of global welfare is due to the exemption of ROW from future emission abatement commitments; that is, in our analysis ROW is always a free rider.

12. Obviously we could have introduced two different types of R&D efforts, contributing, on the one hand, to the growth of an environmental knowledge stock and, on the other, to the growth of a production knowledge stock. Such an undertaking, however, is made difficult by the need to specify variables and calibrate parameters for which there is no immediately available and sound information in the literature.

References

Aumann, R., and J. Drèze. 1974. Cooperative games with coalition structures. *International Journal of Game Theory* 3: 217–37.

Barrett, S. 1994. Self-enforcing international environmental agreements. *Oxford Economic Papers* 46: 878–94.

Barrett, S. 2002. *Environment and Statecraft*. Oxford: Oxford University Press.

Bloch, F. 1995. Endogenous structures of associations in oligopolies. *RAND Journal of Economics* 26: 537–56.

Bloch, F. 1996. Sequential formation of coalitions in games with externalities and fixed payoff division. *Games and Economic Behavior* 14: 90–123.

Bloch, F. 2003. Noncooperative models of coalition formation in games with spillovers. In C. Carraro, ed., *The Endogenous Formation of Economic Coalitions*. Cheltenham: Edward Elgar.

Bosello, F., B. Buchner, and C. Carraro. 2003. Equity, development and climate change control. *Journal of the European Economic Association* 1(2–3): 601–11.

Bosetti, V., C. Carraro, and M. Galeotti. 2005. The dynamics of carbon and energy intensity in a model of endogenous technical change. FEEM Working paper 6.2005; forthcoming in the *Energy Journal*.

Buchner, B., and C. Carraro. 2005a. Economic and environmental effectiveness of a technology-based climate regime. *Climate Policy* 4: 229–48.

Buchner, B., and C. Carraro. 2005b. Modelling climate policy: Perspectives on future negotiations. *Journal of Policy Modeling* 27(6): 711–32.

Buchner, B., and C. Carraro. 2006. US, China and the economics of climate negotiations. *International Environmental Agreements: Law Politics Economics* 6: 63-89.

Buchner, B., C. Carraro, I. Cersosimo, and C. Marchiori. 2005. Back to Kyoto? US participation and the linkage between R&D and climate cooperation. In A. Haurie and L. Viguier, eds., *Coupling Climate and Economic Dynamics*. Dordrecht: Kluwer Academic, pp. 173–204.

Buonanno, P., C. Carraro, and M. Galeotti. 2002. Endogenous induced technical change and the costs of Kyoto. *Resource and Energy Economics* 524: 11–35.

Carraro, C. 1998. Beyond Kyoto: A game theoretic perspective. In the *Proceedings of the OECD Workshop on Climate Change and Economic Modelling. Background Analysis for the Kyoto Protocol*. Paris.

Carraro, C. 1999. The structure of international agreements on climate change. In C. Carraro, ed., *International Environmental Agreements on Climate Change*. Dordrecht: Kluwer Academic.

Carraro, C. ed. 2003. *The Endogenous Formation of Economic Coalitions*. Cheltenham: Edward Elgar.

Carraro, C., J. Eyckmans, and M. Finus. 2006. Optimal transfers and participation decisions in international environmental agreements. *Review of International Organizations*, forthcoming.

Carraro, C., and C. Marchiori. 2003. Stable coalitions. In C. Carraro, ed., *The Endogenous Formation of Economic Coalitions*. Cheltenham: Edward Elgar.

Carraro, C., and D. Siniscalco. 1993. Strategies for the international protection of the environment. *Journal of Public Economics* 52: 309–28.

Chander, P., and H. Tulkens. 1997. The core of an economy with multilateral environmental externalities. *International Journal of Game Theory* 26: 379–401.

Coe, D. T., and E. Helpman. 1995. International R&D spillovers. *European Economic Review* 39: 859–87.

D'Aspremont, C. A., A. Jacquemin, J. J. Gabszewicz, and J. Weymark. 1983. On the stability of collusive price leadership. *Canadian Journal of Economics* 16: 17–25.

Demange, G., D. Ray, and M. Wooders, eds. 2005. *Group Formation in Economics: Networks, Clubs and Coalitions*. Cambridge: Cambridge University Press.

Egenhofer, C., W. Hager, and T. Legge. 2001. Defining Europe's near abroad in climate change: A Russian–EU alliance; sub-global bargaining to further international environmental agreements. CEPS Discussion Paper.

Eyckmans, J., and H. Tulkens. 2003. Simulating coalitionally stable burden sharing agreements for the climate change problem. *Resource and Energy Economics* 25: 299–327.

Finus, M., and B. Rundshagen. 2003. Endogenous coalition formation in global pollution. In C. Carraro, ed., *The Endogenous Formation of Economic Coalitions*. Cheltenham: Edward Elgar.

Goulder, L. H., and K. Mathai. 2000. Optimal CO_2 abatement in the presence of induced technological change. *Journal of Environmental Economics and Management* 39: 1–38.

Griliches, Z. 1979. Issues in assessing the contribution of R&D to productivity growth. *Bell Journal of Economics* 10: 92–116.

Griliches, Z. 1984. *R&D, Patents, and Productivity*. Chicago: University of Chicago Press.

Hoel, M. 1992. International environmental conventions: The case of uniform reductions of emissions. *Environment and Resource Economics* 2: 141–59.

IPCC. 2001. *Third Assessment Report*. Cambridge: Cambridge University Press.

Jensen, J., and M. H. Thelle. 2001. What are the gains from a multi-gas strategy? FEEM Working paper 84.01. Milan.

Manne, A., and R. Richels. 1999. The Kyoto Protocol: A cost-effective strategy for meeting environmental objectives? In J. Weyant, ed., *The Cost of the Kyoto Protocol: A Multi-Model Evaluation*. Special Issue of the *Energy Journal* 23(2): 1–24.

Manne, A. S., and R. G. Richels. 2001. US rejection of the Kyoto Protocol: The impact on compliance costs and CO_2 emissions. Working paper 01-12. AEI-Brookings Joint Center for Regulatory Studies.

Nordhaus, W. D., and Z. Yang. 1996. A regional dynamic general-equilibrium model of alternative climate-change strategies. *American Economic Review* 4: 741–765.

Nordhaus, W. D. 1999. Modelling induced innovation in climate-change policy. Paper presented at the IIASA Workshop on Induced Technological Change and the Environment, Laxenburg, June 21–22.

Ray, D., and R. Vohra. 1997. Equilibrium binding agreements. *Journal of Economic Theory* 73: 30–78.

Ray, D., and R. Vohra. 1999. A theory of endogenous coalition structures. *Games and Economic Behavior* 26: 286–336.

Reinstein, R. A. 2004. A possible way forward on climate change. Mimeo. Reinstein and Associates, Inc.

Romer, P. 1990. Endogenous technological change. *Journal of Political Economy* 94: 1002–37.

Stewart, R., and J. Wiener. 2003. *Reconstructing Climate Policy*. Washington: American Enterprise Institute Press.

The Japan Times. China could help Japan by taking its money and cutting its Kyoto target, January 26, 2002. Available at ⟨www.japantimes.com⟩.

Victor, D. 2007. Fragmented carbon markets and reluctant nations: Implications for the design of effective architectures. In J. E. Aldy and R. N. Stavins, eds., *Architectures for Agreement: Addressing Global Climate Change in the Post-Kyoto World*, Cambridge University Press, Cambridge, pp. 133–60.

Yi, S.-S. 1997. Stable coalition structures with externalities. *Games and Economic Behaviour* 20: 201–23.

Yi, S.-S. 2003. Endogenous formation of economic coalitions: A survey of the partition function approach. In C. Carraro, ed., *Endogenous Formation of Economic Coalitions*, Cheltenham: Edward Elgar, pp. 80–127.

Yi, S.-S., and H. Shin. 1995. Endogenous formation of coalitions: Oligopoly. Mimeo. Department of Economics, Dartmouth College.

Weyant, J. P. 1997. Technological change and climate policy modeling. Paper presented at the IIASA Workshop on Induced Technological Change and the Environment, Laxenburg, June 26–27.

Weyant, J. P., and T. Olavson. 1999. Issues in modelling induced technological change in energy, environmental, and climate policy. *Environmental Modelling and Assessment* 4: 67–85.

8 Cooperation, Stability, and Self-enforcement in International Environmental Agreements: A Conceptual Discussion

Parkash Chander and Henry Tulkens

This chapter is not intended for game theorists—unless they are interested to learn something about how their products are being used. The chapter is intended instead for economists who make use of game theory concepts in analyses that may provide guidance on climate change negotiations.

In 1995 one of us presented a paper[1] that explored the issue of a grand worldwide coalition. The central question then was: Can a grand—worldwide—coalition prevail in climate decisions, or is the problem of such a logical structure that only treaties involving small groups of countries can be signed?

More than ten years later that debate has still not been brought to a close. Is the present exercise more than just a reprise? Has progress been made? During our receiving and selecting presentations for this conference, we observed that clarifications have indeed been made but more is still called for, be that only for ourselves. This state of affairs motivates our present contribution, whose structure should be clear enough from the chapter title.

We begin by introducing some notation. With N the set of all countries of the world, indexed $i = 1, \ldots, n$, let p_i denote the amount (flow[2]) of pollutant emissions in country i, let the value of the increasing function (with an upper bound) $g_i(p_i)$ denote the level of country i's GDP, and let the function $\pi_i(.)$ measure the total cost of damages caused in country i by the aggregate emissions $\sum p_i$.

In this setting we will call a *treaty* a joint choice by several countries of (1) an abatement policy, that is, a level of p_i for each of them, as well as (2) possible transfers of resources among them. In general equilibrium terms, this induces a "state"—or an "allocation"—of the simple international economy specified above.

In the absence of a treaty, we assume that each country chooses the abatement policy that suits it best, given the policies of the other countries, with no transfers. The resulting state of the international economy is the Nash equilibrium of a noncooperative game associated with the above-stated elements of the economy.

Efficiency for a group of countries, be it N or any subset S of N, is a joint policy of the group members that maximizes the group's aggregate welfare W. In the case of N, this objective reads

$$W_N = \sum_N \left[g_i(p_i) - \pi_i \left(\sum_N p_j \right) \right],$$

where all summation signs refer to indexes running from 1 to n. If the group is a (proper) subset S of N, however, the maximand is denoted W_S with the first summation in the expression above including only the members of S, whereas the second summation applies to all countries in N. This difference characteristically makes the problem of an international environmental agreement (IEA) as one of externalities.

8.1 Cooperation

On the notion of cooperation leading to IEAs, we can distinguish between two views. One is economic theoretic, and the other is game theoretic.

8.1.1 Economic Rationale for Cooperation

The economic view finds its justification in the public good or diffuse characteristic of the externality generated by the emissions that cause climate change. Because the public good is global, that is, worldwide, elementary public goods theory (Samuelson 1954) teaches us that efficiency (in the Pareto sense) can be reached only if all concerned parties are involved in the process of resource allocation required to control the externality in question. Thus, getting all parties involved—be it by sharing cost or by revealing preferences, or both, or instead by other means—is an essential requirement for efficiency. Economically the social objective of efficiency entails the necessity of cooperation. Samuelson saw only the state as the appropriate actor for this purpose.

Apart from the public good characteristic just emphasized, that is, the diffuse nature of the externalities under discussion, another economic argument for cooperation in the sense of engaging in bargaining in the presence of externalities is provided by the Coase theorem. IEAs may be viewed as outcomes of voluntary negotiations between generators and recipients of externalities, as described by Coase (1960). His view concludes at an efficient outcome regardless of whether or not the rights to the emissions externality are assigned to the parties. In international environmental affairs polluting countries arrogate to themselves these rights first, and thus prompt negotiations that attempt at an efficient (or at least Pareto superior) outcome. However, the notion of applying the Coasian argument in this context has been recently questioned in a series of papers by Ray and Vohra[3] who conclude that "robust inefficient outcomes" could exist. The issue is discussed further below.

8.1.2 Cooperative Games

The game-theoretic perspective is that offered by the theory of cooperative games. Cooperative game theory flourished in the 1960s and 1970s most prominently within the Jerusalem school of game theory and produced a wealth of "solution concepts" on the outcome of games ("social interactions" is the more recent and more apt expression) where coalitions of players are the objects of analysis. These developments occurred independently of public goods and externality theory.

Surprisingly, it is difficult to find in this literature arguments explaining and justifying the phenomenon of cooperation. Section 8.1 of Myerson's (1991) book, entitled "Noncooperative Foundations of Cooperative Game Theory," should provide an answer, but the author instead remarks on how "subtle" the concept is. The attractive idea of giving a noncooperative foundation to cooperative game theory (the "Nash scheme") strikes at a basic difficulty: the multiplicity of equilibria of the noncooperative games that might support the notion of cooperative solutions. Criteria are discussed at length for explaining how selection from among these equilibria might logically occur (focal arbitration of Schelling, institutions, contracts). These criteria are in one way or another inspired by the notion of efficiency: cooperation finds its raison d'être in the efficiency it allows to be achieved. Thus cooperation can find its starting place in the outcome of some process of bargaining among players.[4]

These arguments hardly explain, however, *how* groups are formed, as admitted by the quoted author. All game theory textbooks, when they come to their cooperative games chapters (if any), take formed groups as given, without inquiring how the groups got formed and why a joint objective of striving for efficiency can be attributed to the group.

8.1.3 Games with Externalities and the Core

Turning to cooperative games for analyzing IEAs can nevertheless be justified. Game theory has provided compelling arguments in support of competitive market exchanges, arguments pointing to strategic stability of market equilibria because they belong to the core of cooperative games associated with market exchanges.

It is thus natural to ask whether the core concept, if applied to international economies with externalities, can offer similar properties for Coasean agreements between generators and recipients. This question was raised in the early 1970s, but it was not clearly dealt with throughout that decade, nor in the 1980s, probably because of imprecise, unrealistic, or ad hoc representations of the externality phenomenon.[5] Typically the core theory in these applications oscillated between results of nonexistence and problems of nonconvexities; moreover the cooperative games as they were formulated were not really bearing on the multilateral and diffuse form of externalities that is commonly used nowadays (and recalled in the introduction to this chapter).

It may be argued that the formulation of "environmental externalities" became standard in the early 1990s because of its appropriateness for dealing with international environmental agreements, and in particular, in its close connection with the public good concept. IEAs appeared to be an ideal field of application, and the externalities analysis started to develop very quickly in the 1990s, after some early contributions such as those of Tulkens (1979) and Mäler (1989). The mentioned formulation of externalities also allowed for game-theoretic concepts to yield results in this field. In addition to Nash equilibrium, used in the two papers just mentioned, the core of a cooperative game was adapted to environmental externalities under the name "γ-core" by Chander and Tulkens (1995, 1997).[6]

Thus at least one major concept of cooperative game theory was imported in the IEA literature. One may wonder why and regret that other such concepts from the Jerusalem school alluded to above—the bargaining set, the kernel, the nucleolus, the Shapley value or the von Neumann-Morgenstern stable sets—have not been similarly more explored in the externalities context.[7]

8.1.4 Coalition Formation

At the early 1990s, however, there appeared in the IEA literature (Carraro and Siniscalco 1993, 1995; Barrett 1994) another category of arguments bearing on the *formation* of coalitions of countries and inspired from earlier cartel formation models.[8] Some authors later called this the noncooperative approach to IEAs.

This theory is built around the idea that a group (coalition) S forms or does not form, depending on whether the payoffs of all players are such that they pass the following two-sided test, called "internal and external stability":

$$\forall i \in S, \quad W_S^i > W_{S \setminus \{i\}}^i \quad \text{(internal stability of } S\text{)} \tag{8.1a}$$

and

$$\forall i \notin S, \quad W_S^i > W_{S \cup \{i\}}^i \quad \text{(external stability of } S\text{),} \tag{8.1b}$$

where for any $i \in N$ and any subset $S \subseteq N$, W_S^i denotes the payoff that i obtains when S forms and i is a member of S as in condition (8.1a), or i is not a member of S as in condition (8.1b).[9]

The 1995 paper mentioned at the outset criticized this concept because the definition, as just stated, does not make precise how the payoff of player i is determined when i is not in S, namely $W_{S \setminus \{i\}}^i$ in (8.1a) and W_S^i in (8.1b). This imprecision was later corrected by specifying that players not in a coalition S are assumed to maximize their individual payoffs $g_i(p_i) - \pi_i(\sum_N p_j)$ (i.e., act as singletons), just as the members of S are assumed to maximize their joint payoffs $\sum_{i \in S}[g_i(p_i) - \pi_i(\sum_N p_j)]$. But this is nothing else than what defines a "partial agreement equilibrium with respect to a coalition S" (PANE wrt S) introduced by Chander and Tulkens (1995–1997; recalled in note 6). Thus internal–external stability of a coalition S appears to be a property of the PANE with respect to that S.

Further progress occurred with the introduction in the IEA literature[10] of the notion of *games in partition function form*. With this tool the all players set N is split into a family of nonoverlapping and collectively exhaustive subsets, which defines what is called a *coalition structure*: each partition is such a structure. A coalition structure is an *equilibrium* coalition structure if it is shown to be a Nash equilibrium between the elements of the partition.[11] Other expressions are "multi-coalitional equilibrium," or even "fragmented equilibrium." If in addition the internal–external (I-E) stability test above is passed by each coalition of the structure, the equilibrium structure is called *I-E stable*. The motivation here is to assert that only coalitions that belong to a stable coalition structure are likely to form.

No analytic conditions[12] ensuring the existence of I-E stable equilibrium coalition structures for the standard IEA model have yet been provided to the best of our knowledge. However, Eyckmans and Finus (2006a, b) have explored the issue by means of numerical simulations with the specific CWS integrated assessment model.[13] They take all conceivable partitions of the set of six regions of the world that the model treats as "countries," compute a multi-coalitional equilibrium for each structure, and check for which of these structures all coalitions they comprise pass the I-E stability test. Similarly Buchner and Carraro (chapter 7 of this volume) examine with simulations on the FEEM-RICE model[14] the I-E stability of some conceivable coalition structures.

Having thus taken stock of the state of the art in coalition formation theory, the following question arises: In a multi-coalitional equilibrium, why is each coalition S assumed to achieve efficiency only within itself, among its members? In the standard IEA model here under scrutiny, it is well known that efficiency at the world level can only be achieved by the grand coalition of all countries. Hence, as each coalition strives for internal efficiency, why should this quest be limited to the members of S? Could coalitions strive as well for external efficiency, that is, contact other coalitions and adopt mutually beneficial and more efficient strategies?

If it can be shown that a resulting merged coalition is not I-E stable, then there is good argument for the merge not to take place. But if the merging coalition happens to be I-E stable, then it should be allowed to form. In the case of multiple equilibria the merged equilibrium coalition would Pareto dominate the equilibrium coalitions it is made of. Giving precedence to the equilibrium with the merge over that without the merge would be based on efficiency domination of the former. So we are driven back to a reasoning on coalition formation essentially led by efficiency considerations.[15]

Finally, it should be clear that, no more than what cooperative game theory has to offer, I-E stability criteria do not teach much on the process of *how* stable coalitions are formed.

8.1.5 An Axiomatic Approach

On the theme of coalition formation in games with externalities, Maskin (2003) has brought about a contribution based on other arguments. He (courageously) tackled the sequential process of discussions between players on whether or not they will act jointly. His analysis is grounded in an explicit axiomatic that bears (in part) on communication among the players. One of these axioms specifies that at some point any player is allowed to break communication lines between himself and (some of) the other players. On this basis, the conclusion is derived that the grand coalition will not form.

But doesn't the axiom in fact contain the conclusion? Apart from this issue, Maskin is to be commended for introducing the important factor of communication among players as a determinant of cooperation. He recognizes,[16] however, that absent this axiom, the grand coalition would form in the public good game of his paper (which is very close to the environmental model we deal with in the IEA literature), and thus efficiency would prevail.

To summarize, for the complementary themes of cooperation and coalition formation in games with externalities we have, on one hand, a γ-core theory that derives cooperation from the nondomination property; we have, on the other hand, (1) an incomplete Nash program that cannot explain cooperation, (2) an I-E stability theory that only explains the nonformation of some coalitions (among these the grand coalition N) and thus only supports partial cooperation, and (3) an axiomatic communication breakdown argument that attempts to explain the nonformation of a grand coalition.

8.2 Stability

8.2.1 Preliminaries

At the outset we need to clarify whether stability refers to coalitions or allocations. In many formulations of the IEA literature the focus has been more on coalitions than on allocations. This may be due to a dependence on the symmetric players assumption[17] systematically used by the authors, which causes them to state their results in terms of a single number, namely the number of signatories, with no mention whatsoever of the ensuing state of the economy or of the environment. The oversimplification of the economic model has caused the object of interest to be lost. When it comes to policy (i.e., normative) statements, such slender analysis cannot provide a strong enough justification for decision making.

Another point to be considered is that in the I-E version, the term stability is used in a conceptually very different sense than that used for at least two decades in cooperative game theory, where the expressions "strategic" and "coalitional" stability are associated with solution concepts such as the core or the bargaining set. So there are now two different concepts of stability that we wish to confront in this section more systematically.

8.2.2 The Alternative Stability Concepts

Let us recall that for a game in general (and an IEA game in particular), the core property of a strategy for all players (resp., of a proposed treaty[18] for the coalition of all countries) is that (1) it be Pareto efficient (in terms of countries' emissions) and (2) that if any individual or group of parties consider deviating from it, the best they can do is less attractive for them than what they gain in the said strategy (resp., in the proposed treaty). In the IEA game, if the first condition is met but the second is not, transfers among countries can be devised[19] to ensure that it is fulfilled.[20] Stability (called strategic or coalitional in this case) is thus a property of robustness of a strategy for all players against the alternatives that any coalition, smaller than N, might look for.

By contrast, I-E stability criteria apply to coalitions of any size, they do not require that the allocation(s) to which they are applied be Pareto efficient, and they bear only on individual deviations from any possible coalition. Stability is in this case a property of lesser scope.

A further basic difference is in the treatment of deviations: when an individual or a group of parties considers deviating, γ-core theory assumes that the other parties abandon any form of cooperation and act to the best of their interest as singletons, whereas I-E stability theory assumes that the nondeviating players keep cooperating among themselves. The rationale for these alternative assumptions will be discussed shortly.

Chapter 8 Cooperation, Stability, Self-enforcement in Agreements   173

Before doing so, we present with the help of diagrams some apparently unnoticed properties of the γ-core solution that are independent of the assumptions just mentioned.

8.2.3 On the Nature of γ-Core Solutions for the IEA Game

In the standard multilateral externality model used for dealing with IEAs, each player (country) i is at the same time a polluter and a pollutee.[21] In an effort to disentangle which roles each one of these two functions plays in the determination of the solution, let us consider the elementary, actually unilateral, forms of the model, successively with two, and then three parties.

In the first instance, we have just one polluting country—the polluter, indexed r, which is not polluted, and one polluted country—the pollutee, indexed e, which is not a polluter. Think of a simple upstream-downstream river pollution situation. In the Edgeworth box-type of diagram appearing in figure 3.1, which one of us introduced in 1975,[22] the core of this two-agents economy consists of all points on the segment A-B if the polluter has the right to pollute (this

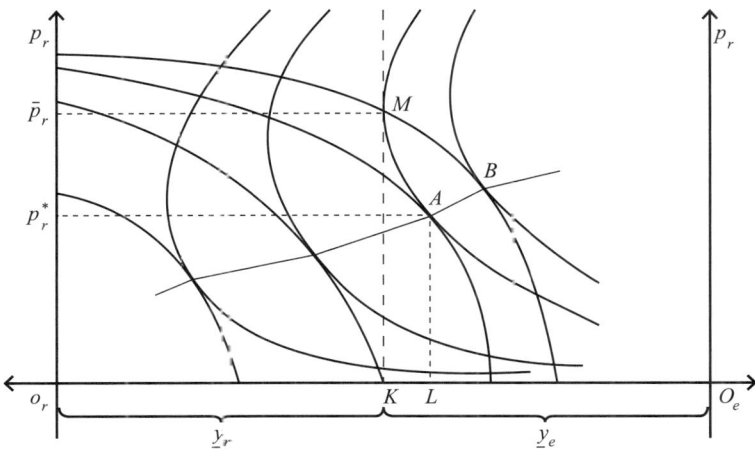

Figure 8.1
A one polluter (r)–one pollutee (e) economy (Source: Tulkens and Schoumaker 1975). Polluter: $u_r(p_r, y_r)$, $y_r \leq \bar{y}_r$ (>0: initial endowment) with $\partial u_r/\partial p_r \geq 0$, $\partial u_r/\partial y_r > 0$. Pollutee: $u_e(p_r, y_e)$, $y_e \leq \bar{y}_e$ (>0: initial endowment) with $\partial u_e/\partial p_r < 0$, $\partial u_e/\partial y_e > 0$. M: Nash equilibrium; A–B: core (with respect to M); A: CT solution (where r receives from e a transfer KL).

proviso implying that point M is the Nash equilibrium in the absence of negotiation); it is also the locus of all allocations that may be reached by Coasean bargaining under the rights allocation just mentioned. This segment A-B is reminiscent of the gain from trade in the exchange interpretations of the Edgeworth box. The figure illustrates how the externality is an object of exchange in this setting, yielding what may be called an "ecological surplus."

Among the core points, the Chander-Tulkens (CT) solution is at point A. It is seen to be a Pareto optimum. It is individually rational with respect to M, and implies a transfer KL such that the polluter is compensated by the pollutee for its abatement cost—but nothing more.

Let us now enlarge this economy by one more pollutee, with the two pollutees being indexed e_1 and e_2 respectively. Figure 8.2 is figure 8.1 with the second pollutee's indifference curve added horizontally to the right of the first pollutee's curve so that at each point along the resulting curve MN the slope of the tangent measures the *sum* of the marginal rates of substitution between environmental pollution and the numéraire y.

Here p_r^* is some Pareto-efficient level of emission for which the line DE is the core relative to M of the economy and the core point D is the CT solution. At this allocation the polluter is compensated just for the

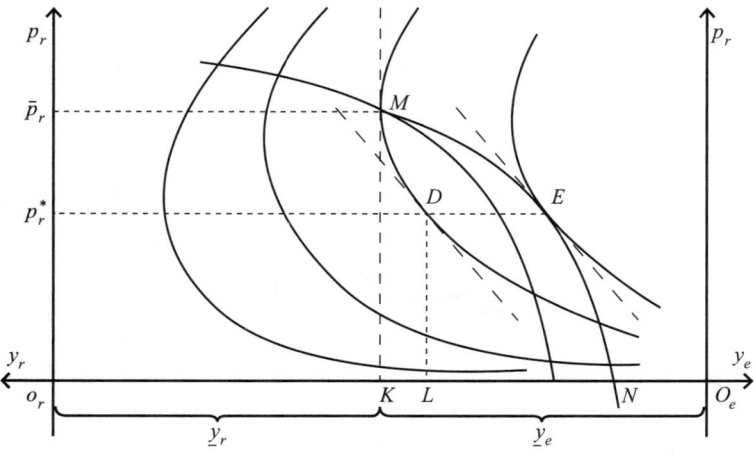

Figure 8.2
One polluter (r) and two pollutees (e_1, e_2). p_r^* is optimal because for that value of p_r the slope at D ($= MRS_r$) = the slope at E ($= \sum_i MRS_{e_i}$). M: *Nash equilibrium*; D–E: *core* allocations; D: *CT solution* (r receives from e_1 and e_2 an aggregate transfer KL; the respective shares of payment by e_1 and e_2 are not shown).

cost of its abatement, and no more. The bargaining gain (DE) is entirely appropriated by the pollutees as if, in the process, both acted as a single party.[23]

This illustration should help make clear:

1. what constitutes the bargaining gain when there are several pollutees, that is, with several recipients of the externality;

2. that the bargaining gain is equal to the core of the game;

3. the particular nature of the CT solution within the core. The polluter is deprived of any pure bargaining gain, which goes entirely to the pollutees; however, all of the polluter's abatement cost is covered.

Note that other core points are conceivable and reachable, all of which are more beneficial to the polluter. They can be reached by the power of bargaining between r and the set of e's. Note that free-riding by r is not an option, for the simple reason that for r, there is nothing to free-ride about!

The relative positions of the players *qua* polluters against *qua* pollutees in the IEA game, in the core, and at the CT solution, can be further demonstrated by diagrams such as those appearing in figures 8.3 and 8.4. These figures show how large the γ-core can be. They also show the strongly pollutees-favoring character of the CT solution: all of the ecological surplus goes to them at that solution. But this result is specific to the CT solution; other outcomes in the core of the game may benefit the polluters, as suggested by point R on figure 8.4 where the two polluters are able to reap some of the bargaining gain (they could reap it all if R were located on the c-d line).

8.2.4 The Rationale for a Game with a Particular Coalition Structure

As we mentioned in section 8.2.3, the γ-characteristic function of an IEA game is a function defined on particular partitions (or coalition structures) of the set N. But why should the partition be limited to just one kind of structure and all conceivable partitions not be considered? Of course, this would transform our IEA game, which is thus far treated as one in characteristic function form, into a game in partition function form.

We have two reasons for not exploring such an extension. One is, as already mentioned, the paucity of results on outcomes, and even of

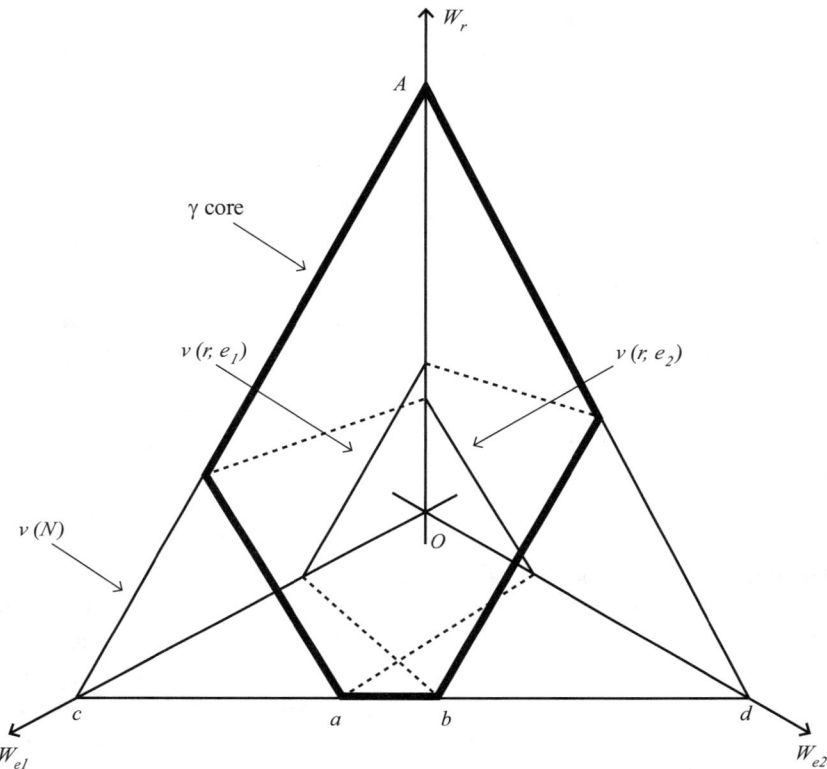

Figure 8.3
The γ-core in payoffs space for any one polluter (r) and two pollutees (e_1, e_2) game. The game is defined by $N = \{r, e_1, e_2\}$ and the characteristic function $v(.)$. Note that $v(e_1, e_2) = 0$. The origin is the welfare levels of players at the Nash equilibrium. The CT solution is one point along the segment $[a, b]$. There all the bargaining gain accrues to the pollutees. Point A belongs to the core illustrates that the (single) polluter can reap all of the bargaining gain.

treatment, of such games in the literature,[24] which limits what we can transpose to our IEA model. The other reason is that not all coalition structures can be considered rational structures, or equally likely to emerge as rational.

Indeed an argument developed in Chander (2007) establishes that when a coalition forms against a proposed γ-core strategy, it is rational *in the sense of an equilibrium strategy for the other players* to break up into singletons, thereby to induce the defectors to accept the proposed γ-core strategy. Thus, instead of considering mechanically all conceivable

Chapter 8 Cooperation, Stability, Self-enforcement in Agreements   177

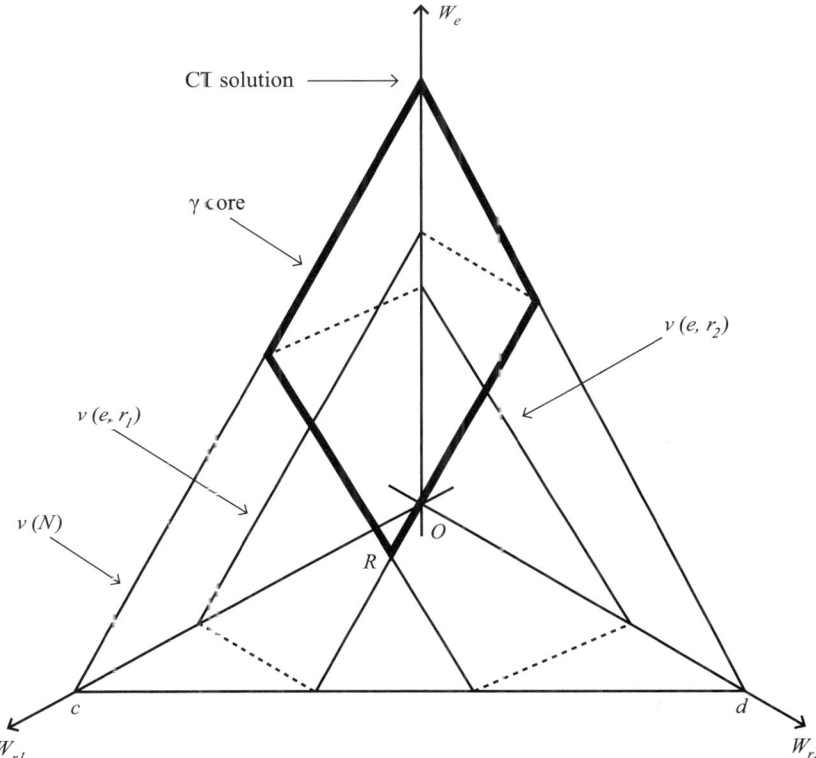

Figure 8.4
The γ-core in payoffs space for any one pollutee (e) and two polluters (r_1, r_2) game. The game is defined by $N = \{e, r_1, r_2\}$ and the characteristic function $v(.)$. Note that $v(r_1, r_2) = 0$. The origin is the welfare levels of players at the Nash equilibrium. The CT solution is one point on the W_e axis. There all the bargaining gain (or ecological surplus) accrues to the (single) pollutee. All other core solutions give some of the gain to the polluters, down to R. In this example, a solution where the two polluters would reap all of the bargaining gain (i.e. a point along $[c,d]$) does not belong to the core. In general, the stronger (weaker) the coalitions between the polluter and one pollutee, the less (more) the pair of polluters can obtain from the bargaining gain.

structures, taking into account the rationality of the collective behavior of the nonmembers of S leads to the selection of a well-justified structure.

The said equilibrium strategy is one of a repeated game of coalition formation. Thus coalition formation theory comes here as a support of the γ-core and the formation of the grand coalition.

8.2.5 Free-Riding and Stability

Two Forms of Free-Riding

Originally the expression free-riding was used by Samuelson (1954) to describe the behavior of economic agents who conceal their preferences with respect to a public good[25] vis-à-vis a single producer—this producer being necessarily the state because of the impossibility of selling the good. On the public good production side, there was no question of leaving or joining coalitions, neither in that paper nor in the following public goods literature—until the international environmental problems were taken up in the late 1960s and early 1970s.

The necessarily voluntary character of public good provision, here abatement of the environmental externality, together with the fact that the externality is multilateral, has shifted the attention from the issue of individual consumers revealing preferences to a planning authority[26] to the problem of having several states participate in international voluntary agreements on a global externality. The expression of free-riding has reappeared not as a preference revelation problem but instead as a way to behave in the face of such agreements.[27]

There are thus two forms of free-riding, which we propose to call "preference revelation (PR) free-riding" and "nonparticipatory (NP) free-riding." While the two forms are not mutually exclusive, we are not aware of any work that treats them together. We will consider here essentially the latter, with occasional allusions to the former.

NP Free-Riding and I-E versus γ-Core Stability

NP free-riding is a special kind of group instability. Depending on the stability concept in use, what such free-riding designates will vary. When a γ-core allocation is declared not to be internally stable, implying that some i prefers to leave the grand coalition, the nonstability statement rests on the assumption that if i leaves N, $N\setminus\{i\}$ remains a coalition (possibly re-optimizing its strategy), and *tolerates* i's free-riding.[28] That is to say, the remaining coalition tolerates the global inefficiency induced by i's defection.

By contrast, the assumption behind γ-core stability is that $N\setminus\{i\}$ *counters* free-riding by reacting, not in an extremely punishing way as would be the case with the α-characteristic function[29] but rather in a rational way that is just sufficient to make the free-rider believe that it will forfeit any benefit from belonging to the grand coalition, as argued in Chander (2007).

The strength of the γ-core concept in dealing with (or even solving) the NP free-rider problem thus lies in the farsighted rationality of the threat it presumes. The weakness of I-E stability is, instead, an inherent myopia that eventually legitimates NP free-riding.

PR Free-Riding and the Particular CT Core Solution
While the CT solution has all the core stability properties just outlined, it allows one in addition to see the effect of a player i joining but incorrectly revealing preferences, through the $\pi_i'^*/\sum_{j\in N}\pi_j'^*$ coefficients in the transfers formula. Understating $\pi_i'^*$ implies a lesser contribution of i to the coverage of the aggregate abatement cost. But that lower value of $\pi_i'^*$ also induces a less than optimal level of aggregate abatement, since the optimality criterion is based on the sum of the π_i's. Thus the CT solution to the IEA game is vulnerable to PR free-riding, at least away from the optimum.[30]

To conclude, we are back again to the motivation behind seeking stability: from a normative point of view, the reason for avoiding free-riding is essentially that it prevents the achievement of efficiency.

8.3 Self-Enforcement

Self-enforcement is an intuitively attractive expression when dealing with international agreements. It evokes the absence of an external authority, which is at the root of the problems raised by international type of agreements. It also contains an implicit reference to incentives. After its introduction by Barrett (1994), the appearance of a book (Barrett 2003) entirely devoted to this idea has positioned the author as its most articulate advocate.

For cooperative game-minded theorists like us, there is a bit of mystery here: it is difficult to find in the standard literature a commonly received definition of self-enforcement. It does usually not appear in the index of game theory textbooks, and when it does (e.g., in Myerson 1991), it is only to refer to a property of occasional interest. More important, in what sense is self-enforcement more desirable than efficiency, or more so than core or I-E stability? Is it an additional concept that we should add to our tool box for IEA analysis?

We feel that while the answer to the last question is definitely yes, the answers to the previous question are difficult to make precise. Self-enforcement is a property of a treaty that "must satisfy three conditions: individual rationality, collective rationality and fairness" (Barrett 2003, pp. xiii–xiv). Apart from the first, which is used in its standard sense, the other expressions are given a special meaning. On the one hand, collective rationality is redefined successively in Barrett's chapters 7 and 11 as a property of a treaty implying not only efficiency for the group under consideration but additionally free-riding deterrence (p. 213),[31] which is given two possible forms (strong and weak collective rationality, respectively; see Barrett, p. 294). A formal definition is offered (in section 11.4) but, unfortunately, with a model of identical players that is hardly convincing. On the other hand, fairness is not formally dealt with, but presented as a requirement that the treaty "be perceived by the parties as being legitimate" (p. xiv).

While potential readers, fond of precise definitions and rigorous developments of sufficiently rich and realistic models, are likely to be sometimes disappointed, the book offers nevertheless a remarkable intellectual challenge to theorists dealing with IEAs.

The one we would like to highlight here is the theme of chapter 11, which describes a possible trade-off between the breadth of international cooperation (in terms of the number of participants in a treaty) and its depth (in terms of the size of the actions agreed upon by the parties): Is a "broad but shallow" treaty better than a "narrow but deep" one?

A shallow treaty would be one that does not achieve full efficiency among the participating countries, such as by abating less than it is optimal; this would be the price, so to speak, for having it signed by many countries. The outcome is called by Barrett a "consensus treaty," asserted elsewhere to be self-enforcing. That this is a better solution than the opposite (deep and narrow) is claimed to be established (p. 302) by means of an ingenious symmetric countries model. But we have already voiced the opinion that such a basis is itself too shallow for transforming this conjecture into scientific truth.

Yet the trade-off brought to light remains an important intellectual challenge: while it surely deserves scrutiny by means of better adapted, and therefore more elaborate, game-theoretic tools, it illustrates once more that before proving an idea to be true, it must be generated. This is a major merit of many ideas in Scott Barrett's book.

Let us close this point with a perhaps timely question: Would the David Bradford scheme presented at this conference be self-enforcing?

8.4 Conclusion

Neither stability nor cooperation is desirable per se. Both are there to achieve efficiency because the welfare of people derives primarily from allocations, not from their stability or from cooperation. The virtues of Barrett's self-enforcement notion eventually point in the same direction, admitting that otherwise, no treaty would be signed at all.

At a less general level, the analysis has revealed that there is much to gain in understanding if one distinguishes more explicitly between the involvement of countries as polluters from their involvement as pollutees.

In fact, this is already done, to some extent, within the Kyoto Protocol: the motivations behind the *aggregate* quotas that have been negotiated are essentially those of the pollutees. The quotas result from country preferences, with the working of flexible mechanisms being of concern essentially for the polluters. What is less clear is how the bargaining gain turns out to be shared among the two categories of parties.

Notes

1. Published thereafter as Tulkens (1998).

2. The specific problems raised by stock externalities will not be considered in this discussion, although such are indeed the externalities generated by greenhouse gas emissions. Our immediate excuse is that they are not dealt with either in the literature we consider. More fundamentally, we think the issues at stake need to be clarified first within flow (static) models before being tackled in the dynamic context required by stock externalities. First extensions in that direction, using dynamic games, have been proposed by Germain et al. (2003) in the γ-core stability perspective and by Rubio and Ulph 2007 in the alternative internal-external stability context.

3. Namely Ray and Vohra (1997, 1999, 2001).

4. Taking this view one step farther, some authors consider cooperative games as normative social science, as opposed to noncooperative games being positive science. This is an oversimplification.

5. See, for instance, how widely different are the formulations of externalities by Shapley and Shubik (1969) in their "lake game" and their "garbage game," or even by Scarf (1972).

6. For easier reference in the developments to follow, we briefly remind the reader that the γ-core of a cooperative game with externalities is defined as the core in the usual sense (that is, a joint strategy of all players that no coalition can improve upon), assuming that the worth of each coalition S is determined by both the joint payoff maximization of the members of S and the payoff maximizing strategies of the other players acting individually, that is, as singletons. This assumption allows one to define for each S what the authors call a "partial agreement Nash equilibrium with respect to coalition S". When the characteristic function form of the game is used, the "γ-characteristic function" is defined over the set of all such partial agreements with respect to a coalition. All details are provided in Chander and Tulkens (1995, 1997).

7. A notable recent exception is to be found in the work of Van Steenberghe (2004) who deals with the nucleolus and the Shapley value of our externality game, using the γ-characteristic function that allowed to define the core.

8. Due to d'Aspremont and Gabszewicz 1986.

9. This is reminiscent of von Neumann and Morgenstern "stable sets," as described by Osborne and Rubinstein (1994, p. 279), but not identical.

10. See Finus (2001, ch. 15).

11. A PANE with respect to any S is an example of an equilibrium coalition structure.

12. In games in partition function form, the partition function plays a role similar to the characteristic function in standard cooperative games. Conditions for results should thus hinge on properties of that function. Notice that the γ-characteristic function of Chander and Tulkens is a special case of a partition function, for which the property of "balancedness" as established by Helm (2001) has allowed the nonemptiness of the core to be confirmed.

13. The CWS model was introduced by Eyckmans and Tulkens (2003).

14. The FEEM-RICE model was introduced in Buonanno, Carraro, and Galeotti (2003).

15. Repeating this reasoning on further mergers might well end up with N as the only coalition!

16. Private communication, after the Coalition Theory Network meeting in Paris, January 2005, where the paper was presented and discussed.

17. As well as the rudimentary description of environmental phenomena; but this is acceptable because no model will ever describe reality entirely.

18. Incidentally, there is no a priori reason to believe that there is only one allocation (or treaty) that can belong to the core of the IEA game. In other words, the core is not a unique point solution concept, neither in general nor in the particular case of IEAs.

19. In the notation of this paper, the Chander-Tulkens (1997) formula for these transfers T_i (>0 if received, <0 if paid) reads

$$T_i = -(g_i(p_i^*) - g_i(p_i^-)) + \frac{\pi_i'^*}{\sum_{j \in N} \pi_j'^*} \left(\sum_{j \in N} g_j(p_j^*) - \sum_{j \in N} g_j(p_j^-) \right),$$

Chapter 8   Cooperation, Stability, Self-enforcement in Agreements   183

where p_i^* and p_i^- are, respectively, the world efficient and the Nash equilibrium emission levels of country i and $\pi_i'^*$ is the derivative of the damage cost function $\pi_i(.)$ at the Pareto efficient point $\sum_{j \in N} p_j^*$.

20. It is important remember (see Chander et al. 2003, sec. 5) that *the same allocation can be achieved with the transfers being substituted by initial allowances of tradable emission permits*, provided that the amounts of these allowances be such that the resulting competitive equilibrium on the permits market induces the γ-core allocation just defined. This point is of major importance when discussing the connection between the theories presently examined and actual treaties such as the Kyoto Protocol where there are no explicit transfers specified. But the treaty's allowances together with the competitive equilibrium they generate play the role of the transfers. For a further and thorough exploration of this substitution, see Van Steenberghe (2004).

21. This is why the term multilateral is used.

22. See Tulkens and Schoumaker (1975, pp. 247ff); the diagram appears to be redrawn in Varian (1990, pp. 539 and 542). The same diagram can be deduced from the simplified version of the IEA model sketched out below the figure. Full details are given in the paper cited.

23. Unfortunately, the picture cannot show how, at the CT solution, the coverage of the polluter's abatement cost KL is shared between the two pollutees e_1 and e_2, and thus how the Coasean gain is shared among them.

24. Thrall and Lucas (1963) is an early source, limited to $n \leq 3$.

25. Correct revelation is necessary for checking whether efficiency is obtained. But if that information is used to determine the individual contributions to the financing of the public good, the contributors will be tempted to understate their preferences and production will be suboptimal; if, however, no connection is made between what contributors reveal and what they have to pay, preferences will be overstated and production will be larger than optimal

26. While Samuelson wrote in 1954 that only a smart game theorist could master the preference revelation problem raised by the type of free-riding behavior he had identified, the challenge was successfully taken up by game-theoretic economists fifteen years later in a series of papers written in the context of decentralized planning procedures, starting with Drèze and de la Vallée Poussin (1971), continued by Roberts (1979), Henry (1979), Groves and Ledyard (1973), and culminating with Champsaur and Laroque (1981). This literature may be regarded as a main source of the mechanism design stream of thought that developed subsequently in wider contexts than just public goods economies.

27. A third form of free-riding has been put forward by Finus (2001), namely the behavior that consists in signing an agreement and then not complying with it. We are not sure this wording is appropriate. Noncompliance is a breach of the agreement. In the two forms described above, free-riding results from either not signing the agreement or signing under conditions of an information bias.

28. Eyckmans and Finus (2004) even reward free-riding, by means of a transfer scheme based on the notion of free-rider payoff that they call "almost ideal." Note that to offer such compensation, the preferences of the free rider must be known. Is there any reason to believe that he/she will reveal truthfully while bargaining on a possible defection?

29. Unlike the γ-characteristic function, the α-characteristic function, which dates back to von Neumann and Morgenstern 1944, is such that for each coalition S the players not in S are assumed to choose the joint strategy which has the worst effects on the members of S.

30. When the optimum is reached, there is an argument due to Drèze and de la Vallée Poussin (1971, sec. 3) that it is a Nash equilibrium of a preference revelation game that all parties reveal correctly their preferences. Away from the optimum, this is no longer the case, but the bias in misrepresentation can be identified (see Roberts 1979).

31. We responded above to Barrett's criticism of the γ-core, in which he introduces his collective rationality concept: we claim that the threat he considers as noncredible is a farsighted rational one, as proved in Chander (2007).

References

Barrett, S. 1994. Self enforcing international environmental agreements. *Oxford Economic Papers* 46: 878–94.

Barrett, S. 2003. *Environment and Statecraft: The strategy of Environmental Treaty-Making.* Oxford: Oxford University Press.

Buonanno, P., C. Carraro, and M. Galeotti. 2003. Endogenous induced technical change and the costs of Kyoto. *Resource and Energy Economics* 25: 11–34.

Carraro, C., and D. Siniscalco. 1993. Strategies for the international protection of the environment. *Journal of Public Economics* 52: 309–28.

Carraro, C., and D. Siniscalco. 1995. International coordination of environmental policies and stability of global environmental agreements. In L. Bovenberg and S. Cnossen, eds., *Public Economics and the Environment in an Imperfect World*. Dordrecht: Kluwer Academic, ch. 13.

Champsaur, P., and G. Laroque. 1982. Strategic behavior in decentralized planning procedures. *Econometrica* 50(2): 325–44.

Chander, P. 2007. The gamma-core and coalition formation. *International Journal of Game Theory* 35: 539-56.

Chander, P., and H. Tulkens. 1995. A core-theoretic solution for the design of cooperative agreements on transfrontier pollution. *International Tax and Public Finance* 2(2): 279–94.

Chander, P., and H. Tulkens. 1997. The core of an economy with multilateral environmental externalities. *International Journal of Game Theory* 26: 379–401.

Chander, P., H. Tulkens, J.-P. van Ypersele, and S. Willems. 2002. The Kyoto Protocol: An economic and game theoretic interpretation. In B. Kriström, P. Dasgupta, and K.-G. Löfgren, eds., *Economic Theory for the Environment: Essays in Honor of Karl-Göran Mäler*. Cheltenham: Edward Elgar, pp. 98–117.

Coase, R. M. 1960. The problem of social cost. *Journal of Law and Economics* (3): 1–44.

d'Aspremont, C., and J. Jaskold Gabszewicz. 1986. On the stability of collusion. In J. Stiglitz and G. F. Mathewson, eds., *New Developments in the Analysis of Market Structure*. Cambridge: MIT Press, pp. 243–64.

Drèze, J., and D. de la Vallée Poussin. 1971. A tâtonnement process for public goods. *Review of Economic Studies* 38: 133–50.

Eyckmans, J., and M. Finus. 2004. An almost ideal sharing scheme for coalition games with externalities. FEEM Working paper 155.04. Revised version 2007. Mimeo.

Eyckmans, J., and M. Finus. 2006a. Coalition formation in a global warming game: How the design of protocols affects the success of environmental treaty-making. *Natural Resource Modeling* 19(3): 223–58.

Eyckmans, J., and M. Finus. 2006b. New roads to international environmental agreements: The case of global warming. *Environmental Economics and Policy Studies* 7: 391–414.

Eyckmans, J., and H. Tulkens. 2003. Simulating coalitionally stable burden sharing agreements for the climate change problem. *Resource and Energy Economics* 25: 299–327.

Finus, M. 2001. *Game Theory and International Environmental Cooperation*. Cheltenham: Edward Elgar.

Finus, M., and B. Runshagen. 2003. Endogenous coalition formation in global pollution control: A partition function approach. In C. Carraro, ed., *Endogenous Formation of Economic Coalitions*. Cheltenham: Edward Elgar, pp. 199–243.

Groves, T., and J. Ledyard. 1977. Optimal allocation of public goods: A solution to the "free rider" problem. *Econometrica* 45(4): 783–810.

Germain, M., Ph. Toint, H. Tulkens, and A. de Zeeuw 2003. Transfers to sustain dynamic core-theoretic cooperation in international stock pollutant control. *Journal of Economic Dynamics and Control* 28: 79–99.

Helm, C. 2001. On the existence of a cooperative solution for a coalitional game with externalities. *International Journal of Game Theory* 30: 141–47.

Henry, C. 1979. On the free rider problem in the M.D.P. procedure. *Review of Economic Studies* 46(2): 293–303.

Mäler, K. G. 1989. The acid rain game. In H. Folmer and E. Van Ierland, eds., *Valuation Methods and Policy Making in Environmental Economics*. Amsterdam: Elsevier, pp. 231–52.

Maskin, E. 2003. Bargaining, coalitions and externalities. Presidential address to the Econometric Society European Meeting, Stockholm. Mimeo (August).

Myerson, R. 1991. *Game Theory: Analysis of Conflict*. Cambridge: Harvard University Press.

Osborne, M. J., and A. Rubinstein. 1994. *A Course in Game Theory*. Cambridge: MIT Press.

Ray, D., and R. Vohra. 1997. Equilibrium binding agreements. *Journal of Economic Theory* 73: 30–78.

Ray, D., and R. Vohra. 1999. A theory of endogenous coalition structures. *Games and Economic Behavior* 26: 286–336.

Ray, D., and R. Vohra. 2001. Coalitional power and public goods. *Journal of Political Economy* 109(6): 1355–84.

Roberts, D. J. 1979. Incentives in planning procedures for the provision of public goods. *Review of Economic Studies* 46(2): 283–92.

Rubio, S., and A. Ulph. 2007. An infinite-horizon model of dynamic membership of international environmental agreements. Mimeo.

Samuelson, P. A. 1954. The pure theory of public expenditure. *Review of Economics and Statistics* 36: 387–89.

Scarf, H. 1971. On the existence of a cooperative solution for a general class of N-person games. *Journal of Economic Theory* 3: 169–81.

Shapley, L., and M. Shubik. 1969. On the core of an economic system with externalities. *American Economic Review* 59: 678–84.

Thrall, R. M., and W. F. Lucas. 1963. n-Person games in partition function form. *Naval Research Logistics Quarterly* 10: 281–98.

Tulkens, H. 1979. An economic model of international negotiations relating to transfrontier pollution. In K. Krippendorff, ed., *Communication and Control in Society*. New York: Gordon and Breach, pp. 199–212.

Tulkens, H. 1998. Cooperation vs. free riding in international environmental affairs: two approaches. In N. Hanley and H. Folmer, eds., *Game Theory and the Environment*. Cheltenham: Edward Elgar, pp. 330–44.

Tulkens, H., and F. Schoumaker. 1975. Stability analysis of an effluent charge and the "polluters pay principle." *Journal of Public Economics* 4: 245–69.

Van Steenberghe, V. 2004. Core-stable and equitable allocations of greenhouse gas emission permits. CORE Discussion paper 2004/75.

Varian, H. R. 1990. *Intermediate Microeconomics: A Modern Approach*, 2nd ed. New York: Norton.

Von Neumann, J., and O. Morgenstern. 1944. *Theory of Games and Economic Behavior*. Princeton: Princeton University Press.

9 Heterogeneity of Countries in Negotiations of International Environmental Agreements: A Joint Discussion of the Buchner-Carraro, Eyckmans-Finus, and Chander-Tulkens Chapters

Sylvie Thoron

An important characteristic of the Kyoto protocol is that the different partners are treated very differently. For the first time an international environmental agreement (IEA) sets not only a global target and a timetable but also very precise country-specific targets. In previous IEAs the objective was to reach a consensual common target as, for example, in the Protocol on the Reduction of Sulphur Emissions by at Least 30 Percent. The Montreal Protocol on Substances That Deplete the Ozone Layer also fixes a uniform target for industrialized countries but introduces a differentiation in the timetable with a ten-year grace period for developing countries. In the case of the Kyoto Protocol the differentiation is much more marked. The developing countries are not committed to any binding target for the first period (until 2012) and nothing is specified for the following period. The global target of a decrease of greenhouse gases by 5.5 percent is supposed to be achieved by assigning specific targets to the different countries that vary from a reduction of 8 percent to an increase of 10 percent. Furthermore in the same Protocol the flexibility mechanisms, the EU bubble, the principles of joint implementation, emissions trading, and the clean development mechanism allow the exploitation of differences and complementarities among countries. Indeed the Kyoto protocol is clearly the result of a negotiation between partners whose differences, far from being ignored in order to reach a consensus on a common target, have occupied a central position in the discussions.

It is not surprising, in this context, that the theorists who wanted to analyze the IEA, in general, and the climate change negotiations, in particular, were not satisfied with a theory of coalition formation that limits its analysis to a symmetric framework. Although I do not want to advocate the latter choice at this point, I will try to give a historical explanation. The integration of externalities and the exploitation of efficiency gains are two important economic justifications for cooperation. Chander and Tulkens explain (in chapter 8 of this volume) how these two arguments have been developed in different literatures. The problem of externalities has been analyzed by economists in the framework of Samuelson's public good theory (1954) or as developments of the Coase theorem (1961). In parallel and exactly during the same period, the cooperative game theory literature was analyzing superadditive games (Shapley 1953, which introduced the Shapley value; Gillies 1959, which developed the concept of the core). This second literature has always been criticized because of its focus on the problem of how to exploit efficiency gains inside a coalition, while leaving aside the incentive problem generated by the persistent externalities between coalitions. Thrall and Lucas (1963) proposed a framework to deal with both problems simultaneously but the cooperative game theory literature did not take up this aspect. Interest in the Thrall and Lucas proposal was revived in the 1990s, with the development of the theory of coalition formation.

It is notable that both literatures then focused on disparities between actors. The Coase theorem explained how polluters and pollutees could reach an efficient outcome by bargaining. The aim of the solution concepts proposed by cooperative game theory was to solve the problem of sharing a coalition's worth between heterogeneous players. Why then, did the theory of coalition formation not deal with the problem of heterogeneity? Because the original aim of this theory was to explain how the existence of these externalities or spillovers could lead to inefficient outcomes. The Coase theorem claimed the emergence of an efficient outcome, and in cooperative game theory efficiency is assumed. By contrast, this new literature focused on the sources of inefficiencies. Noncooperative game theory focused on the problem of compliance to agreements and the theory of coalition formation on the problem of participation in agreements. I will not address the problem of compliance here. After all, to quote Abram Chayes and Antonia Chayes (1991: 311), "International lawyers and others familiar with the operations of international treaties take for granted that most states comply with most of their treaty obligations most of the time."

Conversely, participation is often a serious problem for IEA, and this has been illustrated by the negotiations on climate change. However,

Chapter 9   Heterogeneity in Negotiations of International Agreements   189

now that we know that the incentives are not conducive to the formation of the grand coalition, even in a symmetric framework, several questions remain to be answered. Is the situation better or worse off in a nonsymmetric framework? Did the theory of coalition formation propose solutions to the participation problem? If the answer is in the affirmative, are these solutions relevant for the heterogeneous case? Finally, are there solutions specific to heterogeneous cases? These are the kind of questions addressed in the three papers on international environmental negotiations of this volume.

9.1   Sources of Heterogeneity

In this volume, Eyckmans and Finus (chapter 6), on one hand, and Buchner and Carraro (chapter 7), on the other, use modified versions of the same RICE model (i.e., regional integrated model of climate and the economy). This model, proposed by Nordhaus and Yang in 1996, is a regional, dynamic, general equilibrium model of the economy which integrates economic activity with the sources, emissions, and consequences of greenhouse-gas emissions, and consequences of greenhouse-gas emissions and climate change. It divides the global economy into six different regions: USA (United States), JPN (Japan), EU (European Union), CHN (China), FSU (former Soviet Union), and ROW (rest of the world). The novelty of the RICE model in comparison with previous models of global warming is to allow nations to adopt different strategies.

In the CLIMNEG world simulation model, Eyckmans and Finus introduce differences in discount rates. Discount rates for developing regions are higher than those for developed countries. The countries also differ in energy efficiency. USA, JPN, and EU have steep marginal abatement costs, while CHN and ROW have flat marginal abatement costs. This means that energy-efficient regions face higher costs when cutting back emissions. The countries are also more or less vulnerable or more or less sensitive to climate change. Damage functions are particularly steep in EU and ROW, to a lesser extent in USA and JPN, and relatively flat in FSU and CHN.

The FEEM-RICE model used by Buchner and Carraro incorporates an endogenous technical change that yields endogenous growth and improves the emission or output ratio. The Kyoto Protocol constitutes the starting point: the European Union, Japan, and Russia are committed to complying with their Kyoto targets, and there is a market for pollution permits. The authors present different scenarios and explain in each how the countries can move away from the situation. In each scenario two coalitions are formed. Each coalition can create a market for pollution permits. On each market, the countries draw benefits from their complementarities. Big polluters with stringent targets can buy pollution permits from small or "not yet so big" polluters with mild targets. Despite their mild targets, less energy-efficient regions have an incentive to reduce their emissions. By selling their own emission permits to more energy-efficient regions, they generate funds that can be used to invest in new technologies.

9.2 Transfers and Stability

Eyckmans and Finus (EF) pose the question as to the links between the stability of agreements and transfers among partners. As already discussed above, in most of the literature in economics and cooperative game theory dealing with the formation of agreements, efficiency is a basic criterion. With this as a principle, the outcome should be calculated to satisfy efficiency in a first step. The different partners' contributions are chosen to maximize the sum of their welfares. As a consequence in the framework described above, for example, it may be that energy-inefficient regions facing smaller abatement costs will have to make bigger reductions in pollution. Then, in a second step, different normative criteria can be used to justify transfers between partners. Consent on criteria is necessary because the different partners will not sign an agreement they consider to be unfair. The difficulty is that the definition of fairness is not unique. In particular, much depends on which countries' characteristics are to be considered. There is a whole literature on this delicate question to which Eyckmans and Finus refer.

Eyckmans and Finus define a series of transfer schemes. Each of these guarantees to a partner at least the welfare level it would have had if no agreement were signed. They differ in the way the costs and benefits from partnership are shared: a normative criterion is used to define a country-specific coefficient that is applied to the difference between the sum of welfares before and after the agreement implementation. In one of these schemes, a solution originally proposed by Chander and Tulkens (1997)—the CT solution—each country's contribution to pollution abatement is proportional to the cost that it incurs as a result of the damage from climate change. As a consequence those who care more about climate change, or are more vulnerable to it, must contribute more. From a normative point of view, this is debatable. It depends on what kind of good the climate is. Given an objective that is to determine a contribution for each country, let us consider a parallel with private goods. Suppose that climate can be considered a luxury good: in this case the preference for the environment, in general, and climate stability, in particular, increases with wealth. It may then be justified to ask richer countries that have a higher preference for protecting the climate to contribute more. However, climate is more often considered an inferior good:[1] those who have the highest demand for protecting the climate are those who are not wealthy enough to be able to protect themselves against climate change. In this case it is more difficult to justify, from a normative point of view, the idea that developing countries that are more vulnerable to climate change should contribute more.

However, the CT solution is also attractive because it can be interpreted in a positive framework. In a purely positive view, the stability of an agreement is not guaranteed by its fairness but rather by the fact that each partner has an incentive to sign, given the alternative and considering its own interest. This depends on each country's anticipation about what would be its ex-partners' reactions if that country decided to rescind its agreement. The literature on coalition formation explains how the assumption about the reactions of the ex-partners after a deviation determines the outcome. Suppose that each partner anticipates that a deviation would provoke a complete dismantling of the agreement. Call this anticipation assumption 1. Now suppose that after a deviation, a country anticipates that its ex-partners will continue with the agreement. These countries will just adjust their contributions so that the exiting country's welfare is no longer taken into account. Call this assumption 2. When externalities are positive, as in the case of an abatement game, the dismantling of the original coalition is a bigger threat than a simple adjustment of the coalition's abatement (see Hart and Kurz 1983; Yi 1996; Thoron 2000). As a consequence under assumption 2 it is more difficult to attain stability than under assumption 1.

Eyckmans and Finus propose to test the claim that transfers can increase coalition stability. They consider stability under assumption 2. Each transfer scheme guarantees the partners at least a welfare level that is the same as that in the situation without any agreement. Hence, when transfers are added, the coalitions are stable under assumption 1.

For example, Chander and Tulkens (1997) prove that the CT solution belongs to the core that is also defined under assumption 1. However, there is no direct reason why these coalitions should still be stable under assumption 2, unless the normative criterion used to share the surplus plays a role in stability, which is the case with the CT solution. Indeed the countries that care more about climate change are also those that are least likely to withdraw. If they contribute more, those that care less contribute less and this favors the attainment of stability.

However, this relationship is the exception rather than the rule. Other than by simple coincidence, there is no link between an allocation that satisfies a normative criterion and an allocation that satisfies a stability criterion. I will go even further: there is in fact a certain contradiction between the normative and the positive approaches. On the one hand, the normative approach starts from an efficient outcome and organizes transfers on the basis of a normative criterion. On the other hand, the positive approach describes the outcome of a negotiation, whether or not it is efficient, and considers that the allocation of payoffs is endogenous to this negotiation process. In the first case, the efficiency is a basic criterion, in the second case, it is neither a starting point, nor necessarily an outcome.

I explained above the problem posed by the definition of stability in normal form games of coalition formation. In the literature on coalition formation, other types of models are proposed. Bloch (1996) and Ray and Vohra (2001) used extensive form games to represent the negotiation process. They proved the inefficiency of the outcome when there are positive externalities. In order to focus on the problem of heterogeneity, Maskin (2003) uses these extensive form games to define a Shapley value for games with externalities. Here again, Maskin's conclusion is that this new value does not satisfy the efficiency axiom. This result is due, though, to the introduction of another axiom, namely one that allows for communication breakdown between players, as pointed out by Chander and Tulkens. The other common feature of these positive models is that the payoffs are generated during the negotiation with only the objective of stability. For example, in an extensive form game, when a player makes a proposal to another player, it is just sufficient to convince that player to become a partner.

The conclusion is not that normative criteria can or cannot be used as arguments during a negotiation. Two problems prevent my giving a clear answer to this question. First, it is theoretically difficult to combine the normative and positive approaches. Maybe the reason is that the basis of positive models, the maximization of individual payoffs, is too restrictive. Second, the empirical evidence is lacking, other than simple declarations that normative criteria have some relevance for the outcome of a negotiation.[2]

9.3 Membership Rule and Climate-Blocs

When the internal or external stability criterion is applied in a symmetric framework, it has been proved that equilibrium always exists (d'Aspremont et al. 1983). The agreement generated at equilibrium can be restricted, as in the extreme case it corresponds to the trivial situation where the agreement is signed by a unique player (!), but equilibrium exists. It has also been proved that such an equilibrium is robust against deviations by coalitions (Thoron 1998). This case is no longer true when players are not symmetric. To illustrate this point, consider the following example: it may be that FSU wants to join coalition (EU, JPN) and that USA wants to join (EU, JPN, FSU). Both coalitions are externally unstable. However, it may be that FSU does not want to leave (EU, JPN, FSU) but would want to leave (EU, JPN, FSU, USA). In comparison with the symmetric case, the fact that one country wants to join the coalition no longer means that the extended coalition is internally stable. As a consequence the equilibrium may not exist, and the membership may cycle. Eyckmans and Finus propose to reduce this instability by using an exclusive membership rule.

The difficulty, in the asymmetric case, is that the different countries have different preferences for the type of measures that should be applied to prevent climate change. Indeed economic theory has taught us that two kinds of problems threaten the efficient provision of a public good: the incentive to free ride and the difficulty with choosing the public good for partners that have different preferences for it. While the recent theory of coalition formation has focused on the first problem, which can be studied in a symmetric framework, the second problem, which only occurs when there are asymmetries, is the main object of an older literature on coalition formation: the theory of jurisdiction formation and the theory of clubs. When no agreement is possible with an open membership rule, this means that preferences are too different for all countries to reach a common agreement. Then, as a consequence of what would be the equivalent of a Tiebout mechanism, different agreements may emerge. The countries that manage to reach an agreement will have similar preferences. However, the literature on jurisdiction formation, which tried to model the Tiebout mechanism, showed that the existence of equilibrium cannot be proved in this framework. Furthermore the conclusion that can be drawn from the theory of clubs is that this problem of nonexistence can indeed be solved by using an exclusive membership rule.

Buchner and Carraro in this volume use another argument to justify the formation of climate-blocs: the exploitation of complementarities on markets for pollution permits. We know the advantages of a market for pollution permits: the gains in efficiency generated by the reallocation of the original targets through a market mechanism. These efficiency gains increase with the differences between countries (more precisely the differences in energy efficiencies). As a consequence emissions trading has been proposed as the tool necessary to make the joint implementation and clean development mechanisms work. However, it does not seem to be the case that the same socially beneficial incentives explain why the countries would prefer to form subglobal blocs. In Buchner and Carraro's chapter, if some coalition structures turn out to be better, from a total welfare point of view (see table 8.2), this is due to the assumption that if China and the United States were to enter a coalition, they would accept stringent targets (-10 percent for China). In fact the argument is that countries would have a stronger incentive to accept such targets if they had the possibility to form subglobal blocs. However, in each of the coalition structures the authors consider, there is at least one country in the coalition that is worse off than in the current Kyoto scenario. To be convincing, this kind of argument has to be based on a stability analysis.

Furthermore I am not sure that the possibility for the countries to form parallel climate-blocs is the solution to the participation problem, and this is for two reasons. First, we know that the formation of a coalition structure implies a social loss in comparison with the formation of the grand coalition. Of course, the formation of climate-blocs is only proposed because the grand coalition cannot form. But, and this is the second point, there are arguments in favor of the formation of a unique subglobal bloc. Indeed a country that did not sign or ratify the Kyoto Protocol cannot benefit from the existing market for pollution permits. In other words, the existence of a unique market generates negative externalities and likely incentives to join this subglobal bloc. Even if this perspective is too optimistic, in my opinion it is better than a situation where a country becomes stuck in a collectively suboptimal stable coalition structure, if indeed such a structure exists.

Even in the symmetric case several coalitions can exist simultaneously. When the assumption that only one coalition can be formed is dropped, the driving forces are the same as in the restricted framework. Because the game is superadditive, the countries have an incentive to cooperate; however, they also have an incentive to free-ride, which generates the nonparticipation problem. Now the different coalitions free-ride on each other. Because the externalities are positive, within a given structure the smaller coalitions obtain considerable benefits from the existence of the bigger ones. In Ray and Vohra's model (2001) the coalitions form sequentially. When the number of negotiating partners is intermediate, two coalitions are formed. The biggest coalition forms first, the second coalition forms afterward in order to free-ride on the previous one. When the externalities are negative, the outcome is very different. In this case Yi (1996) showed that the grand coalition is always an equilibrium.

When describing the formation of climate-blocs, Buchner and Carraro draw a parallel with the formation of the World Trade Organization. Indeed coordination and information problems can make the formation of the grand coalition, in one step, difficult. This is why, in the case of the organization of free trade, countries started by signing binary agreements. However, it seems to me that there is a fundamental difference between this last case and negotiations on climate change. Thanks to negative externalities, the incentives made for a move in the right direction to reach the final outcome of the formation of the World Trade Organization. This will not work in the same way when considering negotiations on global warming. In this case, because the externalities are positive, the emergence of several agreements does not necessarily mean that we are moving in the right direction. It may mean that we have reached a suboptimal equilibrium in which one coalition free-rides on the other.

An optimistic point of view would be that there are other driving forces that can create negative externalities between blocks and that markets for permits constitute a source of these externalities. Another force is the emergence of social norms. An argument developed by Finnemore and Sikkink (1998) is that at a certain "tipping point" in a norm's evolution, a "norm cascade" takes place, and then states join the coalition in large numbers because of pressure from other states and nonstate actors. In this case the disparities among countries can help the process, since the countries that are more committed can, in the end, convince the others to follow their lead.

9.4 Conclusion

The theory of coalition formation has been developed essentially in a symmetric framework. It explains that the problem of participation inherent to the negotiations on climate change, is generated by the existence of positive externalities and an incentive to free-ride. A conclusion would then be that the countries involved are unable to reach the social optimum because the incentive to free-ride is too strong. As a result they may form subgroups, the climate-blocs, and reach a suboptimal situation in which one coalition free-rides on the other. However, the forces that can change the externalities can help offset these incentives.

The asymmetries among potential partners introduce new aspects. On the one hand, the differences between the countries' evaluations of climate change costs and abatement benefits produce new difficulties. The impossibility of reaching a consensus among asymmetric parties is another explanation for the emergence of climate blocs of similar countries. On the other hand, Buchner and Carraro argue that technology differences can also be considered as an advantage when they take the form of "complementarities" among countries. In this case they provide an explanation for the formation of climate-blocs of countries that are different. I understand the arguments in favor of a second best in which countries with similar perceptions associate. I am less convinced by a second best generated by collusion on the markets for pollution permits. Furthermore, even in the first case, the difficulty is to disentangle two explanations: free riding or efficient exploitation of similarities.

In conclusion, even if the outcome turns out to involve differential treatment of the countries and, in the extreme case, the formation of several blocs, I would still recommend a global framework for negotiations. In the case of negotiations on climate change, all the countries felt themselves obliged, at least initially, to negotiate together under the guidance of the UN. These lengthy negotiations provided conditions for a more creative approach. Countries had to find other ways to differentiate themselves from each other within a single agreement. In the framework of the Kyoto Protocol they came up with the country-specific targets and the flexibility mechanisms. The advantage of this process is that it is a better guarantee that countries do not free ride but work to increase efficiency when they want to differentiate themselves from each other.

Notes

This work was done while the author was a visitor at the Institute for Advanced Study at Princeton, and she would like to thank the Institute for its hospitality.

1. See, for example, the interpretation given by Schelling (2006) in the chapter What makes greenhouse sense? of his book *Strategies of Commitment*.

2. See the article by Kauppi and Widgren (2004) on the difference between declared and real arguments underlying the European budget allocation.

References

Bloch, F. 1995. Endogenous structures of association in oligopolies. *Rand Journal of Economics* 26: 537–56.

Bloch, F. 1996. Sequential formation of coalitions with externalities and fixed payoff division. *Games and Economic Behavior* 14: 90–123.

Chander, P., and H. Tulkens. 1995. A core-theoretic solution for the design of cooperative agreements on transfrontier pollution. *International Tax and Public Finance* 2(2): 279–94.

Chander, P. and H. Tulkens. 1997. The core of an economy with multilateral environmental externalities. *International Journal of Game Theory* 26: 379–401.

Chayes, A., and A. H. Chayes. 1991. Compliance without enforcement: State regulatory behavior under regulatory treaties. *Negotiation Journal* 7: 311–31.

Maskin, E. 2003. Bargaining, coalitions and externalities. Presidential address to the Econometric Society European Meeting, Stockholm. Mimeo.

Hart and Kurz. 1983. Endogenous formation of coalitions. Econometrica 51(4): 1047–64.

Kauppi, H., and M. Widgrén. 2004. What determines EU decision making? Needs, power or both? *Economic Policy* (July): 221–66.

Lucas, W. F., and J. C. Maceli. 1978. Discrete partition function games. In P. C. Ordeshook, ed., *Game Theory and Political Science*. New York: New York University Press, pp. 191–213.

Ray, D., and R. Vohra. 1999. A theory of endogenous coalition structure. *Games and Economic Behavior* 26: 286–336.

Ray, D., and R. Vohra. 2001. Coalitional power of public goods. *Journal of Political Economy* 109(6).

Roth, A. E. 1998. *The Shapley Value: Essays in Honor of Lloyd S. Shapley*. Cambridge: Cambridge University Press.

Shapley, L. S 1953. A value for n-person games. In Kuhn and Tucker, eds., *Contributions to the Theory of Games*, vol. 2. Princeton: Princeton University Press, pp. 307–17.

Schelling, T. C. 2006. *Strategies of Commitment and Other Essays*. Cambridge: Harvard University Press.

Thoron, S. 1998. Formation of a coalition proof stable cartel. *Canadian Journal of Economics* 1 (February).

Thoron, S. 2000. Conjectures about the reaction of a frustrated coalition. Working paper GREQAM 31A00.

Thrall, R. M., and W. F. Lucas. 1963. n-Person games in partition function form. *Naval Research Logistics Quarterly* 10: 281–98.

Yi, S.-S. 1997. Stable coalition structure with externalities. *Games and Economic Behavior* 20: 201–37.

III Policy Design

10 Economics versus Climate Change

William A. Pizer

There is a tendency in economics to focus on the big picture and key messages. In the arena of climate change these might be: a global externality requires global cooperation, international emissions trading lowers costs for all nations, and emission pricing is the key to the development of new climate-friendly technologies. Emissions trading, in particular, has been a centerpiece of economic thinking in the realm of environmental regulation, allowing policy makers to separate out the choice about who pays while maintaining a cost-effective outcome. Such thinking clearly shaped the design of the Kyoto Protocol, a climate change treaty negotiated by more than 140 nations that establishes a global emissions trading system for greenhouse gases. And even among those who might quibble with the particular targets, timetables, or mechanisms, many would embrace the overall architecture of global cooperation and international emissions trading.

But is this the right message for economists to be bringing to the table? To the extent that economics is fundamentally about informing better public policy decisions, as well as understanding economic (and human) behavior as a means to that end, the discipline must confront three pieces of information in conflict with the earlier message. First, the United States is not part of the Kyoto agreement now and, for many reasons, probably may not ever join a Kyoto-like agreement. Second, developing countries are not lining up behind the Kyoto idea of binding emission limits, a necessity for conventional emissions trading. Third, the kind of technologies we need to solve the long-term climate challenge are not currently available at the prices many nations are willing to pay. For economic insight to be relevant to the climate policy debate, these facts need to be embraced.

In consideration of these observations, this chapter addresses three questions, and in doing so sketches out the economic (and practical) arguments for both a more relaxed international framework, a bottom-up or pledge and review approach as discussed by Bodansky et al. (2004), and flexible domestic architecture involving price mechanisms, technology policy, and vehicles for developing country investment. The three questions are: Is international agreement necessary to initiate action on climate change? Should we pursue international emissions trading, or more generally globally harmonized marginal abatement costs, as a policy goal at this time? And how can domestic and international actions encourage long-term solutions to climate change?

Summarizing, I find that the current state of affairs, whereby the United States has no mandatory emission policy while Europe has already initiated an emissions trading program, provides strong evidence that international agreement is not necessary for national governments to embrace mandatory policies—at least not agreement between the world's two largest economic powers. On the second question, despite the cost effectiveness of international emissions trading, there are easier ways to equalize prices (e.g., national price-based policies) while within-country concerns over equity and climate damages may argue against global price equalization in the first place. On the third question, in addition to considering national price-based policies, I find convincing arguments for explicit technology incentives based both on the existence of technology market failures and the practical desire to complement the "stick" of emission regulation with the "carrot" of incentives. Finally, the real needs on the international front are successful mechanisms to tie national policies to developing country energy investments where the majority of inexpensive global mitigation opportunities exist.[1] Credible international reviews of national actions could also speed up the necessary process of national decisions about future action. My conclusion, therefore, is that international efforts should focus primarily on spurring domestic action through bottom up approaches, creating mechanisms for channeling funds to projects in developing countries, and providing credible reviews of national activities.

10.1 Is International Agreement Necessary to Initiate Action on Climate Change?

While basic economic theory suggests the solution to a global environmental externality like climate change requires global cooperation, recent experience presents challenges to that theory. The Kyoto Protocol has entered into force without the United States—the largest industrialized country responsible for almost one-quarter of the world's emissions. Moreover those countries that have pursued domestic policies to reduce emissions have done so with only partial linkages to their Kyoto obligations and mechanisms. Perhaps cooperation is not so important for a *first* step to address global climate change.

Consider, for example, the hypothesis that the US departure from the Kyoto Protocol, in part, could have encouraged its eventual entry into force. On the one hand, absent the United States, the collective burden of the Kyoto parties was substantially reduced.[2] The United States, because of its rapid population and economic growth since the base year of 1990, as well as its relative size, faced an enormous shortfall to reach its target. Some of this shortfall could have been ameliorated through favorable decisions about sink credits, but the United States would likely have depended on the use of the Protocol's flexibility mechanisms—emissions trading with Russia and Ukraine who possess an allowance surplus, and developing country credits via the Clean Development Mechanism. Absent US participation, these flexibility mechanisms have more capacity to address shortfalls in Europe and Japan.

On the other hand, the US departure put Europe on the spot to prove it was serious about climate change (Grubb 2002). Because the treaty was constructed in a way that would allow it to enter into force without the United States, Europe along with other nations remaining in the Protocol faced the uncomfortable choice of either abandoning the treaty and appearing incapable of action absent the United States, or continuing with the treaty without the world's largest emitter. Conventional economic theory suggests that continuing without the United States makes no sense: How do Kyoto parties convince the United States to take action in the future if they are not bargaining their own collective action in exchange? It is as though the Kyoto parties agree to lose the prisoner's dilemma; the United States benefits from their action and incurs no costs.

Yet, not only did the European Union quickly ratify the Kyoto Protocol despite US inaction, it moved ahead to enact a 2003 directive creating an emissions trading scheme (ETS), partly in parallel the Protocol's trading mechanisms. While their own forecasts suggest the EU-15 will be 8.5 percent above their 1990 target and the EU-23, including new annex-B member states such as Poland and the Czech Republic, 3.6 percent above their target (EEA 2005), the trading program remains a very tangible commitment to reduce emissions—and clear evidence that Europe is not waiting for US cooperation.

Why would the European Union go forward with a trading program, absent a global commitment, with minimal consequences for the environment, and very real consequences for their businesses? One could back up further and ask, why would the annex-I countries agree to emission targets under the Kyoto Protocol, absent a developing country commitment, with minimal consequences for the environment, and (assuming they enact enabling legislation) very real consequences for their economies? Or why did the recent Bingaman-Specter resolution (US Senate, *A resolution to express the sense of the Senate on climate change legislation*, 2005) ask only that US action *encourage* action by other nations—while the previous Byrd-Hagel resolution (US Senate, *A resolution expressing the sense of the Senate regarding the conditions for the United States becoming a signatory to any international agreement on greenhouse gas emissions under the United Nations Framework Convention on Climate Change*, 1997) stipulated "new specific scheduled commitments to limit or reduce greenhouse gas emissions" by developing countries?

The answer is that initial action on climate change need not be entirely cooperative. First, large countries will themselves recognize benefits from their own mitigation, simply not the full, global benefit of that mitigation outside their own borders. Second, even acknowledging these internal benefits, it is unclear what drives some nations to become stronger advocates for action than others.[3] Third, the push for cooperative action may *slow* real action by spending time and resources on the effort to reach an unnecessary agreement.

Meanwhile there are many examples of multilateral issues where actions were initially idiosyncratic or unilateral, and later became cooperative. Trade, disarmament, phase-out of ODS, and other global environmental problems—all of these have (at times) involved one or a few countries taking action. The key is that countries take an initial step in expectation of some level of reciprocation. Countries take one step unilaterally, but not two.

There are at least two distinct advantages to this less cooperative, unilateral approach. First and foremost, it is a chance to determine what can be done domestically. Over the past decade in the United States, we have seen numerous efforts to tackle serious problems that have floundered over domestic/congressional support: health care, social security, major tax reform. There is also an issue of timing: the 1990 Clean Air Act Amendments that created the historic acid rain trading program were in part a product of the successes and failures over the preceding decades (Kete 1993). Finally, the United States has a history of treaty law whereby it does not typically ratify a treaty without implementing legislation in place (CRS 2001). All this suggests that attempting to negotiate an international agreement as a prelude to domestic action creates problems when the time comes to turn the international commitment into domestic action—or even to ratify the international agreement.

Second, focusing on domestic action first means that the international architecture benefits from an initial base of experience. Efforts to address sinks, project-based credits, penalties, and other details under the Kyoto Protocol were arguably hampered by little practical experience at the time provisions were negotiated. More recent plans to incorporate project credits in the Regional Greenhouse Gas Initiative in the United States (spearheaded by Governor George Pataki of New York), for example, have benefited from all the project-based activities that have occurred since 1997. Similarly considerable expansion and improvement in the economic modeling of climate policies has made the near-terms costs of mitigation policy clearer—a significant advantage when setting targets.

From a more subtle economic view, climate change can be viewed as a repeated game. Unlike a static game where players are viewed as simultaneously jumping to the equilibrium, a repeated game allows many more possibilities. This includes the idea that one player will unilaterally follow a cooperative route in an attempt to lead the other player; if the other player does not follow, the cooperative route can be abandoned. And it need not be a particularly cooperative route—simply encouraging an initial round of policies where countries maximize their own net benefits could be a useful first step on the road to maximizing global net benefits. There may even be an advantage to moving first, if the "first-mover" is able to establish architecture and precedent in her favor, or if she develops an expertise that is then valuable.

In concluding this discussion about cooperation, I would note that my use of current action as evidence about the necessity of cooperation (or lack thereof) might seem to raise an obvious concern: Does current action in fact represent an appropriate response given the current scientific understanding? For example, one might agree that the ETS is possible in the European Union without US cooperation but believe that more action is necessary now and that action requires cooperation with the United States. My point, however, is not that current action is too little, too much, or just right, merely that meaningful (i.e., mandatory) action on climate change is possible unilaterally as a first step.

My lack of concern over international cooperation is also only in reference to initiation and first steps. Assuming the arguments for, and willingness to invoke, a stronger climate policy response continue to grow, international coordination and cooperation will become increasingly important.

10.2 Should We Pursue International Emissions Trading Now?

The preceding section has argued that initial mitigation efforts are not only possible but actually occurring without the benefit of cooperation among key parties. But now I want to ask a second question: Suppose we see domestic, market-based policies evolving; should we pursue international emissions trading or alternative efforts to harmonize marginal abatement costs *now*? As the heading suggests, I find convincing evidence that the answer is no. First, economic analysis (and recent experience) suggests price mechanisms ought to be preferred to pure emission caps; introducing price mechanisms makes an international trading program both more difficult and less important. Second, despite arguments for economic efficiency there are a variety of reasons why countries might want to maintain or limit their domestic emissions at a particular level, and therefore want to avoid international trading. Third, the kinds of cash flows associated with a global emissions trading program could create both economic and political challenges. Fourth, global trading in a general sense is unnecessary, assuming the real supply of cheap mitigation opportunities are in developing countries. For efficiency, trading only needs to facilitate North–South movements of mitigation technology to capitalize on those opportunities. All this suggests that a more pragmatic solution might focus on domestic mitigation programs that simultaneously create incentives to pursue opportunities in developing countries.

The first point is ironically supported by both economic theory and political practicality. Most economic analysis shows that price-based mechanisms are much superior to quantity-based approaches in terms of expected welfare (Pizer 2002). This follows from the basic Weitzman (1974) intuition that relatively flat marginal benefits favor price instruments coupled with the extreme persistence of greenhouse gas emissions. That is, cumulative contributions over even a number of years are sufficiently small compared to cumulative contributions since industrialization so as to not affect marginal consequences (Newell and Pizer 2003). Even absent academic studies, debate in the United States suggests that the economic uncertainty associated with an emission cap is problematic (NCEP 2004). Conveniently a "safety valve" can be introduced into an otherwise ordinary emissions trading program to make it behave like a price mechanism: The government stands ready to sell additional allowances at the safety-valve price, mimicking a price policy.

In an international trading program, however, who sells those extra allowances? If national governments do this, the outcome is tricky. A shortage in one country could lead firms to purchase from that government, or import allowances from another country and shift the shortage to the exporting country. Whichever government ends up selling allowances ends up generating revenue for itself—a clear national benefit. If this hazy outcome about who ends up generating revenue is not acceptable, then a single international agency needs to sell the allowances—but again, the question arises, how will the revenue be used? For some, the idea of buying additional emission allowances from a UN-type organization is problematic. But all this is really unnecessary; if a safety valve exists, it can be used to set a common price across countries via domestic programs without the need for trading.

A second question arises when countries have different ideas about what they believe emission reductions are worth and what allowance prices ought to be. Arguably that is the case right now: Allowances in the ETS recently traded in the European Union at nearly €30 per ton of carbon dioxide (Point Carbon 2005) while there is currently no market signal in the United States.[4] If one were to guess about likely prices in a future US trading program, something closer to $5 to $10 is perhaps reasonable.[5] Suppose that this occurs and that two potentially compatible trading systems exist, Europe with a €20 to €30 price and the United States with a $5 to $10 price; would the Europe and the United States really want trading?[6]

Basic economics says yes: Both countries stand to gain from trade. The European Union gets cheap reductions in the United States, and the United States gets to sell reductions at a profit to the European Union. But wait. Does the European Union really want to legitimize what it may believe is an inappropriately weak target in the United States? Not only legitimize, but allow the United States to actually

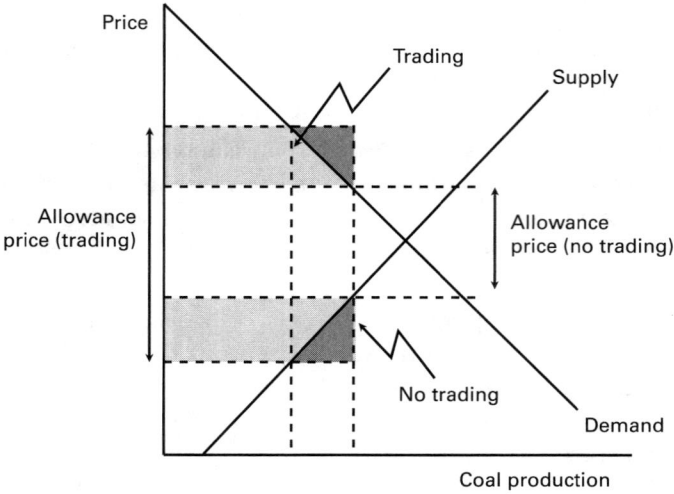

Figure 10.1
Simple analysis of welfare effect of GHG allowance trading in coal market when allowances prices rise

make money off their weak choice? What if the European Union sees €20 to €30 prices as necessary to encourage technological change? And, from the US perspective, the choice of a $5 to $10 price may reflect important domestic interests, such as preserving a particular level of coal production, or limiting energy price increases.[7] That is, trading into the European Union likely means higher allowance prices, large gains for allowance holders, and large losses for those purchasing energy or involved in coal mining, with the net welfare change being a relatively small gain.

Consider figure 10.1, showing a simple welfare analysis for the coal market when allowances prices rise with trading. The net welfare gain is given by the small dark gray triangles. Alongside this gain, however, are transfers from consumers and producers of coal to allowance holders, equal to the much larger light-gray quadrilaterals. If redistribution is difficult, perhaps the gains to trade are less important than preserving a domestic balance.[8]

A third point concerns the more general political and economic practicality of large trade flows in emission allowances. McKibbin et al. (1999) show that trading under the Kyoto Protocol could have had important and adverse effects on capital flows. Indeed the first motivation for a safety-valve mechanism was to avoid these flows, rather than address cost certainty (McKibbin and Wilcoxen 1997). At a more gut level, one has to wonder about the eventual political response to large voluntary cash flows to other industrialized countries based on some negotiated allowance allocation among countries—can such transfers be sustained?[9] It is almost ironic that in order to sustain a trading program, we may need an allocation that does not in fact lead to systematic trading in one direction or another.

Finally, it should be noted that the important trades really need to occur between developing countries, where the very cheapest mitigation options exist, and industrialized countries, where the mitigation burden will (at least initially) be placed, and not *among* industrialized countries. On the plus side, this is an area where we might expect a political acceptance of capital flows—there are already various forms of aid flowing from industrialized to developing countries; allowance purchases would be one more form. Further this flow would likely translate into purchases of mitigation technology from the very industrialized countries initiating the flow. On the minus side, developing countries are both less equipped to establish an emissions trading program—in terms of monitoring and enforcement capacity—and less motivated to adopt a binding cap.

The latter reflects a view that emission caps, however generous at the outset, could eventually be used to limit development and growth, and that climate change is simply not a priority when viewed alongside poverty, hunger, and education (Kopp 2005). I worry that the strength of this view among developing countries may be underestimated among economists theorizing about global trading. If true, it will be critical to put into place alternative effective mechanisms both to channel industrialized country resources into developing countries mitigation activities and to encourage developing country action. It is really this piece of the puzzle that leads me to question the value of an international agenda focused now on international emissions trading and/or equalizing marginal costs of mitigation.

As with the first question concerning the importance of international cooperation, this view does not extend to the longer term if policies evolve toward tighter controls. At such a point in time, concern over international competition as well as emission leakage necessitates some effort to harmonize allowance prices/marginal mitigation costs. Further, tighter controls *without* sufficiently global participation may require trade instruments to address competition and leakage with nonparticipants.

10.3 What Should Domestic Policies and International Action Look Like?

Hopefully none of what I have said so far will be construed to imply that climate change is not an important global issue, or that should it be ignored in international forums. Rather, I see the international stage as a place to re-enforce domestic action in key countries, to challenge those countries to step up and do more where appropriate, but not—now—to focus on global caps and allocation of emission rights, or even specific commitments. I believe it is important to recognize the variety of actions that are going on around the world, focused both on mitigation and technology development. The preceding section suggested that some countries may pursue more expensive mitigation options than others. Some countries will pursue greater technology development; others higher efficiency. Some will embrace nuclear power; others will reject it. Some will pursue emission taxes, others emissions trading, and yet others some form of hybrid policy.[10]

While recognizing the wide variety of potential domestic response, it is worth noting that economics has produced convincing evidence supporting particular policy directions. I have already noted the strong arguments—both economic and practical—favoring price-based mechanisms or some other approach to modulating cost shocks (Pizer 2002; Newell et al. 2005). Others have focused on the need to combine technology and mitigation policies, based on dual market failures (environment and innovation) as well as practical concerns about adequate emissions pricing (Jaffe et al. 2004). This combination also possesses the politically appealing feature of coupling the carrot of technology incentives with the stick of emissions regulation (NCEP 2004). Finally, there is the issue of project-based mechanisms to take advantage of opportunities in developing countries (Fischer 2004).

Lurking behind all of these features is the reality that climate change is a long-term problem where the key is to provide incentives for technological development and deployment, both on the supply side and energy efficiency. While economists have long focused on getting the prices right, a central question is whether nations can sustain those prices in the face of competing forces.[11] Those forces include economic competition with developing countries without similar controls, domestic pressure to preserve the status quo, and the voting public who, while concerned about global warming, rank it relatively low compared to other issues (Krosnick et al. 2000). There is also the concern noted earlier that developing countries may fail to invoke any price on emissions in the foreseeable future. For these reasons the concepts of stable economic costs, linkage with technology policies, and mechanisms for exporting technologies to developing countries are all important in the design of domestic policies.

Despite this reasoning the creation of the ETS within the European Union has obviously created momentum for an architecture with little attention to cost shock modulation or technology incentives, and linkages to developing countries only via the Kyoto Clean Development Mechanism.[12] A key concern, then, is whether the European ETS might falter for exactly these reasons. Experience with the RECLAIM program in California, for example, suggests that simply establishing an emissions trading program does not guarantee its success (Coy et al. 2001).

Many people besides me have ideas about what an effective domestic program should look like, and I believe international forums should be used to encourage domestic activities in whatever form they take. Further they should provide opportunities to describe and evaluate the alternative routes taken by different countries. A key element of the earlier assumption, that domestic action can occur with something less than internationally agreed targets and timetables, is the idea that national governments will willingly take one unilateral step on climate change, but not two. Additionally they are even more likely to take a more meaningful step when they see others doing so. Both of these points suggest an important role for international institutions to both describe current actions, as well as provide multilateral reviews of their consequences (including possibly countervailing policies that might increase fossil energy use). For example, there is no doubt that as Europe moves ahead with its trading program, assuming some measure of success, arguments against such mandatory action in other countries will be weakened. Further, while the decision to take future steps (unilateral or multilateral) inevitably hinges on a country's own assessment about what their allies and competitors are doing, such assessments can be facilitated by multilateral reviews.

Much of what I am describing has been described previously as a bottom-up approach or pledge and review (Bodansky et al. 2004). The chief disadvantage cited in the literature is the potential reluctance of countries to pursue significant reductions absent a firmer commitment—if one believes significant reductions and firmer commitments are both appropriate and possible. Note that even if one believes significant reductions are appropriate, the question of what is possible could still lead one to prefer the bottom-up approach. And note that the pledge step may or may not be important: as nations go through their domestic policy processes, dictated by their idiosyncratic political schedules and cycles, they will inevitably face their choices with a keener focus on what they believe other nations are doing, versus what they might have pledged. Finally, even if an international pledge activity itself leads to more ambitious goals, it is not obvious these goals will translate into different choices when domestic politics is brought to bear. For example, despite more than a decade of international debate focused on historic (1990) emission levels as a policy goal, it is hard to see how that has influenced serious proposals for domestic caps in Europe, the United States, or other countries.[13]

10.4 Conclusions

This chapter began with the hypothesis that basic economic concepts applied to the problem of climate change may have lead us down the wrong path. The ideas that a global environmental externality requires a global intervention, that efficiency requires global trading, and that correctly pricing emissions is the way to bring forward future climate-saving technologies follow from economic principles and are enmeshed in the current Kyoto targets and timetables approach. Yet there are signs of trouble ahead: There is no indication that the United States or developing countries will engage on these terms, few parties to the Protocol are on track to meet their commitments, and we have yet to see the consequences of €15 allowances.[14]

This chapter then looked more closely at the three ideas noted above and at how economic theory and practical issues might reverse such conclusions. Viewing climate negotiations as a repeated activity, where countries can take one step and see how others respond, suggests many more possibilities than a once-off chance at cooperation. Recognizing the within-country distribution of effects from emissions trading provides one of several reasons why equalizing prices across countries may not be desirable even if it lowers aggregate costs. Finally, the relative importance of long-term technology development and deployment over near-term mitigation argues for a program that is broader and more flexible than ordinary emissions trading—providing cost certainty through a safety-valve mechanism, coupling technology-push policies with the demand pull of emissions pricing, and pursuing various approaches to channel resources toward developing country energy investments.

Much of the emphasis of future research and action, then, might be on domestic policy development that later can be ratcheted up as more countries become engaged. This bottom-up approach is not new. What may be new is an economic rationale coupled with a clearer suggestion about what international action could usefully focus on: encouraging developed country action, creating mechanisms to channel funds to projects in developing countries, and providing credible reviews of national activities. This, in turn, calls into question the value of seeking either new agreements on targets or further development of international trading mechanisms, with the exception of funding vehicles for developing country technology investments.

As the first Kyoto compliance period draws near and active discussions are occurring in virtually every industrialized country concerning domestic climate change policies, including those both in and out of the Kyoto Protocol, the time is clearly ripe to debate these issues.

Notes

1. Whether the Clean Development Mechanism under the Kyoto Protocol can become a successful funding vehicle remains to be seen. Alternatives include larger "deals" to bring Russian natural gas to China or encourage more nuclear power in India, as well as crediting for policy reforms (discussed at COP-11 in Montreal). All of this hinges on continued demand from industrialized countries.

2. EIA (2001) estimates that the Kyoto burden of the United States alone exceeded the collective burden of annex I in 2010 (tab. 2, p. 14). See also Nordhaus (2001) and Babiker et al. (2002).

3. For example, Nordhaus (1998) estimates the largest impacts to be in Africa and India—yet those countries are far from leading the charge on the climate change. Interestingly his estimates show Europe to have considerably higher impacts than the United States, Japan about the same, and Russia to observe benefits (all from 2.5°C warming).

4. Although previous analyses of the ETS in the European Union suggested costs of €26 per ton of carbon dioxide, this arose in the later, more stringent Kyoto compliance phase (2008–2012) and only under the assumption of no trading beyond the European Union (Criqui and Kitous 2003). With trading, the cost was estimated to be as low as €5 per ton; presumably the easier targets in the warm-up phase should also result in lower costs.

5. Recent analyses of a US climate proposal by US Senator Jeff Bingaman and the National Commission on Energy Policy suggests costs of around $5 per ton (EIA 2005). This proposal has caught the attention of Pete Domenici, the Republican chairman of the Senate Energy Committee who has indicated his possible support (Fialka et al. 2005). The more favorable analyses of the McCain-Lieberman proposal were also in the $10 per ton range (Paltsev et al. 2003).

6. There could even be a third system: Canada is pursuing a trading program for larger emitters with a limiting price of C$15 per ton of CO_2.

7. Support by the United Mine Workers of America for the NCEP recommendations, for example, hinged partly on the fact that while coal growth slows under their climate recommendations, it remains positive (UMWA 2004).

8. Note that differences in relative prices among countries could be maintained by having nonuniform trading ratios; for example, two US allowances might be required to obtain one EU allowance.

9. See, for example, IEA (2001).

10. New Zealand recently enacted a NZ$15 carbon dioxide tax. Japan recently considered (but did not enact) a ¥2500 to ¥3000 tax per ton of carbon. Canada is pursuing a hybrid emissions trading program with a C$15 per ton CO_2 safety valve.

11. Montgomery and Smith (2005) have argued that emissions pricing is totally unnecessary.

12. As noted at the beginning of this chapter, the effectiveness of the CDM at channeling resources to developing countries remains to be seen.

13. For example, the McCain-Lieberman bill in the United States eventually abandoned any reference to 1990 levels. In Europe, the National Allocation Plans have focused on deviations from current or forecast emission levels more than 1990 levels.

14. Recent attention, for example, has focused on the tight correlation of ETS allowance prices and electricity prices, raising concerns about unfair profits for electric generators who received their allowances for free (IFIEC 2005).

References

Babiker, M. H., H. D. Jacoby, J. M. Reilly, and D. M. Reiner. 2002. The evolution of a climate regime: Kyoto to Marrakech and beyond. *Environmental Science and Policy* 5: 195–206.

Bodansky, D., E. Diringer, J. Pershing, and X. Wang. 2004. *Strawman Elements: Possible Approaches to Advancing International Climate Change Efforts*. Washington: Pew Center on Global Climate Change.

Congressional Research Service. 2001. *Treaties and Other International Agreements: The Role of the United States Senate*. Washington: Government Printing Office.

Coy, C., P. Mueller, D. Luong, S. Tsai, D. Nguyen, and F. Chen. 2001. *White Paper on Stabilization of NO_x RTC Prices*. Diamond Bar, CA: South Coast Air Quality Management District.

Criqui, P., and A. Kitous. 2003. Kyoto Protocol implementation. In *(KPI) Technical Report: Impacts of Linking JI and CDM Credits to the European Emissions Allowance Trading Scheme*. Brussels: CNRS-IEPE and ENER-DATA S.A. for Directorate General Environment, Service Contract No. B4-3040/2001/330760/MAR/E1.

Energy Information Administration (EIA). 2001. *International Energy Outlook 2001*. Washington: Energy Information Administration.

Energy Information Administration (EIA). 2005. *Impacts of Modeled Recommendations of the National Commission on Energy Policy.* Washington: Energy Information Administration.

European Environmental Agency (EEA). 2005. *Analysis of Greenhouse Gas Emission Trends and Projections in Europe 2004.* Copenhagen: European Environmental Agency.

Fialka, J. J., J. D. McKinnon, and J. Ball. 2005. Bush faces heat over global warming. *Wall Street Journal*, June 20, p. A4.

Fischer, C. 2004. Project-based mechanisms for emissions reductions: balancing trade-offs with baselines. *Energy Policy* 33(14): 1807–23.

Grubb, M. 2002. International climate change strategies. Briefing organized by the Environmental and Energy Study Institute, Washington, DC.

Harvey, F. 2005. US agreement paves the way for fresh climate chang talks. *Financial Times*, December 12, p 3.

International Energy Agency (IEA). 2001. *Emissions Trading: From Concept to Reality.* Paris: OECD.

Jaffe, A. B., R. G. Newell, and R. N. Stavins. 2004. *A Tale of Two Market Failures: Technology and Environmental Policy.* Washington: Resources for the Future.

Kete, N. 1993. The politics of markets: The acid rain control policy in the 1990 Clean Air Act amendments. PhD dissertation. Johns Hopkins University, Baltimore.

Kopp, R. 2005. *So Kyoto Is Going into Effect—How Do We Make It Work?* Resources for the Future 2005 (cited February 16, 2005). Available at ⟨http://www.rff.org/rff/News/Features/So-Kyoto-Is-Going-Into-Effect.cfm⟩.

Krosnick, J. A., A. L. Holbrook, and P. S. Visser. 2000. The impact of the fall 1997 debate about global warming on American public opinion. *Public Understanding of Science* 9: 239–60.

McKibbin, W. J., M. Ross, R. Shackleton, and P. Wilcoxen. 1999. Emissions trading, capital flows and the Kyoto Protocol. *Energy Journal* (Special Issue).

McKibbin, W. J., and P. J. Wilcoxen. 1997. *A Better Way to Slow Global Climate Change.* Washington, DC: Brookings Institution.

Montgomery, W. D., and A. E. Smith. 2005. *Price, Quantity, and Technology Strategies for Climate Change Policy.* Washington: Charles River Associates.

National Commission on Energy Policy. 2004. *Ending the Energy Stalemate: A Bipartisan Strategy to Meet America's Energy Challenge.* Washington, DC: National Commission on Energy Policy.

Newell, R., and W. Pizer. 2003. Regulating Stock Externalities Under Uncertainty. *Journal of Environmental Economics and Management* 45: 416–32.

Newell, R., W. Pizer, and J. Zhang. 2005. Managing Permit Markets to Stabilize Prices. *Energy and Resource Economics*.

Nordhaus, W. D. 1998. *New Estimates of the Economic Impacts of Climate Change.* New Haven: Yale University.

Nordhaus, W. D. 2001. Global warming economics. *Science* 294: 1283–84.

Paltsev, S., J. M. Reilly, H. D. Jacoby, A. D. Ellerman, and K. H. Tay. 2003. Emissions trading to reduce greenhouse gas emissions in the United States: The McCain-Lieberman proposal. MIT Joint Center on the Science and Policy of Global Change, Cambridge.

Pizer, W. A. 2002. Combining price and quantity controls to mitigate global climate change. *Journal of Public Economics* 85(3): 409–34.

Point Carbon. 2005. Midday Market Update. *Carbon Market News*, July 8.

United States Senate. 1997. *A resolution expressing the sense of the Senate regarding the conditions for the United States becoming a signatory to any international agreement on greenhouse gas emissions under the United Nations Framework Convention on Climate Change.* 105, S 98.

United States Senate. 2005. *A resolution to express the sense of the Senate on climate change legislation.* 109, S Amdt 866.

United Mine Workers of America. 2004. United Mine Workers of America Commends Newly Released Report by the National Commission On Energy Policy. Washington: United Mine Workers of America.

Weitzman, M. L. 1974. Prices vs. quantities. *Review of Economic Studies* 41(4): 477–91.

11 Economics versus Climate Change: A Comment

Richard S. J. Tol

On matters of climate policy and economics, I usually agree with Billy Pizer. The chapter he contributed to this book is no exception. My commentary is complementary to it rather than critical. I follow the structure of his discussion, remarking on international cooperation, international permit, and technological change, followed by some brief conclusions.

11.1 International Cooperation

Pizer's first point is that an international agreement is not needed for greenhouse gas emission abatement. A survey of domestic action shows that it is impossible to disagree. In fact this point may be stronger than Pizer suggests: international agreements may hinder abatement. One reason is that a risk-averse country may be more conservative if negotiating legally binding targets than if pledging voluntary action. Another reason is that international treaties often find the lowest common denominator, and so provide support for the conservative opposition in progressive countries.

Pizer alludes to the different positions that international treaties have in domestic law. In some countries (Australia, the United States), "legally binding" means just that. In other countries (the European continent), the government is above civil law and can flaunt targets at will. The EU Stability and Growth Pact is a recent example; the quota for milk production is an older but more pertinent example. A treaty that means different things to different parties is hard to negotiate, and perhaps should not be negotiated.

Nonetheless, some form of international cooperation is needed, since domestic policy is not isolated. Emission abatement implies higher energy prices. If a country has more stringent targets than other countries, it will become less competitive, first on the commodity market and then on the capital market. Its emissions would leak to other countries, and abatement would be more expensive (Kuik and Gerlagh 2003). Domestic policy should therefore not deviate too much from the policies of the main trading partners and competitors. This effect would be stronger in smaller and more open countries; it is probably small for the United States. International cooperation is about information exchange and confidence building. As trade and investment are crucial issues, one may wonder whether trade bodies (ASEAN, EU, MERCUSOR, NAFTA) could adopt this role rather than the UN FCCC and its bodies.

At two points, I disagree with Pizer. He asserts that the US nonratification of the Kyoto Protocol induced the European Union to ratify. The European Union would have ratified anyway. One of the aims of European climate policy is to demonstrate moral superiority to the filthy Americans. The Bush administration's lack of ambition in climate policy has made that rather easy, and may well have slowed European climate policy.

I also disagree with Pizer on the European Trading System. Although functioning, it is too early to assess its success. It is clear, however, that only a small fraction of the planned and promised emission reduction will be by permit trade; the initial allocation of emission permits is simply too generous to have much impact. Besides, permit trade did not replace preexisting regulation but was simply placed on top. This takes away many of the potential benefits (Pearce 2006).

11.2 International Permit Trade

International permit trade would equalize marginal abatement costs among trading partners and guarantee the least-cost solution to emission reduction. However, this is true under rather restrictive conditions only (Baumol and Bradford 1970), which are unlikely to be met in the case of climate change.

Pizer points out that countries would likely disagree about what the price of the permits should be. This is particularly pronounced if there are safety valves. Indeed the lowest safety valve would be valid in the global market. The country with the lowest safety valve would effectively set international climate policy. It would flood the market with exported permits. The revenues may induce a veritable race to the bottom.

This is true for an unregulated market only. Rehdanz and Tol (chapter 13 in this volume) propose one way around the problem: Quantity or price instrument may be used to regulate international trade in emission permit. A country may impose import restrictions, either on total imports or on imports from a particular country. The World Trade Organization has no say in this matter, as permits do not fall under its regulations. Unlike with commodities, permits have arbitrary units. The European Union may simply declare that there are two American tonnes in one European tonne of carbon. Alternatively, the European Union may put a tariff on imports of American permits.

Bradford (chapter 12 in this volume) proposes an alternative solution. Countries would pursue domestic abatement policies, as in Pizer, but would also contribute funds to a "Global Climate Facility" that would use the money to purchase emission reduction where that is cheapest. This would probably not equalize marginal abatement costs, but there is a mechanism of convergence for marginal costs—which is lacking in Pizer's suggestions.

11.3 Technological Change

The key to a zero-carbon economy is to accelerate and re-direct technological change. While "sticks" such as taxes and permits would contribute to this, "carrots" have a place too and are underrepresented in current climate policies. I fully agree with Pizer. Carrots would include subsidies on R&D. However, there are plenty of blueprints for a carbon-free energy sector already (Pacala and Socolow 2004). The key is commercialization rather than invention. Governments may be good at fundamental research (through its support for university and research laboratories), but their track record on applied research is less impressive (Gomulka 1990). With subsidies, the attraction of "picking winners" is too great, particularly as "winners" may be picked based on lobbying rather than merit. Prizes, guaranteed purchases, and temporary monopolies may work better, particularly if based on performance rather than for a specific technology. The long-term credibility of the regulator is crucial, however.

At this point Pizer could have returned to international cooperation. Energy is big. Smaller countries would be unable to make much of an impact on international technological trends. Some challenges are too large for even the biggest countries. Nuclear fusion is an example. In some industries (e.g., cars), a few countries dominate the international market. Negotiations with a few parties are much easier than with many partners. If such countries would agree on performance standards, the rest of the world has little choice but to follow.

11.4 Conclusion

In sum, I largely agree with the arguments of Billy Pizer: There is little hope for an international climate treaty; and there is less hope that climate policy will be run according to textbook economics. That does not imply, however, that there is no hope for climate policy. And, as any fool can read a textbook, it also implies that skilled economists are even harder needed to successfully negotiate the subtle and treacherous waters of real-life climate policies.

References

Baumol, W. J., and D. F. Bradford. 1970. Optimal departures from marginal cost pricing. *American Economic Review* 60(3): 265–83.

Bradford, D. F. 2004. A no-cap-but-trade proposal for international climate policy. Mimeo. Available at ⟨http://www.wws.princeton.edu/~bradford/globalpublicghgcontrol01.pdf⟩.

Gomulka, S. 1990. *The Theory of Technological Change and Economic Growth*. London: Routledge.

Kuik, O. J., and R. Gerlagh. 2003. Trade liberalisation and carbon leakage. *Energy Journal* 24(3): 97–120.

Pacala, S. W., and R. H. Socolow. 2004. Stabilization wedges: Solving the climate problem for the next 50 years with current technology. *Science* 305(5685): 968–72.

Pearce, D. W. 2006. The political economy of an energy tax: The United Kingdom's climate change levy. *Energy Economics* (in press).

Rehdanz, K., and R. S. J. Tol. 2005. Unilateral regulation of bilateral trade in greenhouse gas emission permits. *Ecological Economics* 54: 397–416.

12 Absolute versus Intensity Limits for CO_2 Emission Control: Performance under Uncertainty

Ian Sue Wing, A. Denny Ellerman, and Jaemin Song

Prediction is very hard ... particularly of the future ...
—Niels Bohr

A simple and fundamental environmental policy question is: If a country makes a commitment to constrain or keep emissions at or below a target level in some future time period, does it make a difference if the commitment is expressed as a limit on the absolute level of emissions or on the intensity of emissions? Nowhere is this question more relevant than in the design of policies to mitigate the emissions of greenhouse gases (GHGs). Already the widespread concern is that attempts to cut GHG emissions will cause significant increases in energy prices and reductions in economic output and welfare. The GHG emission limits negotiated under the Kyoto Protocol have been criticized as contributing to this unfavorable outcome because they are expressed as fixed caps on countries' ability to emit. The absolute character of these caps, it has been argued, fails to account for the possibility that economies and their emissions might grow more quickly than was expected at the time the targets were negotiated, and that larger-than-anticipated economic losses would be inflicted on the Kyoto signatories.

Several generic proposals have been advanced in response to these concerns. A "safety valve" would set an upper bound on the marginal costs of abatement and thereby truncate the upper end of the distribution of outcomes (Kopp et al. 2000; Jacoby and Ellerman 2002; Philibert 2005). We do not engage in further discussion of these proposals here but focus instead on an alternative: intensity limits. Although rare in the domain of GHG emissions control, limits on the pollution intensity of output are by far the more common method of constraining emissions in the field of environmental regulation.[1] Nevertheless, "relative" or intensity-based targets have been adopted as a component of climate policy in the UK Emissions Trading Scheme (UK DEFRA 2001),[2] and in 2001 the Bush administration proposed a voluntary target of 18 percent reduction by 2012 in GHG emissions intensity for the United States.

Implicit in the adoption of all these measures is the recognition of the general principle that the pollution from a source can be limited by specifying either an absolute cap on the quantity of emissions that it generates or by setting a maximum allowable intensity of emissions relative to some measure of output or input. Examples are the units of output or the amount of energy input required by some production processes at the firm level, and the volume or value of commodities purchased by consumers at the level of an economic sector, or even GDP at the national level. Such an intensity limit can be imposed either directly as an emissions rate limit or as an efficiency standard, or indirectly by means of technology mandates that have the same effect.

The choice of intensity targets is not without controversy, however. An often-heard environmentalist critique of intensity caps is that indexing an emission limit to GDP allows GHGs to exceed an ex ante equivalent absolute cap if economic growth is more rapid than expected. But this criticism overlooks the symmetric opposite case, of which there has been comparatively little discussion, where intensity caps require *more* abatement than an ex-ante equivalent absolute cap if economic growth is less than expected. In this chapter we elucidate the differences between absolute and intensity limits under uncertainty. Our guiding assumption is that the variance in the intended environmental and economic effects of an emission constraint is a key factor in deciding how it is to be implemented. Both impacts derive from the reduction in emissions relative to no-policy levels; therefore we examine the divergence between actual and expected *abatement*, and treat higher and lower than expected economic growth outcomes as symmetric. For simplicity, our discussion will focus on the economic costs of this choice, without consideration of the associated environmental benefits. As well, we restrict our analysis to the setting of a single economy, and leave the interaction of absolute and intensity limits in an international emission trading system to future research. Also left to future investigation is a rigorous comparison of the merits of intensity limits relative to safety valves or intertemporal banking and borrowing as means to reduce the variance of outcomes. We concentrate instead on laying the groundwork for such assessments by elaborating the conceptual and theoretical foundation introduced in Ellerman and Sue Wing (2003).

These sacrifices in terms of scope allow us to make three contributions. First, we demonstrate that an emissions constraint can be expressed equivalently as an absolute or intensity limit on emissions when there is no uncertainty about the future. At face value this point may appear trivial, but there seems to be much misunderstanding in policy circles on the issue of intensity limits, and we are hard-pressed to find analyses that rigorously address this basic fact.

Second, we demonstrate the conditions under which an absolute or indexed intensity limit would be preferred, which we model with relation to reduced variance and also discuss the characteristics of an optimal degree of indexing where an intensity limit would produce less variance.

Third, we explore the policy implications of these conditions using time series data on different countries' actual CO_2 emissions and GDP, as well as historical forecasts of these variables. We do this by conducting a backcasting analysis that considers an alternate state of the world in which countries decided to limit their emissions of CO_2 in earlier decades. This allows us to investigate what would have been the optimal choice for the form of a country's emissions cap.

12.1 Literature Review

A recent and diverse literature has developed concerning the use of intensity-based and indexed caps in the context of climate policy.[3] The nearly uniform motivation is the widespread perception that developing countries would not accept absolute caps because of the perceived limit on economic growth. The proposal by Argentina in November 1999 at the Fifth Conference of the Parties to the Kyoto Protocol first drew official attention to this subject (Argentina 1999; Barros and Conte Grand 2002). Shortly thereafter, one of President Clinton's economic advisors proposed indexing GHG emission targets to GDP growth as a means of making Kyoto-type caps more acceptable to developing countries (Frankel 1999). Key early papers by Baumert et al. (1999) and especially Lutter (2000) introduced the idea of intensity targets as a hedge against uncertainty—in particular, their potential to mitigate excess abatement costs incurred by higher than expected business-as-usual (BAU) emissions. Subsequently the Bush administration's announcement of a target to reduce GHG intensity in the United States by 18 percent by 2012 (White House 2002), and its advocacy of intensity limits for developing countries prompted a spate of analyses concerned both with the adequacy of the US target and with the more general merits of intensity-based caps.[4]

While analysts appear united in finding that the target set by the Bush administration is indistinguishable from BAU emissions, opinion on the attractiveness of intensity limits is less uniform. Gielen, Koutstaal, and Vollebergh (2002) and Fischer (2003) draw on an old literature in environmental economics, going back to Spulber (1985) and Helfand (1991), that criticizes intensity limits because of the incentive they give producers to use larger quantities of the input or output in which the intensity index is denominated. Compared to absolute limits, intensity caps are a "subsidy" to firms' use of the denominated input or to their production of the denominated output, thereby giving rise to an inefficient allocation of resources.

The output subsidy critique of intensity limits applies only in so far as the limit is faced by individual producers. It does not apply at the country level because the indexation variable, aggregate output, does not figure in firm-level decisions. Within the participating country, producers could be expected to take into account the fact that an indexed emissions cap would be adjusted upward (or downward), but the practical incentive they would face is a lower (or higher) cost for the use of allowances. Individual firms would not face any greater or lesser constraint as a result of variations in the output of, or the inputs to, their production processes.

Another persistent critique of intensity limits is that relative to an absolute ceiling on emissions that is fixed ex ante, an intensity cap creates the potential for an environmentally adverse outcome if GDP is higher than expected since the cap would be adjusted upward, making the target less stringent in absolute terms (see Dudek and Golub 2003). Comparisons between Kyoto's absolute emission targets and intensity limits that characterize the latter as economically advantageous while being environmentally disadvantageous reflect this criticism. What happens if GDP growth declines or is *less* vigorous than expected is, however, rarely noted. The level of an intensity cap will adjust downward, making the target *more* stringent than an unchanging absolute cap. In this case an intensity cap is environmentally advantageous and economically disadvantageous. Ellerman and Sue Wing (2003) argue intuitively that an intensity limit trades off less stringent control of emissions in a state of the world with higher than expected economic growth for more stringent control in a state of the world with lower than expected growth. Mirroring this ex post divergence in stringency and environmental outcome will be ex post divergence in the quantity and cost of abatement. For this reason, the presumption that intensity-based limits are inherently less stringent is wrong.

This presumption is rife among the negative reactions to the Bush climate change plan, which uniformly argue that indexing future emission constraints to GDP would allow GHG emissions to continue to rise when GDP is increasing, as it is generally expected to do.[5] Such criticism belies confusion of the *stringency* of the target with the *form* of the instrument employed in its execution. Despite the fact that these are two separate issues, the unstated implication appears to be that intensity limits allow emissions to continue growing unabated while absolute caps do not. The flaw in this argument is that it ignores the counterfactual no-policy path of emissions, which can as easily be higher than that of an indexed cap. An intensity-based cap would allow emissions to increase over time, as emissions can be higher than that under an absolute cap, which has been shown by the experience of Russia and the East European countries under the Kyoto Protocol. The intensity cap would nevertheless produce real reductions and the indexed cap would impose no constraint despite being an absolute limit. A country's decision to set an absolute cap on emissions is invariably informed by a sense of the limit's expected effects, which typically incorporate a forecast of GDP in the future period when that instrument is slated to enter into force. Given this set of expectations, there are numerous schemes for specifying GDP-indexed emission targets that are entirely equivalent to the absolute limit, a point that we demonstrate in section 12.2 and in an appendix.

The essential caveat to this equivalence is uncertainty about the future. Of principal concern is the ex post level of the emission limit that results from imposing either instrument ex ante. Different instruments whose effects are predicted to be equivalent based on ex ante expectations of GDP may turn out not to hold if actual GDP in the target period diverges from its expected level. In particular, the level of an intensity-based cap will fluctuate in proportion to the ratio of actual to expected GDP.

A third critique of intensity limits can be found in Müller and Müller-Furstenberg (2003), who in addition to the preceding arguments cite problems of implementation in choosing appropriate indices and avoiding biases in these indexes. While these concerns are legitimate, they are also typical of many forms of indexing that are commonly accepted such as indexing wages and benefits, and more recently, inflation-protected bonds.

On the other side of this debate, Baumert (1999), Lutter (2000), Kim and Baumert (2002), Strachan (2007), and Kolstad (2005) all find merit in the concept of intensity-based caps because of the reduction in uncertainty in the economic outcome gained by indexing the cap to GDP and, crucially, the effect on the willingness of countries to participate in international agreements. Baumert (1999), Lutter (2000), and Lisowski (2002) also see intensity limits as a means to avoid the "hot air" resulting from overly generous absolute caps that might be needed to reassure acceding countries that the emissions limit would not place undue costs on them in the event of greater than expected economic growth. Along these lines, Jotzo and Pezzey (2007) provide a theoretical analysis and simulations of binding absolute and intensity caps in which parties are assumed to posses varying degrees of risk aversion to unexpectedly high-cost outcomes with particular attention given to developing country participation. They find that intensity-based caps are superior to absolute caps for circumstances where all parties to a treaty place some positive value on global abatement, face positive abatement costs, and are risk averse in varying degrees to high-cost outcomes. For individually varying but positive valuations on global abatement, parties would be willing to embrace tighter binding targets in return for the removal of some of the uncertainty relating to high-cost outcomes.

Our own contribution to this debate (Ellerman and Sue Wing 2003) treats absolute and intensity limits with equanimity while focusing squarely on the nature of the relevant uncertainties. Under conditions of certainty, equivalent absolute and intensity-based caps would have identical effects, and the outcomes between the two forms differ only to the extent that realized values for GDP or other indexes diverge from expectation. Our aim in this chapter is to develop the implications of uncertainty in baseline emissions and GDP for policy makers' choice between an absolute and an intensity cap. In particular, we will establish the conditions under which one or the other form of emission limit will give rise to smaller variance in cost outcomes, and we test which form of the limit would have produced less variance in abatement cost using historical and forecast data.

All of the analyses focusing on the merits of intensity limits as a means of reducing uncertainty assume that emissions and GDP are positively correlated. Like Jotzo and Pezzey we find that the positive correlation between emissions and GDP is often large enough that intensity caps reduce the variance of cost outcomes. However, we also demonstrate that this result has failed to hold for some countries over varying periods of time.

An important assumption in our analysis is that policy makers care about variance in outcomes. As we have stressed, if expectation were all that mattered, the form of the limit could be treated with indifference so long as the limits being compared are ex ante equivalent. Accordingly the concern is whether by choosing one form or the other, the policy maker can reduce or even minimize the expected variance in outcomes.

At least two different motivations can be offered for seeking to minimize variance, which can be characterized as preserving initial expectations and avoiding undue adjustment costs. Since policy makers will tend to set the level of an emission constraint based on their expectations of the economic and environmental conditions that will prevail when that target enters into force, they might seek a limit that would result in less deviation from the initially expected environmental and economic outcomes as a result of the inevitable changes that will occur over time. Also in a non–putty-putty world in which investments cannot be instantly made and undone to ensure optimal responses to the constraint, the form of the limit enacted would reduce the adjustment costs associated with over- or underinvestment in abatement capability. Both of these motivations would lead to an interest in minimizing variance.

12.2 Absolute and Intensity Limits: Equivalence under Certainty

Our first task is to establish the equivalence of absolute and intensity limits under certainty. Our analytical approach builds on Ellerman and Sue Wing (2003). We consider a country that commits to limit its emissions but is undecided whether to express this limit as a constraint on the absolute level of emissions or on the intensity of emissions indexed to GDP.

Let Q denote emissions, Y denote GDP, and the emission intensity of the economy:

$$\gamma = \frac{Q}{Y}. \qquad (12.1)$$

Suppose that the country chooses to limit its emissions to an absolute level, $\underline{Q}$. We assume that this decision is made conditional on an initial information set, θ, which we represent using the conditional expectation operator, E_θ. With expected baseline emissions $E_\theta[Q^{BAU}]$, if the country chooses a binding absolute cap on emissions, the level of abatement (A^A) is, in expectation:

$$E_\theta[A^A] = E_\theta[Q^{BAU}] - \underline{Q} > 0. \qquad (12.2)$$

Equation (12.1) implies that this fixed limit can be transformed into an emission intensity cap according to the expectation of GDP, $E_\theta[Y]$. If the emission target is expected to bind, there exists a corresponding ceiling on emission intensity:

$$\underline{\gamma} = \frac{\underline{Q}}{E_\theta[Y]} < E_\theta[\gamma^{BAU}]. \qquad (12.3)$$

Therefore, under stable expectations, the expected level of abatement with an intensity limit (A^I), is the same as in (12.2):

$$E_\theta[A^I] = E_\theta[Q^{BAU}] - \underline{\gamma} E_\theta[Y] = E_\theta[A^A]. \qquad (12.4)$$

The condition expressed by equation (12.4) expresses what we later refer to as ex ante equivalence. Thus, given an abatement cost schedule $C(A)$, which we assume is positive, monotonic increasing, and known with certainty, the expectation at time zero of the cost of reducing under either instrument is $C(E_\theta[A^A]) = C(E_\theta[A^I])$. The policy maker would be indifferent between the two, and the form of the emissions limit would be irrelevant if expectation were all that mattered.

12.3 Choice between Absolute and Intensity Limits under Uncertainty

In keeping with the motivation of this chapter, we imagine a policy maker who is concerned about variance in abatement and cost outcomes. The actual levels of abatement and cost under the two forms would correspond to

$$A^A = Q^{BAU} - \underline{Q}, \qquad (12.5)$$

$$A^I = Q^{BAU} - \underline{\gamma}Y. \tag{12.6}$$

Since it will be generally true that $E_\theta[Q] \neq Q$, and $E_\theta[Y] \neq Y$, different levels of abatement and cost will be associated with the two limits. Since the emissions target expressed by the intensity limit adjusts to changes in GDP, whereas that determined by absolute cap does not, the difference in actual abatement will be

$$A^I - A^A = \underline{\gamma}Y - \underline{Q}. \tag{12.7}$$

Any rational policy maker would know that things will change. Not knowing the future changes, he or she might well ask whether some form of the emissions constraint might reduce variance in outcomes so that the actual outcomes not deviate too much from the initial set of expectations.

To evaluate variance more formally, we use the "hybrid" GDP-indexed limit introduced by Ellerman and Sue Wing (2003), which specifies the indexed cap on emissions, $\tilde{Q}$, as the convex combination of a fixed cap and a pure intensity target:

$$\tilde{Q} = (1-\eta)\underline{Q} + \eta\underline{\gamma}Y. \tag{12.8}$$

The form of the emission limit combines an absolute limit with an intensity target specified by the product of the intensity limit in equation (12.3) and actual GDP. The coefficient $\eta \in [0,1]$ is an indexation parameter which represents the degree to which the limit accommodates changes in GDP from its expected level, and it is under the policy maker's control. When $\eta = 0$ the limit is absolute, and when $\eta = 1$, it is a pure intensity limit that adjusts fully to the change in GDP. The result is a more general form of equation (12.6):

$$\tilde{A}^I = Q^{BAU} - \tilde{Q} = Q^{BAU} - (1-\eta)\underline{Q} - \eta\underline{\gamma}Y. \tag{12.6'}$$

In keeping with the result of the previous section, if $\underline{Q}$ and $\underline{\gamma}$ are set initially to be ex ante equivalent, such that $\underline{Q} = \underline{\gamma}E_\theta[Y]$, and it be further assumed that $E_\theta[Y] = Y$, then $\tilde{Q} = \underline{Q}$ regardless of the value of η. Further results, with different forms of the emissions limit, are provided in the appendix.

12.3.1 Indexed Limits

We now establish the conditions under which an indexed limit will be preferred. Our criterion in making this determination is minimization of the variance in the cost of abatement. Given the monotone increasing character of the cost function C, it therefore suffices to demonstrate which instrument generates the smaller variance in abatement.

From equation (12.5) the variance of abatement under the absolute cap is simply

$$\text{var}[A^A] = \text{var}[Q^{BAU}], \tag{12.9}$$

while (12.6′) implies that the variance of an indexed intensity cap ($\eta > 0$) is

$$\text{var}[\tilde{A}^I] = \text{var}[Q^{BAU}] + (\eta\underline{\gamma})^2 \text{var}[Y] - 2\eta\underline{\gamma} \text{cov}[Q^{BAU}, Y]. \tag{12.10}$$

The key question is whether the variance in the expected effect of the latter instrument is less than that of the former. This can be determined by subtracting (12.9) from (12.10) and rearranging. The variance of expected abatement and cost is smaller for the indexed intensity limit if

$$\frac{\eta\underline{\gamma}}{2} < \frac{\text{cov}[Q^{BAU}, Y]}{\text{var}[Y]}.$$

The intuition behind this expression becomes clearer if we multiply both sides by $E_\theta[Y]/E_\theta[Q^{BAU}]$ to express the target, covariance and variances in normalized form:

$$\frac{\eta}{2}\left(\frac{\underline{Q}}{\bar{Q}}\right) < \frac{\rho v[Q]}{v[Y]} = Z, \tag{12.11}$$

where $\bar{Q} = E_\theta[Q^{BAU}]$, ρ is the correlation between BAU emissions and GDP, $v[Q]$ and $v[Y]$ are the coefficients of variation of baseline emissions and GDP, and $\underline{Q}/\bar{Q}$ expresses the ex ante equivalent absolute limit as a fraction of expected BAU emissions. The left-hand side is the product of two important policy variables: the form of the limit, given by the value of indexation parameter, and its stringency, expressed as the ratio of the constrained emissions to expected BAU emissions. By contrast, the quantity on the right-hand side, Z, is a function solely of stochastic properties of the economy, none of which are subject to manipulation by the policy maker.

Equation (12.11) is the main result of the chapter, and it provides a mathematical statement of the conditions under which an intensity limit indexed by η would result in less variation of outcomes. The implication of equation (12.11) is that the conditions under which an indexed intensity limit better preserves initial expectations about the level of actual abatement and cost are more likely to obtain the higher the correlation is between Q and Y and the greater the variation in Q is relative to that in Y. For a given emission target, consider first the case of a fully indexed intensity cap ($\eta = 1$). Since the left-hand side of (12.11) is always positive, as are the coefficients of variation, a necessary condition for the intensity limit to exhibit less variation is that the correlation between emissions and GDP be positive. This is, however, not a sufficient condition. If the variation in emissions were very small relative to the variation in GDP, indexing to GDP would produce more variance than an absolute limit. Therefore the sufficient condition is that either the degree of indexation or the level of the emission limit (or both) be small enough for the inequality to hold.

Also, if emissions and GDP are perfectly correlated and have similar degrees of variability, so that the right-hand side of (12.11) equals unity, then any indexed intensity limit will always exhibit less variability and be preferred, since the left-hand side will always be less than half of unity, or 0.5. For any value of $Z < 0.5$ it is possible that an absolute cap might generate less variability, and be preferred. More generally, where there is a sufficiently weak positive correlation between emissions and GDP ($0 < \rho < 1$) or the volatility of emissions is sufficiently small relative to GDP ($v[Q]/v[Y] < 1$), Z may be small enough that $\eta Q/\bar{Q} > 2\rho v[Q]/v[Y]$, in which case an absolute cap would produce less variance in outcomes and be preferred to an intensity limit. Obviously for any nonpositive correlation (and therefore a nonpositive value of Z) an absolute limit would always be preferred, since the left-hand-side variables cannot be negative.

The intuition behind these results can grasped by considering first the case of negative correlation. If emissions decline when GDP increases (and vice versa), any amount of indexing to GDP will cause the emissions constraint to vary inversely with deviations in emissions, and will thereby produce greater variance in abatement and cost than an absolute limit would. Alternatively, if correlation is positive, it is still possible that indexing would produce more variance. For instance, if there were no variation in emissions ($v[Q] = 0$) but variation in Y, any amount of indexation to Y would create variation in abatement and cost where an absolute limit would produce none. Where there is variation in Q, the choice of form of the limit depends on both the magnitude of its fluctuations and the correlation between Q and Y. Where either of these is sufficiently small, an absolute limit can exhibit less variance than an indexed cap.

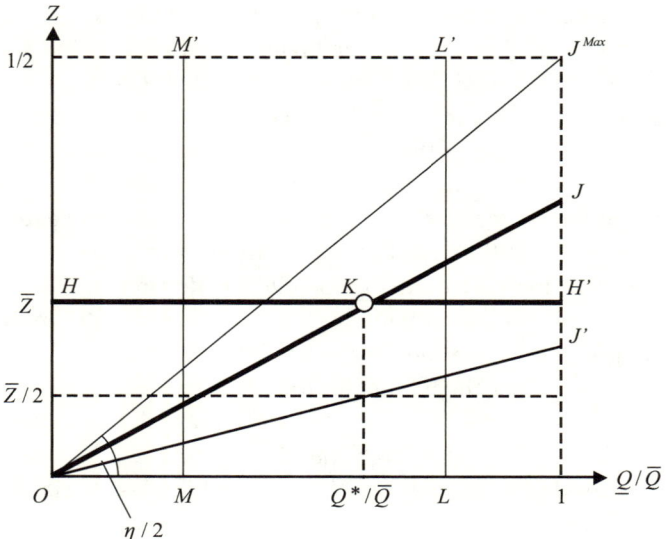

Figure 12.1
Trade-off between absolute and intensity limits

The relationships between the stochastic properties of Q and Y, the desired emission constraint, and the degree of indexation to GDP are illustrated in figure 12.1. The target's fraction of initially expected baseline emissions ranges from zero to one, and is plotted on the horizontal axis. The value of Z is given on the vertical axis. The horizontal line HH' indicates the value of Z for the economy in question ($\bar{Z}$), which represents the boundary between the regions where less variance in outcomes is produced by an absolute or indexed intensity limit ($Z > \bar{Z}$ and $Z < \bar{Z}$, respectively). The diagonal ray OJ gives the locus of values of $\frac{1}{2}\eta Q/\bar{Q}$ over the range of possible reduction fractions for some value of η. Its maximum slope, which is attained when $\eta = 1$, corresponds to the ray OJ^{Max}, which intersects the BAU emission level (i.e., $Q/\bar{Q} = 1$) at $Z = 0.5$ on the vertical axis.

The point K where HH' and OJ intersect represents the equality of both sides of equation (12.11), and defines the level of an emission target Q^* below (above) which an intensity limit will exhibit the lower (higher) variance, and thus will (will not) be the preferred instrument. For example, if the degree of indexation is as indicated by the ray OJ and the emission target is given by the vertical line LL', the intensity cap would be associated with higher variance and would not be preferred. Conversely, if the constrained level of emissions were much lower, say at MM', an intensity limit with the degree of indexing implied by ray OJ would generate less variance, and be preferred.

12.3.2 Optimal Indexation

The preceding section identifies, and figure 12.1 illustrates, the relationships between variables that are under the control of policy makers and those that are fundamentally exogenous. If we assume that the stringency of the emissions constraint is determined exogenously without regard to concerns about variance, and that the policy maker desires to minimize variance of outcomes, then the degree of indexation can be used to achieve this goal. It can be easily shown that as long as there is any degree of positive correlation between emissions and GDP ($\rho > 0$) and any variation in emissions ($v[Q] > 0$), there exists a partially indexed cap that will *always* generate less variance in abatement, and will therefore be preferred to an absolute limit, since it is always possible to choose a sufficiently small value for η to shift the sign of the inequality in equation (12.11) so that $\eta Q/\bar{Q} < 2Z$.

For a given Q the value of the index that minimizes the variance in abatement can be found by differentiating equation (12.10) with respect to η and solving the first-order condition to yield

$$\eta^{Opt} = \frac{E_\theta[Y]\,\text{cov}[Q^{BAU}, Y]}{Q\,\text{var}[Y]} = \frac{Z}{Q/\bar{Q}}. \tag{12.12}$$

Substituting this expression into equation (12.10) yields the minimized variance of abatement:

$$\text{var}[\tilde{A}^I] = (1 - \rho^2)\,\text{var}[Q]$$

for the optimally indexed limit. Any nonzero value of ρ creates the possibility of an indexed limit that will exhibit less variance than the absolute limit. The optimal index will have the same sign as Z, which is to say, ρ, so that for feasible values of the indexation parameter, namely $0 < \eta^{Opt} \leq 1$, indexing will be attractive only in the presence of positive correlation, as was previously demonstrated.

Equation (12.12) implies that for any emission target Q there is a level of indexation given by $\eta^* = 2\eta^{Opt}$ that equalizes the variance of the indexed and absolute forms of the limit, making the policy maker indifferent between them. However, whether η^* or η^{Opt} lies between zero and one depends on the particular values of Z and Q. This outcome is also captured by figure 12.1, where $\eta = \eta^* \in [0,1]$ is indicated by the OJ locus passing through the intersection of HH' and $Q/\bar{Q}$. The corresponding optimal value of η is associated with the ray OJ' that would intersect $Q/\bar{Q}$ at $\bar{Z}/2$. For any $Z \in (0, 0.5]$, both η^* and η^{Opt} will fall within the range $[0,1]$ so long as $2Z \leq Q/\bar{Q} \leq 1$. For $Z \leq 0$, there is no point in considering an indexed limit, while for $Z > 0.5$, a fully indexed limit will always exhibit less variance than an absolute cap and the minimum variance will be achieved by a partially indexed cap so long as $Z < Q/\bar{Q} < 1$.

12.3.3 Measurement Issues and Their Implications for Instrument Choice

An unstated assumption that underlies the foregoing results is that policy makers have the capability to choose among instruments based on the true moments of the distribution of GDP and BAU emissions. But the true moments are not observed; rather, they are inferred from a finite sample of data. Consequently, to give empirical content to the results obtained thus far, the *population* variances and covariance in (12.9) and (12.10) should be replaced by their *sample* counterparts, which are conditional on θ. Then the right-hand side of (12.11) becomes

$$Z_\theta = \frac{\rho_\theta v_\theta[Q]}{v_\theta[Y]}. \tag{12.13}$$

To clarify the implications of expression (12.13), consider the effect of new information about the indexing conditions on output and emissions. The latter represents a shift in the information set to θ' (say). This might not only induce a revision of the expectations that led to the setting of the emission target (i.e., the denominators of $v_\theta[Q]$ and $v_\theta[Y]$), a sufficiently large structural shift can affect policy makers' estimates of the variances of these quantities, as well as the perceived correlation between emissions and GDP. The conclusion is that the quantity Z is not immutable; rather, its value evolves as conditions change.

Two consequences follow: First, the optimal degree of indexation will no longer be constant, as by equation (12.12) even an arbitrarily small shift $Z_\theta \to Z_{\theta'}$ will induce a change $\eta_\theta^{Opt} \to \eta_{\theta'}^{Opt}$ for any emission target Q. Second, a large enough shift in Z can switch the direction of the inequality in (12.11), with the result that the even the binary choice between an absolute and a fully indexed intensity cap would not remain constant over time. Policy makers should be concerned about such outcomes because of the often substantial errors that creep into forecasts of emissions (Lutter 2000) and the CO_2 intensity of GDP (Strachan 2007; Philibert 2005) and can lead to drastic revisions of expectations. We undertake an assessment of this issue in the following section.

12.4 Empirical Tests

We illustrate the practical importance of the preceding theoretical results by examining the properties of Q and Y, and their implications for the choice of the form of an emission target, using two different approaches. In the first, we make assumptions about the character of policy makers' information set and the procedures they follow in using such information to estimate future values of Z. We employ historical data on emissions and GDP for a large number of countries, and define the information set on the assumption that policy makers will invariably make projections of Z based on recently available data. Our second approach attempts to proxy for the conditional moments of Z directly by using a sequence of forecasts of emissions and GDP for a fixed future year. In both approaches the changes in the projections of Z yield insights into whether countries will tend to stick with an absolute or an intensity limit or will more likely shift back and forth between the two instruments as circumstances change.

12.4.1 Using Historical Time Series Data

Our first experiment focuses on observed values of Q and Y, for which there is an abundance of data. Using statistics on carbon emissions from Marland et al. (2003) and real GDP from the Penn World Table 6.1 (Heston et al. 2002), we compile a dataset of 30 developed and developing countries over period 1950 to 2000, from which we compute the value of Z.

Our use of these historical statistics attempts to recreate the kind of prospective assessment and data availability lags that are characteristic of climate policy. We therefore assume that a constraint that is in effect in a particular year (e.g., year t') is determined based on data that become available with a five-year lag and are observed over the course

Table 12.1
Empirical results: Historical data

	Year	ρ	$v[Q]$	$v[Y]$	Z	Q^* (MTC)	$E[Q^{BAU}]$ (MTC)	Q^{BAU} (MTC)	Q^*/Q^{BAU}
Developed countries									
United States	1980	0.929	0.064	0.070	0.847	2470	1458	1300	1.900
	1990	−0.276	0.031	0.063	−0.136	−393	1444	1374	**−0.286**
	1999	0.979	0.041	0.061	0.668	2241	1676	1567	1.430
Japan	1980	0.994	0.163	0.113	1.442	819	284	251	3.259
	1990	−0.046	0.026	0.071	−0.017	−10	286	292	**−0.033**
	1999	0.970	0.069	0.080	0.831	592	356	315	1.877
United Kingdom	1980	0.365	0.026	0.050	0.186	75	203	158	0.476
	1990	−0.583	0.053	0.044	−0.698	−242	173	155	**−1.555**
	1999	0.082	0.044	0.055	0.066	22	171	147	0.153
Canada	1980	0.944	0.092	0.094	0.916	234	128	115	2.041
	1990	0.119	0.029	0.067	0.051	13	124	113	0.111
	1999	0.860	0.041	0.049	0.719	207	144	120	1.725
Italy	1980	0.987	0.110	0.080	1.358	294	108	102	2.893
	1990	0.095	0.027	0.055	0.046	10	112	106	0.097
	1999	0.934	0.042	0.063	0.622	151	122	116	1.311
France	1980	0.954	0.083	0.090	0.886	261	147	132	1.984
	1990	−0.813	0.103	0.051	−1.658	−393	118	99	**−3.988**
	1999	−0.170	0.045	0.058	−0.131	−27	104	100	**−0.274**

Chapter 12 Absolute versus Intensity Limits for CO_2 Emission Control

Country	Year								
Australia	1980	0.976	0.075	0.083	0.893	100	56	55	*1.801*
	1990	0.885	0.059	0.060	0.866	121	70	73	*1.670*
	1999	0.969	0.069	0.068	0.979	178	91	94	*1.896*
Spain	1980	0.979	0.155	0.101	1.510	170	56	55	*3.114*
	1990	−0.142	0.032	0.022	−0.209	−25	60	58	**−0.431**
	1999	0.869	0.073	0.073	0.866	118	68	75	*1.567*
Developing countries									
China	1980	0.983	0.180	0.092	1.909	1447	379	403	*3.590*
	1990	0.939	0.097	0.119	0.764	935	612	655	*1.428*
	1999	0.963	0.093	0.117	0.764	1424	932	771	*1.847*
India	1980	0.896	0.084	0.075	1.010	172	85	95	*1.816*
	1990	0.990	0.122	0.094	1.285	394	153	184	*2.138*
	1999	0.992	0.126	0.093	1.353	733	271	294	*2.494*
South Korea	1980	0.992	0.173	0.119	1.436	75	26	34	*2.202*
	1990	0.952	0.127	0.107	1.128	120	53	66	*1.817*
	1999	0.981	0.156	0.110	1.399	302	108	107	*2.813*
Mexico	1980	0.994	0.126	0.104	1.200	119	49	69	*1.721*
	1990	0.971	0.143	0.105	1.325	227	86	102	*2.212*
	1999	0.891	0.125	0.072	1.559	381	122	113	*3.383*
South Africa	1980	0.957	0.088	0.103	0.816	99	60	58	*1.711*
	1990	0.929	0.107	0.072	1.381	241	87	78	*3.087*
	1999	0.652	0.030	0.030	0.668	132	98	91	*1.449*
Brazil	1980	0.996	0.156	0.126	1.233	114	46	50	*2.285*
	1990	0.334	0.049	0.067	0.243	27	55	55	*0.482*
	1999	0.880	0.062	0.038	1.426	211	74	83	*2.551*

Note: Bold text indicates that absolute caps are unambiguously preferable; italics indicate that intensity caps are unambiguously preferable.

of a decade—namely the interval $(t' - 15, t' - 5]$. Moreover, since at t' current emissions and GDP are not observed, we approximate the denominators of $v[Q_{t'}]$ and $v[Y_{t'}]$ using forecasted quantities, which we estimate based on the growth rates of these variables over the lagged observation period.[6] Thus, for a constraint that is assumed to take effect in 1990, we use the data from 1975 to 1985 to determine the value of Z, and so provide these values for 14 countries for the 1990 experiment as well as for constraints in 1980 and 1999 (where we use data from 1965 to 1975 and 1986 to 1994, respectively).

The most striking feature of table 12.1 is the strong positive correlation between emissions and GDP for developing countries over the length of the entire sample period, and for developed countries before 1975 and after 1985. By contrast, OECD nations exhibit weak or even negative emissions–GDP correlation throughout the decade of high energy prices. The coefficients of variation of emissions and GDP are an order of magnitude smaller and similar in size, and show no trend in the dominance of one type of volatility over the other.[7] The values of Z mostly exceed 0.5. Of the 42 data points in the table, 31 indicate an unambiguous preference for an intensity limit, 6 indicate an unambiguous preference for an absolute limit, and the remaining 5 instances can go either way depending on the stringency of the emission target and the degree of indexation. The unambiguous choice of an intensity cap is far more characteristic of the developing countries than the developed countries due mostly to the consistently high temporal correlations between emissions and GDP. We find that intensity caps are unequivocally preferable for developing countries and may be generally preferable for developed countries. The qualification to the latter conclusion arises from the potential for rapid energy price increases to decouple emissions and GDP.

We conduct a more systematic exploration of these outcomes by computing annual values for the indifference point $Q^*/\bar{Q} = 2Z$ over the period 1965 to 1999 on a rolling basis for a sample of 22 developed and 7 developing countries.[8] Figure 12.2 presents these results as probability density functions (PDFs). In both panels the shaded region corresponds to the range of values in which the choice of an absolute or indexed limit depends on the values of η and $Q^*/\bar{Q}$. In panel A the bulk of the probability masses of both developed and developing countries lie to the right of this range (which we henceforth refer to as the equivocal region).[9] In terms of the geometry this means that the

Chapter 12 Absolute versus Intensity Limits for CO_2 Emission Control   239

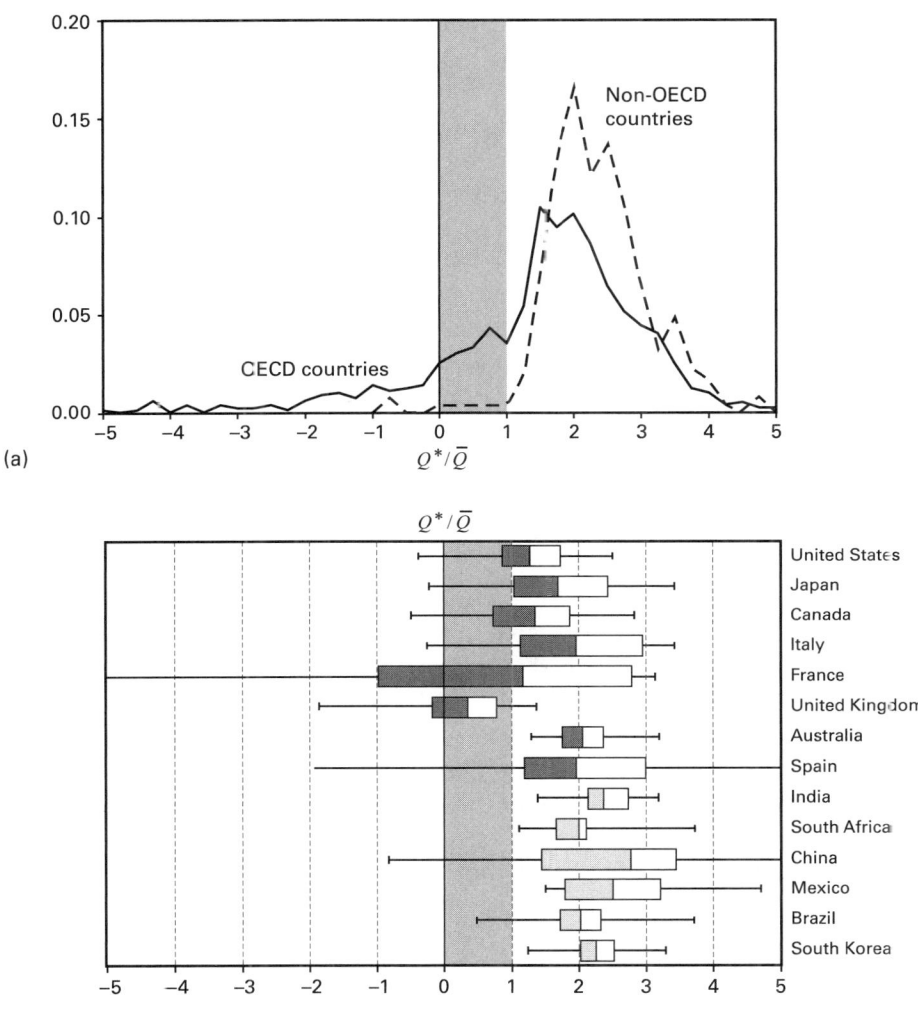

Figure 12.2
Probability density functions for choosing a fully indexed intensity limit. (a) Global aggregates; (b) high-emitting countries

point K lies completely to the right of the 0–1 scale, so binding emission limits will tend to be positioned to K's left. These results echo our previous findings, and imply a clear preference for the use of an indexed intensity limit, especially in developing countries.

The box plot in panel B illustrates the substantial intercountry heterogeneity that underlies the foregoing conclusion. While the entire PDFs of $2Z$ for India, South Africa, Mexico, and Korea lie to the right of the equivocal region, portions of the first quartiles of the distributions for Brazil and especially China overlap with the feasible region, indicating that in some (albeit rare) circumstances these countries might prefer an absolute cap. Even among developed countries the medians of the distributions of the indifference point almost always exceed unity, again indicating a preference for intensity limits. Nevertheless, their lower quartiles intersect the equivocal region and the negative orthant to a greater degree than is the case for the developing countries, indicating that there are more occasions when an absolute cap might be preferred, especially in countries such as France and the United Kingdom.

For each of the observations of countries in a given time period, we also calculate the optimal degree of indexation for emission targets set at 95 and 75 percent of BAU levels using equation (12.12). The box plots in figure 12.3 give the PDFs of the corresponding values of η^{Opt} for each country. There are broad similarities with the results for the indifference levels of the emission target, with slight differences for individual countries. The bulk of the probability masses for large non-OECD emitters lies to the right of the range of allowed values of η^{Opt} (denoted by the shaded area), indicating that fully indexed intensity limits would produce the least variance in outcomes for these countries. Although the PDFs of OECD countries overlap the shaded region to a greater degree, the results for some of these countries, such as Australia and Spain, are similar to the non-OECD patterns.

12.4.2 Using Historical Forecasts

While historical data are plentiful, for our purposes the data suffer from the defect of assuming that policy makers are purely extrapolative in their expectations and that they would not incorporate expected changes from past experience in their set of expectations. Historical forecasts would remedy this problem, but there is a dearth of projections on emissions and GDPs. Nevertheless, we were able to use the

Chapter 12 Absolute versus Intensity Limits for CO_2 Emission Control   241

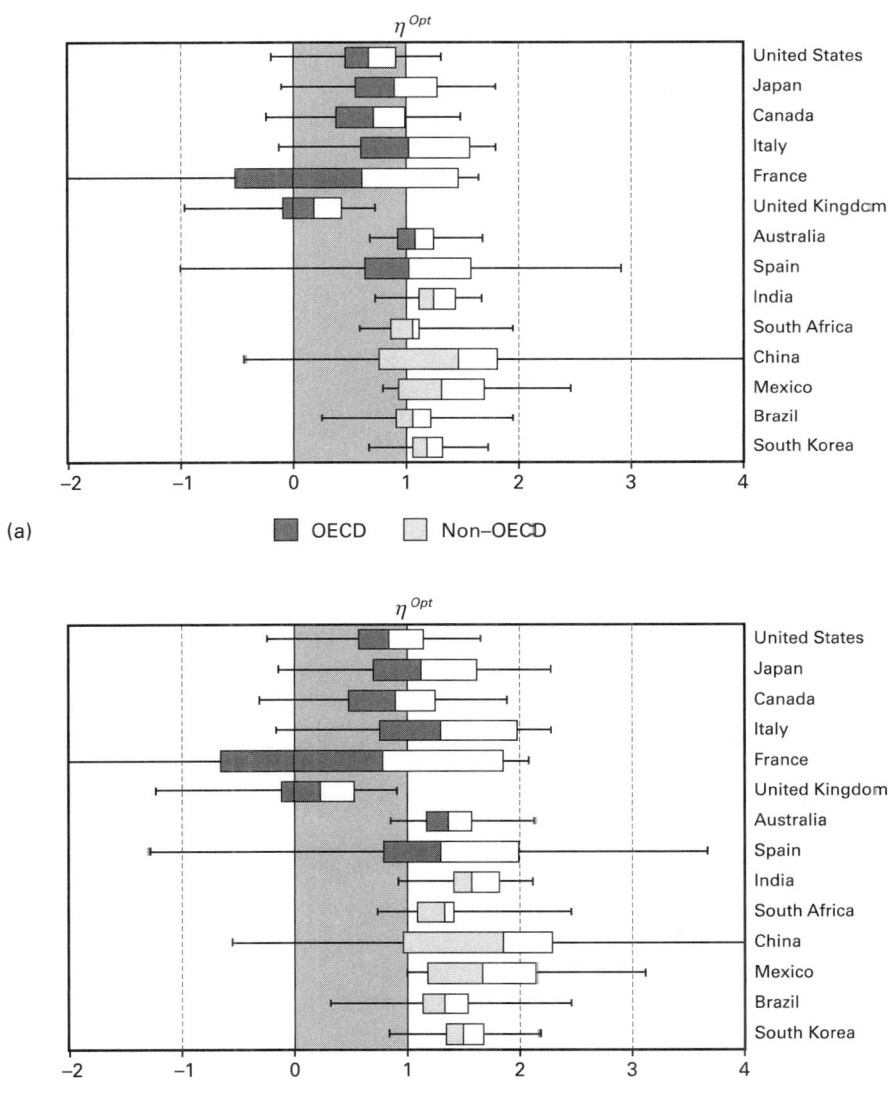

Figure 12.3
PDFs of the optimal index conditional on the level of the emission constraint. (a) $\underline{Q}/\bar{Q} = 0.95$; (b) $\underline{Q}/\bar{Q} = 0.75$

Table 12.2
Empirical results: Forecast data

	USA	OECD Europe	Japan	Canada	Former USSR	China	Mexico
A. In year $T = 2000$							
ρ	0.297	0.233	0.135	0.158	0.313	0.644	0.192
$v[Q]$	0.026	0.090	0.127	0.037	0.446	0.061	0.101
$v[Y]$	0.023	0.031	0.175	0.029	0.147	0.148	0.056
Z	0.336	0.670	0.097	0.205	0.947	0.267	0.349
$E_0[Q^{BAU}]$[a,b]	1491	672	284	144	786	840	115
$E[Q^{BAU}]$ range[a]	1471–1605	672–1235	273–401	143–161	249–810	840–1031	97–123
Q^{*a}	1001	902	55	59	1487	449	80
Q^* range[a]	987–1077	901–1655	53–78	58–65	471–1533	449–551	67–85
$Q^{BAU a}$	1619	787	323	119	185	762	116
Q^*/Q^{BAU} range	0.61–0.67	1.15–2.1	0.16–0.24	0.49–0.55	2.55–8.29	0.59–0.72	0.58–0.73
$\eta^{Opt}(\underline{Q}/\bar{Q} = 0.95)$	0.35	0.71	0.10	0.22	1.00	0.28	0.37
$\eta^{Opt}(\underline{Q}/\bar{Q} = 0.75)$	0.45	0.89	0.13	0.27	1.26	0.36	0.47
B. In year $T = 2010$							
ρ	0.597	0.409	−0.042	0.140	0.837	0.575	−0.021
$v[Q]$	0.036	0.111	0.127	0.047	0.205	0.135	0.091
$v[Y]$	0.082	0.035	0.209	0.047	0.125	0.208	0.153
Z	0.260	1.306	−0.026	0.138	1.379	0.373	−0.012
$E_0[Q^{BAU}]$[a,b]	1819	1101	309	168	1265	944	133
$E[Q^{BAU}]$ range[a]	1621–1835	982–1385	309–466	160–186	666–1265	944–1523	127–164
Q^{*a}	944	2877	−16	46	3490	703	−3

Q^* range[a,c]	842–953	2566–3619	(24)–(16)	44–51	1837–3490	703–1135
Q^{BAU}[a,c]	0.48–0.54	2.22–3.14	(0.07)–(0.04)	0.26–0.30	2.17–4.12	0.55–0.89
$Q/E[Y]$ Kyoto range[d]	0.68–0.77	0.64–0.91	0.55–0.83	0.63–0.74	0.78–1.49	—
η^{Opt} ($\underline{Q}/\bar{Q} = 0.95$)	0.27	1.37	−0.03	0.15	1.45	0.39
η^{Opt} ($\underline{Q}/\bar{Q} = 0.75$)	0.35	1.74	−0.03	0.18	1.84	0.50

a. Megatons of carbon.
b. $E_0[Q^{BAU}]$ = initial emission forecast.
c. Figures in parentheses indicate negative values.
d. Kyoto emission targets as specified in DOE/EIA (1998, tab. 8).

forecasts made annually for a small number of regions for the years 2000 and 2010 by the DOE/EIA for the International Energy Outlook.

We focus first on the year 2000, for which there are the longest series of comparable historical forecasts over the broadest range of countries. EIA prepared forecasts of emissions and GDPs in this year for four developed economies (United States, Japan, Canada, and OECD Europe), one economy in transition (the former Soviet Union), and two industrializing economies (China and Mexico).[10] We used these country series to compute values for ρ, $v[Q]$, $v[Y]$, and Z, for whose source of variability were the changes in expectations captured by the revisions to the DOE/EIA forecasts of the relevant variables. The results for 2000 are shown in panel A of figure 12.3. A first result to note is that none of the values of Z are negative and two of the seven economies exceed 0.5. Therefore in no region does an absolute limit generate less variance than a partially indexed limit, and in only two regions (OECD Europe and the former Soviet Union) does an indexed limit unambiguously generate less variance than an absolute cap. For the remaining countries, the question of which limit exhibits less variance depends on the emissions target and the degree of indexation. The last two rows of panel A provide the optimum values η^* for emissions constraints of 0.95 and 0.75. A fully indexed limit is indicated only for the former Soviet Union; for all others, a partially indexed limit would minimize variance.[11]

The defining characteristic of this result is not so much the values of the correlations between Q and Y (which, except for China, are all comparable in magnitude and small), but the variability of emission forecasts relative to that of GDP forecasts. For OECD Europe and the former Soviet Union, the variability of emission projections exceeds that of GDP forecasts by a factor of three, so that a high degree of indexation is warranted despite a relatively low Q to Y correlation. By contrast, China is an example of a case where indexation has *less* of a tendency to reduce variance despite the high correlation. This appears to be because the variability in emissions forecasts is so much less than that for of GDP forecasts.

To test the robustness of these findings, we computed the values again, using the forecast data for the year 2010, projections for which are available from 1990 onward. The results, shown in panel B, exhibit some interesting differences but the conclusions are broadly the same. The values of Z for OECD Europe and the former Soviet Union exceed

0.5, which continues to argue unequivocally for an intensity cap. However, there are now two countries, Japan and Mexico, with negative values of Z, which points unequivocally to the use of an absolute cap. The remaining regions fall in the interval $0 < Z < 0.5$, for which the choice of instrument can go either way. For the five regions for which indexing is indicated, OECD Europe has joined the former Soviet Union as regions that would choose a fully indexed intensity cap to reduce variance because the correlation of emissions and GDPs is considerably stronger for the 2010 forecasts than for those for 2000. For the remaining three regions, a partially indexed intensity cap would be optimal.

12.4.3 Comparing the Two Sets of Experiments

The results from the forecast tend to support the historical data, namely conditions that suggest a general preference for indexed intensity limits. But they also provide clear evidence that these conditions are far from universal. More important, the results highlight the dependence of the choice between an absolute and an intensity cap of the expected statistical relationships between emissions and the GDP. The much larger sample for the historical data might allow one to argue for placing more confidence in those results than the few instances of actual repeated forecasts of emissions and GDPs that we could find. But even though the forecasts are restricted to a single source and a fairly narrow period of time, they do indicate how actual expectations evolve, whereas the experiments based on historical data suffer from the assumption of extrapolative expectations that remain constant as conditions change moving forward in time.

Moreover, for any given region, what may be preferred for one interval of time may not be for another period. For instance, for many of the developed countries, an intensity limit would have been the wrong choice for late 1970s and early 1980s, but then would have returned to being the right choice when energy prices declined after 1985. Thus a policy maker faced with such a choice of limit would need to pay close attention to factors that might shift the historical relationship between Q and Y. For instance, at the time of this writing, when energy prices are once again at high levels and are expected to remain there, intensity limit might not be as strongly preferred as past data from the low-energy-price 1990s might suggest.

12.5 Conclusion

In this chapter we have sought to elucidate the differences under uncertainty between absolute and intensity-based limits as they may be applied to CO_2 emissions. We demonstrated that the two are identical when there is no uncertainty about the future, and we analyzed the choices between them on the assumption that the policy maker would want to reduce the variance in environmental and economic outcomes from the application of the limit. This analysis consisted of identifying the conditions under which an intensity-based limit would be preferred to an absolute limit and of specifying the optimal index when an intensity-based limit is preferred. We also investigated the frequency of the conditions for preferring an intensity-based limit using historical data and past forecasts, and then the distribution of the optimal level of indexing conditional on the emissions constraint.

The main result of the mathematical analysis is that positive correlation between emissions and GDPs (or whatever other index is chosen) is a necessary but not sufficient condition for an intensity limit to be preferred. In addition the variability of emissions relative to income must be sufficient to make indexation variance-reducing. Otherwise, intensity-based limits will increase the variance of outcomes. Alternatively, there are conditions under which absolute limits would minimize variance and be preferred. The empirical part of the chapter shows that conditions favoring intensity-based limits predominate but that the conditions in which absolute limits would be variance-reducing cannot be dismissed. Moreover the choice of the optimal index, as well as the binary choice between an absolute or intensity-based limit, can change over time as conditions and expectations change.

In this chapter we did not wish to suggest that other means of limiting variance in outcomes are not available. Safety valves and temporal trading (banking and borrowing) have similar, although not identical, advantages in avoiding undesirable outcomes. Our purpose has been to clarify the differences between absolute and intensity-based emission limits that are often discussed as if used in pure form. An important underlying assumption of the chapter is that the reduction of the variance in intended outcomes is an important consideration in policy choices. If policy makers are concerned mostly with expected effects, the form of the limit is not as important so long as the two are ex ante equivalent.

Appendix: Further Results on the Equivalence of Absolute and Intensity Limits

We consider a situation where GDP and emissions are known with certainty at a particular reference point in time, given by t, and policy makers commit to an emission target Q, which is to take effect in some future period $t+k$. We assume that expectations are conditioned on data on the economy in the reference period, and use the subscript t as a shorthand to represent the information set $\theta(t)$. In this setting the projected emission intensity of the economy under the cap is given by the analogue of equation (12.3):

$$\underline{\gamma}_{t+k} = \frac{\underline{\zeta}_{t+k}}{E_t[Y_{t+k}]}. \tag{12A.1}$$

An Emission Target Based on the Growth of GDP

An intensity cap may be expressed in terms of the rate of growth of emissions. In particular, policy makers may choose to limit the growth of emissions to some maximum allowable fraction, $\bar{\omega}$, of the expected growth of GDP over the period t and $t+k$:

$$\left(\frac{Q_{t+k}}{Q_t} - 1\right) = \bar{\omega}\left(\frac{E_t[Y_{t+k}]}{Y_t} - 1\right). \tag{12A.2}$$

For the indexed limit in equation (12.8) to behave similarly to the growth target specified above, it must be the case that

$$\left(\frac{\tilde{Q}_{t+k}}{Q_t} - 1\right) = \tilde{\omega}\left(\frac{E_t[Y_{t+k}]}{Y_t} - 1\right), \tag{12A.3}$$

where $\tilde{\omega}$ specifies the fraction of the rate of GDP growth at which emissions are allowed to increase. It is obvious that $\tilde{Q}_{t+k} = Q_{t+k}$ if $\tilde{\omega} = \bar{\omega}$, implying that emissions are allowed to grow by the same fraction of GDP under both the absolute and the intensity cap, so the two instruments are ex ante equivalent.

This result does not generally hold under uncertainty. Using (12A.1), (12A.2), and (12A.3) to substitute for γ, Q, and $\tilde{Q}$ in (12.8) allows us to solve for $\tilde{\omega}$ as follows:

$$\tilde{\omega} = \frac{1}{E_t[g_Y]}\left\{\left[(1-\eta) + \eta\frac{Y_{t+k}}{E_t[Y_{t+k}]}\right](1 + \bar{\omega}E_t[g_Y]) - 1\right\},$$

where $E_t[g_Y] = E_t[Y_{t+k}]/Y_t - 1$ is the projected rate of GDP growth between t and $t + k$. This expression makes clear that $\tilde{\omega}$ will diverge from $\bar{\omega}$ as GDP at $t + k$ differs from its forecast value, and the gap between these parameters will increase the more accommodation is made for fluctuations in GDP (i.e., as $\eta \to 1$).

An Emission Target Based on the Growth of Emission Intensity

An intensity cap may also be expressed as an upper bound on the future rate of decline in the economy's emission intensity. Denoting this maximum rate by $\bar{\phi}$, we have

$$\bar{\phi} = \frac{Q_{t+k}/E_t[Y_{t+k}]}{\gamma_t} - 1. \tag{12A.4}$$

For our indexed cap to behave in the same way, it must be the case that

$$\tilde{\phi} = \frac{\tilde{Q}_{t+k}/E_t[Y_{t+k}]}{\gamma_{t+k}} - 1, \tag{12A.5}$$

where $\tilde{\phi}$ specifies the rate of decline in the emissions intensity of the economy. As before, once $\tilde{Q}_{t+k} = Q_{t+k}$, the limits produce identical effects if $\tilde{\phi} = \bar{\phi}$, thus proving their equivalence under certainty.

To illustrate the effect of uncertainty, we use (12A.1), (12A.4), and (12A.5) to substitute for γ, Q, and $\tilde{Q}$ in (12.8) and then solve for $\tilde{\phi}$ to obtain

$$\tilde{\phi} = \left[(1 - \eta) + \eta \frac{Y_{t+k}}{E_t[Y_{t+k}]} \right] (1 + \bar{\phi}) - 1.$$

As before, $\bar{\phi}$ and $\tilde{\phi}$ will diverge if actual GDP differs from its forecast value, and the gap between the two policy variables increases as $\eta \to 1$.

Notes

ADE and JMS were supported by the MIT Joint Program on the Science and Policy of Global Change, funded through a government-industry partnership including US Department of Energy Office of Science (BER) Grant DE-FG02-94ER61937, US Environmental Protection Agency Cooperative Agreement XA-83042801-0, and a group of corporate sponsors from the United States and other countries. This chapter has benefited from suggestions by two anonymous referees, as well as helpful comments by Jake Jacoby, John Reilly, John Parsons, Dick Eckaus, Philippe Quirion, and participants at the

Cambridge-MIT Electricity Policy Forum Spring Research Seminar, the EPRI Global Climate Change Research Seminar, and the David Bradford Memorial Conference on the Design of Climate Policy.

1. Familiar examples of intensity limits are the emissions rate limits imposed on nearly all sources under state implementation plans in the United States, best available control technology mandates, such as in the US New Source Performance Standards or the EU Large Combustion Plant Directive, and the Corporate Average Fuel Economy standards in the United States and similar programs in Europe. Although many of the latter do not explicitly specify an emissions rate, the effect of these programs is to reduce emissions (or energy) intensity and to allow emissions to vary with the level of output. However, absolute emissions caps can also be found in several programs controlling conventional pollutants, for example, the SO_2 trading (acid rain), RECLAIM, and the Northeastern NO_x Budget programs in the United States. Rosenzweig and Varilek (2003) review experience with these and other rate-based emission regulations.

2. The UK Emissions Trading Scheme is unique in having two sectors, an absolute sector containing firms with absolute limits on GHG emissions and a relative sector containing firms with intensity limits, and allowing trading (with some restrictions) between the two sectors.

3. The latter has been the focus of studies by Quirion (2005), Jotzo and Pezzey (2007), and Newell and Pizer (2006), who use a cost-benefit framework to analyze the performance of intensity limits relative to other instruments.

4. We use the terms intensity-based and indexed as virtually interchangeable. A conventional intensity limit is automatically indexed to whatever is the denominator by which the intensity is stated. By the same token, an otherwise fixed absolute cap can be indexed to vary the level of allowed emissions according to movements of some denominated quantity, such as output or GDP.

5. The stated intent of the Bush administration's espousal of intensity targets is to take future economic growth into account: "This new approach focuses on reducing the growth of GHG emissions, *while sustaining the economic growth* needed to finance investment in new, clean energy technologies" (White House 2002) [our emphasis]. The stringency of the Bush target (or lack thereof) is a legitimate concern. The 18 percent reduction in the GHG emission intensity of the US economy by 2012 is to be compared with the contemporary DOE/EIA (2004) forecast that projects a decline in the CO_2–GDP ratio of 15 percent by 2010. By contrast, the reduction in the CO_2 emissions *intensity* over the same period implied by the US Kyoto target is greater than 40 percent. Moreover the Bush target is specified not as a legally binding limit but as a goal to be achieved through an array of voluntary actions, creating the potential for little or no abatement to take place.

6. See, for example, "Blowing Smoke," *Economist*, February 14, 2002, p. 27.

7. Formally,

$$v_{t'}[\zeta] = \left\{ \frac{\sigma_\zeta}{\zeta_{t'-5} \cdot \exp(5 \cdot E[g_\zeta])} \middle| \zeta \in (\zeta_{t'-15}, \zeta_{t'-5}] \right\},$$

where $\zeta = Q, Y$, and σ_ζ and g_ζ denote the historical standard deviation and historical average annual growth rate of each of these quantities.

8. The exceptions are India, South Korea, and Mexico, whose emissions are persistently more variable than their GDP.

9. The OECD country panel ($N = 790$) is made up of Australia, Austria, Belgium, Canada, Denmark, Finland, France, Greece, Iceland, Ireland, Italy, Japan, Luxembourg, Netherlands, New Zealand, Norway, Portugal, Spain, Sweden, Switzerland, United Kingdom, and United States. The developing country panel ($N = 247$) is made up of Brazil, China, India, Mexico, South Korea, South Africa, and Turkey.

10. The probability of the indifference point falling in the range $[0, 1]$ is less than unity is 28 percent for OECD countries and only 3 percent for non-OECD countries, while the probability of it being negative is 14 percent for OECD countries and only 1.2 percent for non-OECD countries. As in figure 12.3, the long lower tail of the distribution for OECD countries reflects the influence of the period of high energy prices from 1974 to 1984, and the consequent negative correlation between emissions and GDP over this period.

11. The date of the last forecast is 1999 for all of these regions, but the date of the first forecast differs by region. Complete data were available for Canada, Europe, Japan, and the United States from 1987, for China from 1990, for the former Soviet Union from 1994, and for Mexico from 1995.

References

Argentina. 1999. Revision of the First National Communication to the UN Framework Convention on Climate Change. ARG/COM/2 B.

Barros, V., and M. Conte Grand. 2002. Implications of a dynamic target of greenhouse gas emission reduction: The case of Argentina. *Environment and Development Economics* 7(3): 547–69.

Baumert, K., R. Bhandari, and N. Kete. 1999. *What Might a Developing Country Commitment Look Like? Climate Notes.* Washington: World Resources Institute.

DOE/EIA. 1998. International Energy Outlook, DOE/EIA-0484(98). Washington, DC.

DOE/EIA. 2004. Annual Energy Outlook, DOE/EIA-0383(2005). Washington, DC.

Dudek, D., and A. Golub. 2003. "Intensity" targets: pathway or roadblock to preventing climate change while enhancing economic growth? *Climate Policy* 3(suppl 2): S21–S28.

Ellerman, A. D., and I. S. Wing. 2003. Absolute vs. intensity-based emission caps. *Climate Policy* 3(suppl 2): S7–S20.

Fisher, A. C. 2001. Uncertainty, irreversibility, and the timing of climate policy. Presented before Conference on the Timing of Climate Change Policies, Pew Center on Global Climate Change, Washington, DC, October 10–12.

Fischer, C. 2003. Combining rate-based and cap-and-trade emissions policies. *Climate Policy* 3(suppl 2): S89–S103.

Frankel, J. 1999. Greenhouse gas emissions. Policy brief 52. Brookings Institution, Washington, DC.

Gielen, A. M., P. R. Koutstaal, and H. R. J. Vollebergh. 2002. Comparing emission trading with absolute and relative targets. Presented at the 2nd CATEP Workshop on the Design and Integration of National Tradable Permit Schemes for Environmental Protection, London, March 25–26.

Chapter 12 Absolute versus Intensity Limits for CO_2 Emission Control

Hahn, R. W., and R. N. Stavins. 1999. *What Has the Kyoto Protocol Wrought?: The Real Architecture of International Tradable Permit Markets.* Washington, DC: American Enterprise Institute Press.

Helfand, G. E. 1991. Standards versus standards: The effects of different pollution restrictions. *American Economic Review* 81: 622–34.

Heston, A., R. Summers, and B. Aten. 2002. Penn World Table Version 6.1. Center for International Comparisons at the University of Pennsylvania (CICUP).

Jacoby, H. D., and A. D. Ellerman. 2002. The safety valve and climate policy. *Energy Policy* 32(4): 481–91.

Jacoby, H. D., and I. Sue Wing. 1999. Adjustment time, capital malleability, and policy cost. *Energy Journal* (special issue): 3–92.

Jotzo, F., and J. Pezzey 2007. Optimal intensity targets for greenhouse emissions trading under uncertainty. *Environmental and Resource Economics* 38(2): 259–84.

Kolstad, C. D. 2005. The simple analytics of greenhouse gas emission intensity reduction targets. *Energy Policy* 33: 2231–36.

Kim, Y.-G., and K. A. Baumert. 2002. Reducing uncertainty through dual-intensity targets. In K. A. Baumert et al., eds., *Building on the Kyoto Protocol: Options for Protecting the Climate.* Washington: World Resources Institute, pp. 109–34.

Kopp, R., R. D. Morgenstern, and W. Pizer. 2000. Limiting cost, assuring effort, and Eencouraging ratification: Compliance under the Kyoto Protocol. CIRED/RFF Workshop on Compliance and Supplemental Framework ⟨http://www.weathervane.rff.org/features/parisconf0721/summary.html⟩.

Lisowski, M. 2002. The emperor's new clothes: redressing the Kyoto Protocol. *Climate Policy* 2(2/3): 161–77.

Lutter, R. 2000. Developing countries' greenhouse emissions: Uncertainty and implications for participation in the Kyoto Protocol. *Energy Journal* 21: 93–120.

Marland, G., T. A. Boden, and R. J. Andres. 2002. Global, regional, and national fossil fuel CO_2 emissions, in trends: A compendium of data on global change. Carbon Dioxide Information Analysis Center, Oak Ridge National Laboratory, US Department of Energy, Oak Ridge, TN.

Müller, B., and G. Müller-Fürstenberger. 2003. Price-related sensitivities of greenhouse gas intensity targets. *Climate Policy* 3(suppl 2): S59–S74.

Newell, R. G., and W. A. Pizer. 2006. Indexed regulation, resources for the future. Discussion paper 06-32. Washington, DC.

Philibert, C. 2005. New commitment options: Compatibility with emissions trading. Annex I Expert Group on the UNFCCC. Presented before United Nations Climate Change Conference, Montreal, December 5. COM/ENV/EPOC/IEA/SLT(2005)9.

Quirion, P. 2005. Does uncertainty justify intensity emission caps? *Resource and Energy Economics* 27: 343–53.

Rosenzweig, R., and M. Varilek. 2003. Key issues to be considered in the development of rate-based emissions trading programs: Lessons learned from past programs. Presented at the EPRI Workshop, April 29, Vancouver BC.

Spulber, D. F. 1985. Effluent regulation and long-run optimality. *Journal of Environmental Economics and Management* 12: 103–16.

Strachan, N. 2007. Setting greenhouse gas emission targets under baseline uncertainty: The Bush Climate Change Initiative. *Mitigation and Adaptation Strategies for Global Change* 12(4): 455–70.

UK Department for Environment, Food and Rural Affairs. 2001. Framework for the UK emissions trading scheme. Available at ⟨http://www.defra.gov.uk/environment/climatechange/trading/pdf/trading-full.pdf⟩.

White House, US Office of the President. 2002. *US Climate Strategy: A New Approach*. Policy Briefing Book. Washington, DC: Government Printing Office.

13 On Multi-period Allocation of Tradable Emission Permits

Katrin Rehdanz and Richard S. J. Tol

Much has been said and written about markets of emission permits, and this is particularly true for the initial allocation of emission permits (e.g., see Woerdman 2000; Harrison and Radov 2002). In a sense, emission permits are property rights, and introducing them has implications for the distribution of wealth. However, the attention to the allocation of permits in later periods has not been apace with large attention given to the *initial* allocation, and there is even less on the effect of emission permits on the behavior of the market in the early periods. This chapter seeks to fill this gap.

We set up an analytically tractable model of an emission permit market. In particular, we refer to the carbon dioxide market and contrast three dynamic allocation rules and two mechanisms that could be used to steer overall emission reduction. The first dynamic allocation rule is actually static: allocations depend on emissions in the period before emission reduction. The next two rules are entirely dynamic: in one, allocations depend on actual emissions in the previous period, and in the other, allocations depend on the emission reduction effort in the previous period. We compare the case where the regulator sets the overall emission target with the case where the regulator sets the price of emission reduction. In a static model, the two strategies are equivalent (under perfect information but not under uncertainty; see Weitzman 1974). In a dynamic model, they are not. We investigate this difference further by way of the effect that banking and borrowing might have on the two stategies.

Various authors have addressed the problem of the initial allocation of emission permits. Woerdman (2000), for example, analyzes the issue of permit allocation as a major political barrier to establishing an (inter)national emissions trading scheme. Holmes and Friedman (2000) present alternative designs for a domestic trading scheme in the United

States. Viguier (2001) discusses different allocations of emission allowances across the member states of the European Union. Woerdman (2001) considers under which conditions differing European domestic permit allocation procedures will lead to competitive distortions (see also OECD 1999; Zhang 1999) and result in state aid. Cramton and Kerr (2002) analyze the distributional implications of allocating CO_2 permits through auctions. They argue that because of the political economy differences auctioning is superior to grandfathering. Harrison and Radov (2002) consider different allocation mechanisms for the European Union.

The numerical simulations used to evaluate allocation procedures for emission permits have largely centered on static approaches. Edwards and Hutton (2001) use a computable general equilibrium model to assess different allocation methods for the United Kingdom. Burtraw et al. (2002) apply an electricity market simulation model to compare three different allocation mechanisms for the United States. Other studies include preexisting distortions in their analyses for the United States, for example, Goulder et al. (1999).

All studies find that auctioning the permits and using the revenue to reduce distorting taxes is less costly than grandfathering or than output-based allocations. Burtraw et al. (2002) show how auctioning might be favored by owners of existing generation assets. Nevertheless, Stavins (1998) observes that wherever tradable permits have been adopted, the initial allocation of permits has always been through grandfathering rather than through other methods. These findings are supported by Schwarze and Zapfel (2000) who stress the trade-off between efficiency and political acceptability. They evaluate the design strategies of the two most prominent US cap and trade programs: US EPA Sulfur Allowance Trading Program and southern California's Regional Clean Air Incentives Market (RECLAIM). As for the European effort, the EU Directive (EU Commission 2003) calls for establishing a Community wide emissions trading scheme whereby 95 percent of the allowances are to be allocated free of charge for the first commitment period.

Our approach takes a different turn. We investigate the effects of various types of *dynamic* permit allocations on market behavior. In our approach, static allocation is based on emissions before regulation (period 0 in our model). By contrast, dynamic allocation is based on emissions in a previous period;[1] that is to say, the allocation in period 1 is based on period 0 emissions, in period 2 on period 1 emissions. The allocation is dynamic in the sense that regulation in the far future (period 2) depends on regulation in the near future (period 1). We consider three forms of dynamic allocation: one where allocations depend on previous allocation (this is essentially static), one where allocations depend on previous emissions as they are, and one where allocations depend on previous emissions as they could have been. That is to say, dynamic allocation is based on emissions being traded over time—analogous to banking and borrowing—which we model explicitly.

To our knowledge, only two other studies take into account dynamic effects of permit allocation. Böhringer and Lange (2003) show that dynamic allocation schemes as discussed here cannot be optimal, but they also derive conditions for second-best dynamic allocations. They consider emission-based allocations (which correspond to our rolling grandfathering) and output-based allocations (which correspond to our technology standards). Böhringer and Lange do not investigate the permit market in detail as we do in this chapter. Jensen and Rasmussen (2000) use a dynamic multisectoral model of the Danish economy to investigate the effects of different allocations on welfare, CO_2 leakage, employment, and stranded costs. They compare auctioning to grandfathering and use an output-based allocation scheme based on a company's market share. Neither Jensen and Rasmussen (2000) nor Böhringer and Lange (2003) consider an intertemporal transfer of permits. We compare two alternative dynamic allocation approaches (rolling grandfathering and technology standards) that include intertemporal transfers of permits.

Although banking and/or borrowing are an integral part of most policy programs, it is only recently that economists have started to formally investigate the connection to emission allocation schemes. A theoretical analysis by Cronshaw and Kruse (1996), for example, considers a competitive intertemporal model for bankable emission permits. Rubin (1996) uses a continuous time model of banking and borrowing and derives permit prices and emission paths. Kling and Rubin (1997) use a similar framework to examine the efficiency properties of a permit banking system. The results indicate that allowing banking reduces the costs of emission reduction. However, Kling and Rubin (1997) find that in a system allowing for banking and borrowing, firms could choose to produce excessive damage and output levels in the early period. Leiby and Rubin (2001) extend the analysis to include stock pollutants and show that environmental regulation can achieve a socially optimal level of emissions. Godby et al. (1997) analyze the extent of uncertainty in the control of emissions. In their experimental setting they find price stability to be improved substantially by a banking scheme. Steenberghe (2005) applies the effect of banking to the Kyoto Protocol on world emissions, abatements costs, and the permit price using different scenarios. Hagem and Westskog (1998) explore the optimal design of an intertemporal trading system with banking and borrowing under market imperfections. None of the above-mentioned studies combine dynamic permit allocation approaches with intertemporal transfer of permits.

13.1 The Market

Let us begin with a two-period market for tradable permits with I companies. Permits cannot be transferred between periods. Emission reduction costs C are quadratic (by this restrictive assumption, the model becomes analytically tractable). Each company solves the problem

$$\min_{R_{i1},R_{i2},P_{i1},P_{i2}} C_i = \alpha_{i,1} R_{i,1}^2 + \frac{\alpha_{i,2} R_{i,2}^2}{1+\delta} + \pi_1 P_{i,1} + \frac{\pi_2 P_{i,2}}{1+\delta}$$

subject to

$$R_{i,1} + P_{i,1} \geq E_{i,1} - A_{i,1}; \quad R_{i,2} + P_{i,2} \geq E_{i,2} - A_{i,2}, \tag{13.1}$$

where R is emission reduction, α is a parameter, δ is the discount rate, P denotes the amount of emission permits bought or sold, π is the emission permit price, E are the uncontrolled emissions, and A are the allocated emission permits. Because we assume here a perfect market, all companies face the same price: if a company emits more than has been allocated, $E > A$, it will have to reduce emissions or buy permits on the market. $E - A$ is the emission reduction target; $A + P$ are the actual or controlled emissions, which are below (above) the allocation if permits are sold (bought), that is, $P < 0$ ($P > 0$).

The first-order conditions of (13.1) are

$$\frac{2\alpha_{i,t} R_{i,t}}{(1+\delta)^{t-1}} - \lambda_{i,t} = 0, \quad i = 1,2,\ldots,I,\, t = 1,2, \tag{13.2a}$$

$$\frac{\pi_t}{(1+\delta)^{t-1}} - \lambda_{i,t} = 0, \quad i = 1,2,\ldots,I,\, t = 1,2, \tag{13.2b}$$

$$R_{i,t} + P_{i,t} - E_{i,t} + A_{i,t} = 0, \quad i = 1,2,\ldots,I;\, t = 1,2, \tag{13.2c}$$

Chapter 13 On Multi-period Allocation of Tradable Emission Permits   257

where λ denotes the LaGrange multiplier. This is a system with six I equations and six $I+2$ unknowns. However, we also have

$$\sum_{i=1}^{I} P_{i,t} = 0, \quad t = 1, 2. \tag{13.2d}$$

Equations (13.2a) through (13.2d) solve as[2]

$$\pi_t = \lambda_{i,t} = \frac{\sum_{i=1}^{I}(E_{i,t} - A_{i,t})}{\sum_{i=1}^{I} 1/2\alpha_{i,t}}, \tag{13.3a}$$

$$R_{i,t} = \frac{\pi_t}{2\alpha_{i,t}}, \tag{13.3b}$$

$$P_{i,t} = E_{i,t} - A_{i,t} - \frac{\pi_t}{2\alpha_{i,t}}. \tag{13.3c}$$

So the permit price goes up if the emission reduction obligation increases or if the costs of emission reduction increase. All companies face the same marginal costs of emission reduction, and the trade-off between reducing emissions in-house and buying or selling permits is driven by the ratio of marginal emission reduction costs and the permit price. As there is no banking and borrowing, the markets in the two periods are independent and the discount rate does not influence the result. Rehdanz and Tol (2005) consider the special case $I = 2$ for one period only.

13.2 Dynamic Allocation

13.2.1 Alternative Allocations

Let us assume that the emission reduction obligations in the first period are based on grandfathering.[3] For example, all companies should reduce a fixed percentage τ (with $0 < \tau < 1$) of their emissions E in period 0, $A_{i,1} = E_{i,0} - (1 - \tau_1)E_{i,0} = \tau_1 E_{i,0}$.[4] This is what is mostly found in the existing trading schemes in Europe and also in the SO_2 trading scheme in the United States.

The second period is more interesting. In (13.1) we assume that the emission allocation in period 2 is independent of what happens in period 1. For instance, the allocation may be based on the emissions of period 0, $A_{i,2} = \tau_2 E_{i,0}$. In the long run a system of "fixed grandfathering" (or static allocation) based on the period before emission reduction policies can lead to substantial redistributions as the emission allocation gets more and more out of step with actual emissions.[5] It is therefore likely that the emission allocation in the second period somehow reflects the reality of period 1 rather than period 0.

One way for this to happen is through a "rolling grandfathering" (or "updated grandfathering") scheme whereby the emission allocation in period 2 is based on the actual (residual) emissions in period 1. That is, $A_{i,2} = \tau_2(E_{i,1} - R_{i,1})$.[6] In words, the emission reduction obligation in period 2 falls with emission reduction in period 1, or the more a region reduces now, the more that region has to reduce in the future. There would be less of an incentive to reduce emissions as it would only reduce the amount of permits receiving in the future. This was recently discussed by Harrison and Radov (2002).

Alternatively, emission allocations could shift away from grandfathering to technology standards. Such standards would be based on the emission intensity of companies. Suppose that the emission allocation is based on the best available, commercially proven technology. That is, emission allocations are based on some fixed percentage of potential emissions (those emissions that would have been had a company used the best technology), except for the technology leader, whose allocation is based on actual emissions.[7] Without loss of generality, assume that in period 0 all companies are the same size and have a turnover of unity; further assume that they all have equal uncontrolled emissions (and hence emission intensities) but different abatement costs. The emission allocation might then be something like $A_{i,2} = \tau_2(E_{i,1} - R_{\max,1} + R_{i,1})$. In words, the emission reduction obligation in period 2 falls with emission reduction in period 1, or the more a company reduces now, the less the company has to reduce in the future.

For rolling grandfathering, (13.1) changes to

$$\min_{R_{i1}, R_{i2}, P_{i1}, P_{i2}} C_i = \alpha_{i,1} R_{i,1}^2 + \frac{\alpha_{i,2} R_{i,2}^2}{1+\delta} + \pi_1 P_{i,1} + \frac{\pi_2 P_{i,2}}{1+\delta}$$

subject to

$$R_{i,1} + P_{i,1} \geq E_{i,1} - A_{i,1}; \quad R_{i,2} + P_{i,2} \geq E_{i,2} - A_{i,2} \tag{13.4}$$

with

$$\sum_{i=1}^{I} P_{i,t} = 0, \quad t = 1, 2$$

and

$$A_{i,2} = \tau_2(E_{i,1} - R_{i,1}).$$

The first-order conditions are

$$2\alpha_{i,1}R_{i,1} - \lambda_{i,1} + \tau_2\lambda_{i,2} = 0, \quad i = 1, 2, \ldots I, \tag{13.5a}$$

$$\frac{2\alpha_{i,2}R_{i,2}}{1+\delta} - \lambda_{i,2} = 0 \quad i = 1, 2, \ldots, I, \tag{13.5b}$$

$$\pi_1 - \lambda_{i,1} = 0, \quad i = 1, 2, \ldots, I, \tag{13.5c}$$

$$\frac{\pi_2}{1+\delta} - \lambda_{i,2} = 0, \quad i = 1, 2, \ldots, I, \tag{13.5d}$$

$$R_{i,t} + P_{i,t} - E_{i,t} + A_{i,t} = 0, \quad i = 1, 2, \ldots, I, \tag{13.5e}$$

$$\sum_{i=1}^{I} P_{i,t} = 0, \quad t = 1, 2. \tag{13.5f}$$

Note that for period 2, the first-order conditions are the same as (13.2). In period 2, the target is fixed, so the problem is identical to (13.1) and solved as in (13.3). Substituting this and (13.5c) in (13.5a) gives

$$2\alpha_{i,1}R_{i,1} = \pi_1 - \frac{\tau_2\pi_2}{1+\delta}. \tag{13.5a'}$$

Substituting this in (13.5e), solving for P, and substituting this in (13.5f) gives

$$\pi_1 = \frac{\sum_{i=1}^{I}(E_{i,1} - A_{i,1})}{\sum_{i=1}^{I} 1/2\alpha_{i,1}} + \frac{\tau_2\pi_2}{1+\delta}. \tag{13.6}$$

Note that solution (13.6) is identical to (13.3a) for $\tau_2 = 0$.

For technology standards (13.1) changes to

$$\min_{R_{i1}, R_{i2}, P_{i1}, P_{i2}} C_i = \alpha_{i,1}R_{i,1}^2 + \frac{\alpha_{i,2}R_{i,2}^2}{1+\delta} + \pi_1 P_{i,1} + \frac{\pi_2 P_{i,2}}{1+\delta}$$

subject to

$$R_{i,1} + P_{i,1} \geq E_{i,1} - A_{i,1}; \quad R_{i,2} + P_{i,2} \geq E_{i,2} - A_{i,2} \tag{13.7}$$

with

$$\sum_{i=1}^{I} P_{i,t} = 0, \quad t = 1, 2$$

and

$$A_{i,2} = \tau_2(E_{i,1} - R_{\max,1} + R_{i,1}).$$

The first-order conditions are (13.5b) through (13.5f), while (13.5a) changes to

$$2\alpha_{i,1} R_{i,1} = \pi_1 + \frac{\tau_2 \pi_2}{1+\delta}. \tag{13.5a''}$$

Equation (13.6) changes to

$$\pi_1 = \frac{\sum_{i=1}^{I}(E_{i,1} - A_{i,1})}{\sum_{i=1}^{I} 1/2\alpha_{i,1}} - \frac{\tau_2 \pi_2}{1+\delta}. \tag{13.6'}$$

Again, (13.6') is identical to (13.3a) for $\tau_2 = 0$. The first element at the right-hand sides of (13.6) and (13.6') is identical to (13.3a), so we see that rolling grandfathering (technology standards) increases (decrease) the price of carbon permits. This is because there is a penalty (premium) for selling permits. However, the price increase is exactly compensated by the second element at the right-hand sides of (13.5a') and (13.5a''). For every company, emission reduction and therefore net permit trade in period 1 is unaffected. In effect, the trade in emission permits is a zero-sum game.

Figure 13.1 illustrates this.[8] With rolling grandfathering, a company would be prepared to pay more for emission permits, as this would increase its emission allotment in the second period; the demand curve shifts upward. At the same time, a company could demand a higher price for permits sold, as this would decrease its emission allocation in the second period; the supply curve would shift upward too. The result is that the same amount is traded in period 1 but at a higher price. The reverse happens with technology standards. Both supply and demand curves shift downward, the quantity traded is the same, but the permit price is lower.

Something similar happens in the second period. The permit price only depends on the total emission allocation of all companies put together. The emission reductions of each company only depend on the price. So, if rolling grandfathering and technology standards lead to the same total allocation of emission permits, the emission reduction

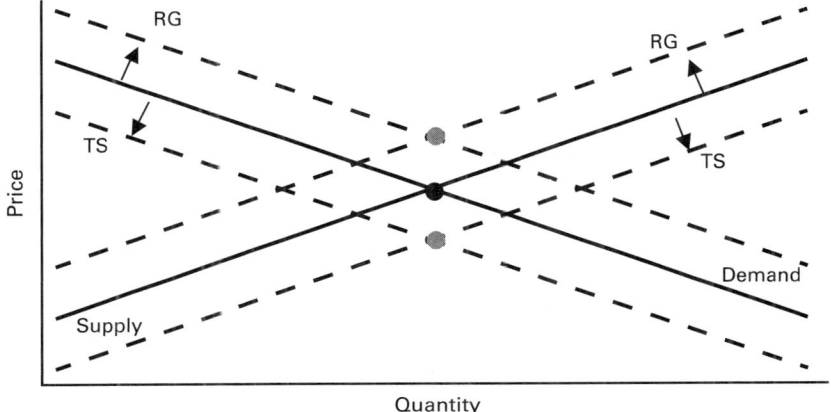

Figure 13.1
Demand and supply of permits in period 1 for a static allocation (solid lines) and two alternative forms of dynamic allocation: rolling grandfathering (RG) and technology standards (TS)

of each company is unaffected. However, as the initial allocation is different, emission permit trade is affected. As a result the costs to the companies are different in both periods, although the total costs are again unaffected. Rolling grandfathering and technology standards have a distributional effect only. This is in agreement with the Coase (1960) theorem.

13.2.2 A Numerical Illustration

The points made above can be better illustrated by a numerical analysis. Let us assume that there are five companies of equal size. Each company emits 20 tC in period 1 and period 2. In period 1, the emission allocation is 19 tC; in period 2, the emission allocation is 18.50 tC. That is, emission reduction is 5 percent in the first period and 7.25 percent in the second (compared to period 0). The firms differ in emission reduction costs. For firm i, $\alpha_{i,1} = 0.01(1+i)$. In the second period, $\alpha_{i,2} = 0.01i$.

Figure 13.2 shows emission reductions in the first period, without and with trade. Recall that allocation in period 1 is based on emissions in the previous period. This is grandfathering. The firms are ordered by emission reduction costs. Firm 1 has the lowest abatement costs and firm 5 the highest. Trade in emission permits makes that firms

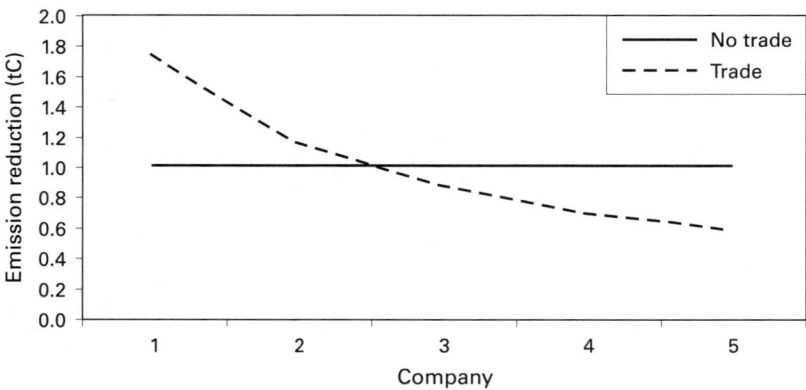

Figure 13.2
Emission reduction with and without trade in the first period

with low (high) abatement costs do more (less). The results are as expected.

Figure 13.3 shows emission reductions in the second period, without and with trade. Under the static allocation, all firms have the same obligations, as they had identical emissions in period 0. Under the rolling grandfathering scheme, the firms with high marginal cost that bought permits in the first period have a higher allocation (a lower emission reduction obligation). Under a technology standard scheme, the firms with low marginal cost that sold permits have a higher allocation. The total emission reduction is the same under the three rules. After trade, emission reduction efforts are the same regardless of the allocation mechanism.

Figure 13.4 shows the net present value of the emission reduction costs, with a 5 percent discount rate. Trade reduces the costs for all firms under all three allocation rules. As expected, companies with very high or low abatement costs benefit most from trade under all allocation rules. The net present costs per company are identical for the three allocation rules.[9] The changes in the permit price in the first period along with the number of traded permits traded in the second period and the emission reduction obligations in the second period offset each other. Compared to the static allocation, under the rolling grandfathering scheme (technology standard scheme), the permit price in the first period is higher (lower), the total number of traded permits is lower (higher), and the emission reduction obligations for companies with low abatement costs is higher (lower). As more periods are

Chapter 13 On Multi-period Allocation of Tradable Emission Permits   263

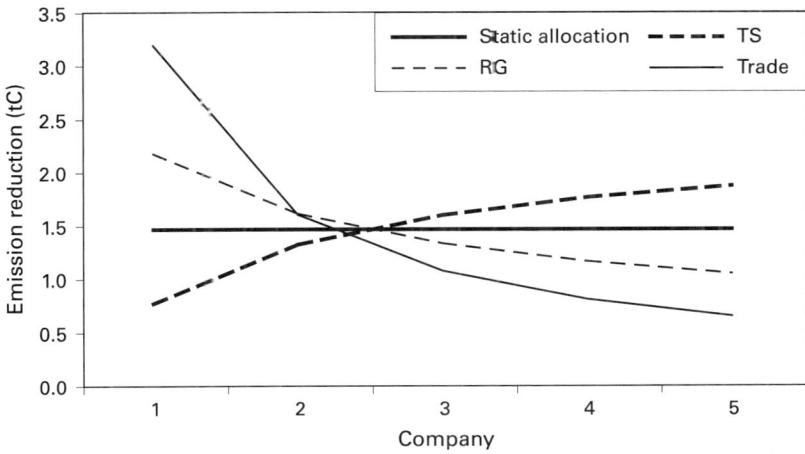

Figure 13.3
Allocation of emission reduction obligations in the second period according to three alternative rules (static allocation, rolling grandfathering, technology standard), and emission reduction effort after trade

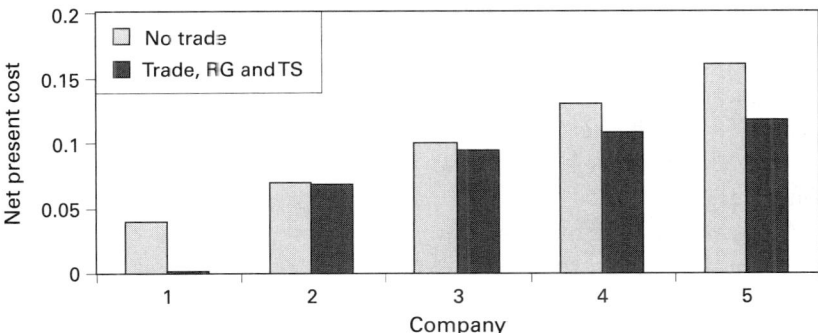

Figure 13.4
Net present costs of emission reduction without and with trade

added, the net present costs diverge for the different allocation rules. A technology standard reduces costs most for the technology leader by contrast to the static allocation and rolling grandfathering schemes. For companies with high abatement costs, the permit buyers on the market, costs would be lowest under rolling grandfathering. Because the net present costs are the same, the intuition is that the total emission reduction effort is the same, and all companies behave in an optimal way both between and within periods.

13.3 Banking and Borrowing

Banking and borrowing allow a company to transfer abatement activities over time, forward in time through banking and backward through borrowing. Excess emission rights can be saved for future use or present emissions can be extended for future abatement. As a result the permit price becomes arbitraged over time. Whether companies choose to bank or borrow permits depends on the price in the first period compared to a later period. If, for example, the permit price in the first period is expected to be lower compared to the second period, companies would bank permits, increasing the price in the first period and decreasing the price in the second period, until the permit price is balanced over the two periods. This is beneficial to all companies regardless of whether a company is planning on selling or buying a permit. The cost-saving potential is highest for high abatement cost companies.

We assume that companies can freely choose to bank or borrow permits, but there is a carbon interest factor $\beta \geq 1$. The interest factor captures that permits borrowed in the first period are worth less in the second period. In a market with permit banking and borrowing and static allocation, equation (13.1) changes to

$$\min_{R_{i1}, R_{i2}, P_{i1}, P_{i2}, B_{i,1}, B_{i,2}} C_i = \alpha_{i,1} R_{i,1}^2 + \frac{\alpha_{i,2} R_{i,2}^2}{1+\delta} + \pi_1 P_{i,1} + \frac{\pi_2 P_{i,2}}{1+\delta}$$

subject to

$$R_{i,1} + P_{i,1} + B_{i,1} \geq E_{i,1} - A_{i,1},$$

$$R_{i,2} + P_{i,2} - \beta B_{i,2} \geq E_{i,2} - A_{i,2}, \tag{13.8}$$

with

Chapter 13 On Multi-period Allocation of Tradable Emission Permits

$$\sum_{i=1}^{I} P_{i,t} = 0, \qquad t = 1, 2,$$

and $B_{i,1} = B_{i,2}$. $B_{i,1}$ is borrowing; if $B_{i,1}$ is negative it is banking. First-order conditions (13.9) are (13.2a), (13.2b), and (13.2d), while (13.2c) changes to

$$R_{i,1} + P_{i,1} + B_{i,1} - E_{i,1} + A_{i,1} = 0, \qquad i = 1, 2, \ldots, I,$$
$$R_{i,2} + P_{i,2} - \beta B_{i,2} - E_{i,2} + A_{i,2} = 0, \qquad i = 1, 2, \ldots, I, \qquad (13.9a)$$

and

$$-\lambda_1 + \beta \lambda_2 = 0 \qquad (13.9b)$$

is added.

Equation (13.9) solves as (13.3b), while (13.3a) changes to

$$\pi_1 = \lambda_1 = \frac{\beta \pi_2}{(1+\delta)} = \frac{\sum_{i=1}^{I}(E_{i,1} - A_{i,1})}{\sum_{i=1}^{I} 1/2\alpha_{i,1}} - \frac{\sum_{i=1}^{I} B_{i,1}}{\sum_{i=1}^{I} 1/2\alpha_{i,1}};$$

$$\pi_2 = \frac{\lambda_2}{1+\delta} = \frac{\sum_{i=1}^{I}(E_{i,2} - A_{i,2})}{\sum_{i=1}^{I} 1/2\alpha_{i,2}} + \frac{\beta \sum_{i=1}^{I} B_{i,2}}{\sum_{i=1}^{I} 1/2\alpha_{i,2}}. \qquad (13.10a)$$

At the company level, permits borrowed $E_{i,t}$ and permits bought $P_{i,t}$ are perfect substitutes, and therefore not determined. However, the total amount borrowed is

$$\sum_{i=1}^{I} B_{i,1} = \frac{(1+\delta) \sum_{i=1}^{I} 1/2\alpha_{i,1} \sum_{i=1}^{I} 1/2\alpha_{i,2}}{\beta^2 \sum_{i=1}^{I} 1/2\alpha_{i,1} + (1+\delta) \sum_{i=1}^{I} 1/2\alpha_{i,2}}$$

$$\times \left[\frac{\sum_{i=1}^{I}(E_{i,1} - A_{i,1})}{\sum_{i=1}^{I} 1/2\alpha_{i,1}} - \frac{\beta}{1+\delta} \frac{\sum_{i=1}^{I}(E_{i,2} - A_{i,2})}{\sum_{i=1}^{I} 1/2\alpha_{i,2}} \right]. \qquad (13.10e)$$

As expected, equation (13.10a) says that if there is net borrowing (banking), the price in the first period is lower (higher), and the price in the second period is higher (lower). Equation (13.10e) says that there is net borrowing $\Sigma E_{i,1} > 0$ (banking $\Sigma B_{i,1} < 0$) if the emission reduction obligation, normalized by the emission reduction costs, in the first (second) period is large relative to the emission reduction in the second (first) period, corrected for the discount factor and the interest rate on the borrowed carbon. This is as expected. However, if, for example, emission reduction obligations are the same in the two periods, but emission reduction costs decrease by technological progress (lower value for α in period 2), companies would do well to borrow permits in the first period instead. In general, however, both banking and borrowing would occur.

In the rolling grandfathering or technology standards scheme the result would be different and depend on the emission reduction obligations in later periods introduced by τ. For the rolling grandfathering and intertemporal transfer of permits equation (13.1) is rewritten as[10]

$$\min_{R_{i1},R_{i2},P_{i1},P_{i2},B_{i,1},B_{i,2}} C_i = \alpha_{i,1}R_{i,1}^2 + \frac{\alpha_{i,2}R_{i,2}^2}{1+\delta} + \pi_1 P_{i,1} + \frac{\pi_2 P_{i,2}}{1+\delta}$$

subject to

$$R_{i,1} + P_{i,1} + B_i \geq E_{i,1} - A_{i,1},$$

$$R_{i,2} + P_{i,2} - \beta B_i \geq E_{i,2} - A_{i,2}, \qquad (13.11)$$

with

$$\sum_{i=1}^{I} P_{i,t} = 0, \qquad t = 1, 2,$$

$$B_{i,1} = B_{i,2}, \qquad \beta \geq 1,$$

$$A_{i,2} = \tau_2 (E_{i,1} - R_{i,1}).$$

Equation (13.10e) changes to

$$\sum_{i=1}^{I} B_{i,1} = \left[\frac{(1+\delta)(1-\tau_2/\beta)\sum_{i=1}^{I} 1/2\alpha_{i,1} \sum_{i=1}^{I} 1/2\alpha_{i,2}}{(\beta^2 - 2\beta\tau_2 + \tau_2^2)\sum_{i=1}^{I} 1/2\alpha_{i,1} + (1+\delta)\sum_{i=1}^{I} 1/2\alpha_{i,2}} \right]$$

$$\times \left[\frac{1}{(1-\tau_2/\beta)} \frac{\sum_{i=1}^{I}(E_{i,1} - A_{i,1})}{\sum_{i=1}^{I} 1/2\alpha_{i,1}} - \frac{\beta}{(1+\delta)} \frac{\sum_{i=1}^{I} E_{i,2}}{\sum_{i=1}^{I} 1/2\alpha_{i,2}} \right.$$

$$\left. + \frac{\beta\tau_2}{(1+\delta)} \frac{\sum_{i=1}^{I} A_{i,1}}{\sum_{i=1}^{I} 1/2\alpha_{i,2}} \right]. \qquad (13.12e)$$

The same as in the static allocation scheme of (13.10e), equation (13.12e) says that there is net banking if emission reduction obligations in later periods are substantially larger ($0 < \tau_2 < 1$) relative to the first period. There is net borrowing, if emission reduction obligations in later periods are large but not too large. This can be seen from the second term of equation (13.12e). If τ_2, for example, is close to zero (substantial reductions in the second period), the first term becomes smaller than the last two together and the whole second term becomes negative. The opposite happens if τ_2 is close to one with emission reduction obligations similar to those of the previous period. However, if there is technological progress (lower value for α in period two), there is still borrowing for smaller values of τ_2. For technology standards and intertemporal transfer of permits equation (13.1) is rewritten as[11]

$$\min_{R_{i1}, R_{i2}, P_{i1}, P_{i2}, B_{i,1}, B_{i,2}} C = \alpha_{i,1} R_{i,1}^2 + \frac{\alpha_{i,2} R_{i,2}^2}{1+\delta} + \pi_1 P_{i,1} + \frac{\pi_2 P_{i,2}}{1+\delta}$$

subject to

$$R_{i,1} + P_{i,1} + B_i \geq E_{i,1} - A_{i,1},$$

$$R_{i,2} + P_{i,2} - \beta B_i \geq E_{i,2} - A_{i,2}, \qquad (13.13)$$

with

$$\sum_{i=1}^{I} P_{i,t} = 0, \qquad t = 1, 2,$$

$$B_{i,1} = B_{i,2}, \qquad \beta \geq 1,$$

$$A_{i,2} = \tau_2(E_{i,1} - R_{\max,1} + R_{i,1}).$$

Equation (13.10e) becomes

$$\sum_{i=1}^{I} B_{i,1} = \left[\frac{(1+\delta)(1+\tau_2/\beta) \sum_{i=1}^{I} 1/2\alpha_{i,1} \sum_{i=1}^{I} 1/2\alpha_{i,2}}{(\beta^2 + 2\beta\tau_2 + \tau_2^2) \sum_{i=1}^{I} 1/2\alpha_{i,1} + (1+\delta) \sum_{i=1}^{I} 1/2\alpha_{i,2}} \right]$$

$$\times \left[\frac{1}{(1+\tau_2/\beta)} \frac{\sum_{i=1}^{I}(E_{i,1} - A_{i,1})}{\sum_{i=1}^{I} 1/2\alpha_{i,1}} - \frac{\beta}{(1+\delta)} \left[\frac{\sum_{i=1}^{I} E_{i,2}}{\sum_{i=1}^{I} 1/2\alpha_{i,2}} \right. \right.$$

$$\left. \left. + \frac{\tau_2 R_{\max,1}}{\sum_{i=1}^{I} 1/2\alpha_{i,2}} - \frac{\tau_2 \sum_{i=1}^{I} E_{i,1}}{\sum_{i=1}^{I} 1/2\alpha_{i,2}} - \frac{\tau_2 \sum_{i=1}^{I}(E_{i,1} - A_{i,1})}{\sum_{i=1}^{I} 1/2\alpha_{i,2}} \right] \right].$$

(13.12e′)

The result is analogous to that of the rolling grandfathering scheme in (13.12e): equation (13.12e′) says that there is net banking if emission reduction obligations in later periods are substantially larger $(0 < \tau_2 < 1)$ relative to the first period. Unlike (13.12e), the total amount of net banked permits is greater as the emission reduction of the technology leader $(R_{\max,1})$ are subtracted. Also, in the last term of the second half of equation (13.12e′) more coefficients are multiplied by τ_2. Compared to rolling grandfathering, there is little room for borrowing.

Note that under any allocation scheme, banking of permits allows companies to reduce their emission reduction obligations in future periods. The total emission reduction achievement will be lower than the previously defined target. To achieve the same emission reduction target, the regulator has to lower τ_2 and hence increase emission reduction obligations. This has repercussions for the intertemporal transfer of permits.

Using the numerical example of section 13.2.2 and setting τ_2 such that total emission reduction obligations of period 1 plus 2 are the same for all approaches, we find the intertemporal transfer of permits under static allocation to be almost zero (0.4 tC).[12] For rolling grandfathering, borrowing permits are, in general, more attractive. This is because the total emission reduction obligations in period 2 are not ambitious enough to make net banking beneficial ($\tau_2 = 0.93$). The total amount of borrowed permits is 4.5 tC. This amount is almost identical to the total emissions that have to be reduced (5 tC) in period 1. In general, net permit borrowing becomes less attractive the lower the regulator sets τ_2 and the more ambitious the emission reduction obligations of future periods are. If τ_2 takes a certain value, net banking becomes more favorable (as discussed above). Under a system of technology standards, banking is more favorable. The calculated total amount of banked permits is 1.5 tC. The total emission reduction is not ambitious but also not so small as to make net borrowing more attractive ($\tau_2 = 0.88$).

At the company level the implications might be different. Depending on the dynamic allocation scheme, a company's abatement costs and the permit price banking or borrowing might be beneficial. To analyze this at the company level, the dynamic allocation scheme and intertemporal transfer of permits need to be solved simultaneously. This is impossible in our model, as both arbitrage the permit price over time and buying or borrowing permits are perfect substitutes. A solution would be to constrain the amount of intertemporally transferred permits.[13] The results of this model (not shown) confirm that depending on the allocation approach, a company's abatement costs and the emission reduction target banking or borrowing might decrease costs. Intertemporal transfer of permits is beneficial especially for companies with low abatement costs, as these are the permit sellers on the market. Interestingly the two dynamic approaches create opposite incentives. The same company would borrow permits under a rolling grandfathering scheme, and bank permits under a technology standard scheme.

13.4 Discussion and Conclusion

For the dynamic aspects of allocating greenhouse gas emission permits, we have examined three different allocation approaches: one is static while the other two are dynamic. We have extended the analysis to investigate two different mechanisms that could be used for emission reduction: the regulator could in either instance set the price of emission reduction or the overall emission reduction target.

As we showed above, these approaches present different strategic incentives for companies. In the absence of intertemporal transfer rolling grandfathering appears to be best for companies with high abatement costs with respect to emission reduction efforts. Companies with low abatement costs might prefer a technology standard scheme. We found that in expanding the model to include intertemporal transfer of permits through banking and borrowing, the model can become difficult to handle. The challenge is to solve a system whereby the permit price becomes arbitraged over time by different factors in a dynamic allocation scheme that involves the intertemporal transfer of permits. As we restricted the two dynamic approaches to banking and borrowing we could better see the tension in the incentives. The EU plan on letting the member states decide on how to allocate permits nationally (EU Commission 2003) is likely to affect markets outside their national borders.

Likely the analysis we presented could use extensions in at least two directions. First, emission reduction in period 1 could lead to lower emissions in a later period. This situation might result if investments in emission saving technology have a longer life span than a policy period. For example, power plants can last 30 to 50 years, whereas the UNFCCC commitment periods are 10 to 15 years. Our model only implies that the effective emission reduction in period 2 is lower because of emission reduction in period 1. Second, more periods than two could be considered. Third, the modeling of banking and borrowing could be more fully developed. We were able to derive results only for a corner solution. Fourth, the interactions between dynamic target setting and incentives to invest in research and development were neglected. These tasks we are deferring to future research.

Notes

The CEC DG research through the NEMESIS/ETC project (ENG2-CT01-000538), the US National Science Foundation through the Center for Integrated Study of the Human Dimensions of Global Change (SBR-9521914), the Hamburg Ministry for Science and Research, and the Michael Otto Foundation provided welcome financial support. All errors and opinions are ours.

1. Allocations can also be based on (expectations of) the current period, and even future periods. We omit this.

2. Equations (13.3a) through (13.3c) hold for all i's and t's.

3. Note that grandfathering has two connotations. First, it implies that emission permits are given to polluting companies *for free*. Second, it implies that emission permits are given *based on past emissions*. In this chapter all permits are allocated *for free*; the three alternative emission allocation schemes discussed here are grandfathering schemes in this sense of the word. Our three alternative schemes are each based on past emissions; however, we refine the definition of grandfathering as *proportional to past emissions*.

4. Emission allocations could also be based on a company's share in the total emissions cap. If the total cap is also based on emissions in period 0, τ would be replaced by another constant, leaving the analysis unaffected.

5. This goes for uncontrolled as well as controlled emissions. For uncontrolled emissions, a fixed allocation based on some historical year would only work if companies grow at approximately the same rate. For controlled emissions, the situation gets more complicated as different companies would prefer a different mix of in-house emission reduction and permit trade.

6. Emission allocations could also be based on a company's share in the total emissions cap. The τ would then not be constant, but a function of total emission reduction in period 1 (which is a constant) and the company's contribution to that (which is a decision variable). This would complicate the notation and the analysis without adding much insight; we in fact suspect that the two cases are equivalent.

7. Note that we could have set the allocation for the technology leader to a fraction of potential emissions too. However, the results would not change.

8. Note that this is an illustrative example only, used to clarify the market mechanisms; RG (TS) leads to higher (lower) permit prices than in a static allocation. The supply and demand curves are not directly based on the mathematical model above.

9. This is a numerical result that is independent of the parameter and target choices. We have not been able to demonstrate this analytically. The problem is that optimal emission

reduction in period 1 depends on the permit prices in both periods, while the permit price in the first period depends on the permit price in the second period, and the permit price in the second period depends on the sum of emission reductions in the first period. In order to make sure that the total emission reduction of the two periods is equal to that of the static allocation. τ_2 also depends on the emission reduction in the first period. Substituting all this in the equation for, say, optimal emission reduction in period 1 results in a system of simultaneous quadratic equations in R_1, with very elaborate constants. There is no general solution, and the equations are too complicated for us to glean insights.

10. The first-order conditions and the solutions are available on request.

11. The first-order conditions and the solutions are available on request.

12. We set $f = 1$.

13. More detailed insight from a model with constraints on intertemporal transfer can be obtained from the authors on request.

References

Böhringer, C., and A. Lange. 2003. On the design of optimal grandfathering schemes for emission allowances. ZEW Discussion paper 03-08. Mannheim.

Burtraw, D., K. Palmer, R. Bharvirkar, and A. Paul. 2002. The effect on asset values of the allocation of carbon dioxide emission allowances. *Electricity Journal* 15(5): 51–62.

Coase, R. 1960. The problem of social cost. *Journal of Law and Economics* 3: 1–44.

Cramton, P., and S. Kerr. 2002. Tradable carbon permit auctions—How and why to auction not grandfather. *Energy Policy* 30: 333–45.

Cronshaw, M. B., and J. B. Kruse. 1996. Regulated firms in pollution permit markets with banking. *Journal of Regulatory Economics* 9: 179–89.

Edwards, T. H., and J. F. Hutton. 2001. Allocation of carbon permits within a country: A general equilibrium analysis of the United Kingdom. *Energy Economics* 23: 371–86.

EU Commission. 2003. Directive 2003/87/EC of the European Parliament and of the Council of 13 October 2003 establishing a scheme for greenhouse gas emission allowance trading within the Community and amending Council Directive 96/61/EC. *Official Journal of the European Union* 50(275): 32–46.

Godby, R. W., S. Mestelman, R. A. Muller, and J. D. Welland. 1997. Emissions trading with shares and coupons when control over discharges is uncertain. *Journal of Environmental Economics and Management* 32: 359–81.

Goulder, L. H., I. W. H. Parry, R. C. Williams III, and D. Burtraw. 1999. The cost-effectiveness of alternative instruments for environmental protection in a second-best setting. *Journal of Public Economics* 72(1999): 329–60.

Hagem, C., and H. Westskog. 1998. The design of a dynamic tradable quota system under market imperfections. *Journal of Environmental Economics and Management* 36: 89–107.

Harrison, D., Jr., and D. B. Radov. 2002. Evaluation of alternative initial allocation mechanisms in a European Union greenhouse gas emissions allowance trading scheme. Report prepared for DG Environment, European Commission.

Holmes, K. J., and R. M. Friedman. 2000. Design alternatives for a domestic carbon trading scheme in the United States. *Global Environmental Change* 10: 273–88.

Jacoby, H. D., and A. D. Ellerman. 2004. The "safety valve" and climate policy. *Energy Policy* 32: 481–91.

Jensen, J., and T. N. Rasmussen. 2000. Allocation of CO_2 emissions permits: A general equilibrium analysis of policy instruments. *Journal of Environmental Economics and Management* 40: 111–36.

Kling, C., and J. Rubin. 1997. Bankable permits for the control of environmental pollution. *Journal of Public Economics* 64: 101–15.

Leiby, P., and J. Rubin. 2001. Intertemporal permit trading for the control of greenhouse gas emissions. *Environmental and Resource Economics* 19: 229–56.

McKibbin, W. J., and P. J. Wilcoxen. 1997. A better way to slow global climate change. Policy brief 17. Brookings Institution, Washington, DC.

OECD. 1999. *Permit Allocation Methods, Greenhouse Gases, and Competitiveness.* ENV/EPOC/GEEI(99)1/Final, Paris: OECD.

Pizer, W. A. 1999. The optimal choice of climate change policy in the presence of uncertainty. *Resource and Energy Economics* 21: 255–87.

Pizer, W. A. 2002. Combining price and quantity controls to mitigate global climate change. *Journal of Public Economics* 85: 409–34.

Rehdanz, K., and R. S. J. Tol. 2005. Unilateral regulation of bilateral trade in greenhouse gas emission permits. *Ecological Economics* 54: 397–416.

Rubin, J. D. 1996. A model of intertemporal emission trading, banking, and borrowing. *Journal of Environmental Economics and Management* 31: 269–86.

Schwarze, R., and P. Zapfel. 2000. Sulfur allowance trading and the regional clean air incentives market: A comparative design analysis of two cap-and-trade permit programs? *Environmental and Resource Economics* 17: 279–98.

Stavins, R. N. 1998. What can we learn from the grand policy experiment? Lessons from SO_2 allowance trading. *Journal of Economic Perspectives* 12: 69–88.

van Steenberghe, V. 2005. Carbon dioxide abatement costs and permit price: Exploring the impact of banking and the role of future commitments. *Environmental Economics and Policy Studies* 7: 75–107.

Viguier, L. L. 2001. Fair trade and harmonization of climate change policies in Europe. *Energy Policy* 29: 749–53.

Weitzman, M. L. 1974. Prices vs. quantities. *Review of Economic Studies* 41: 477–91.

Woerdman, E. 2000. Organizing emissions trading: The barrier of domestic permit allocation. *Energy Policy* 28: 613–23.

Woerdman, E. 2001. Developing a European carbon trading market: Will permit allocation distort competition and lead to state aid? FEEM Working paper 51.01. Milan.

Zhang, Z. X. 1999. Should the rules of allocating emissions permits be harmonised? *Ecological Economics* 31: 11–18.

14

Optimal Sequestration Policy with a Ceiling on the Stock of Carbon in the Atmosphere

Gilles Lafforgue, Bertrand Magné, and Michel Moreaux

In this chapter we characterize optimal time paths for two energy resources. One is that of depletable, carbon-based fossil fuels that have been contributing to climate change (coal, oil or gas), and the other is that of renewable, clean, and nonbiological,[1] such as solar energy, which is directly converted via photovoltaic cells, or indirectly converted as in the case of wind energy. In determining the optimality of the energy paths of our study, we weigh two important features: a critical threshold that should not be exceeded by the cumulative atmospheric pollution stock, as above this threshold, the induced environmental damage cost would be unbearable, and the carbon sink that the pollutant emissions produce through the use of fossil fuel but that can be reduced at source and stockpiled in a natural reservoir.[2]

In 2005 IPCC (Intergovernmental Panel on Climate Change) addressed the capture and sequestration of carbon (CCS) to reduce anthropogenic CO_2 emissions to the atmosphere. IPCC recommended CO_2 fluxes be filtered at the source of emission, that is, in fossil energy-fueled power plants, by use of scrubbers installed near the top of chimney stacks. The carbon would be sequestered in geological reservoirs. Among the reservoirs discussed were coal mines, depleted oil and gas reservoirs,[3] deep saline aquifers, and even the oceans (IPCC 2005). The potential capacities of such carbon sinks and their efficiency are still under assessment.[4]

To our knowledge, there are no analytical studies that clearly demonstrate the trade-offs involved in the management of an exhaustible fossil resource, the accumulation of the related pollution, and sequestration of limited capacity. Nevertheless, the possibility of sequestering some fraction of the pollution has motivated a number of empirical studies, via complex integrated assessment models (see McFarland et al. 2003; Edmonds et al. 2004; Kurosawa 2004; Gitz et al. 2005).

These studies generally favor the early introduction of sequestration as implementation of this measure can lead to a substantial decrease in the cost of environmental externality.

While a generic abatement option can take several forms, such as sequestration by forests or pollution reduction at the source, in this study we are mainly concerned with the capture and direct disposal of carbon, although we take into regard the size and access cost of the reservoir. We use the standard Hotelling model of exhaustion but introduce three features. The first is the possibility a cleaner abundant energy flow serving as a backstop. Following an optimal path in our hypothetical economy, the resource price has two components: its marginal extraction cost and the Hotelling rent, which necessarily grows at a rate equal to the interest rate and thus drives the prices up over time. When this price reaches the backstop marginal cost, which is assumed to be constant, only the renewable source is used. In effect, the two energy sources are used one after the other and the backstop is only relevant when the fossil resource has been exhausted.

The second feature is the ceiling placed on cumulative carbon emissions from consumption of a fossil energy resource. The changes in fossil fuel consumption drive the dynamics of the transition from pollution accumulation to cleaner technology, in that each extraction trajectory generates a cumulative emission trajectory. The ceiling constraint thus adds a third component to the expression of the fossil fuel price: the externality cost associated with the accumulation of pollution in the atmosphere. In this regard, a certain amount of natural regeneration reduces the level of pollution, thus allowing some use of the fossil fuel to continue at the ceiling. We then show that the optimal consumption path consists of four phases. In the first phase, only the fossil fuel resource supplies the economy. During this phase the scarcity rent of the resource and the shadow cost of carbon emissions are both increasing. Hence the energy price increases and the fossil fuel consumption (i.e., the emission flow) decreases, if we assume demand function to be stationary over time.[5] However, the amount of pollutants in the atmosphere increases because the emission flow is larger than the regenerated flow. At the ceiling the fossil fuel consumption is limited by the natural regeneration process. The pollution stock at its ceiling level—induced by natural regeneration flow—is in balance with the emission flow. Fossil fuel consumption and energy price both remain constant. Because at the ceiling rate the Hotelling rent increases (in current value), the shadow cost of the constraint on carbon emissions decreases. The second phase ends with the shadow cost dropping finally to zero. In the third phase, the price increases once again, driven only by the increase in the scarcity rent. Price rises until it is equal to the marginal cost of the backstop, where the fossil fuel supply is exhausted. Nevertheless, emissions keep on decreasing; the ceiling constraint is no longer a limiting factor and the path reverts to the benchmark Hotelling level. Last, in the fourth phase, the backstop supplies the whole demand. This scheme holds as long as the energy price, defined at the ceiling where only fossil fuel is used, is higher than the marginal cost of the clean substitute. If this marginal cost is lower and the clean resource is abundant, clean energy will be substituted for the fossil fuel at the precise moment the ceiling is attained. At that point the full marginal cost of carbon (i.e., the sum of the marginal extraction cost, the Hotelling rent, and the shadow cost of the carbon constraint) will equal the marginal cost of the clean technology. At the ceiling, then, emissions are balanced by the regeneration process, some part of the supply is provided by the renewable resource, and the energy price is equal to the marginal cost of the renewable resource as shown in Chakravorty et al. (2006a). Both resources have to be exploited simultaneously because, at this price, even if the renewable is competitive, the fossil resource remains less costly (excluding the cost of externality) and thus must be used jointly. At the ceiling the fossil fuel supply is indirectly restricted by the regeneration flow. Because in the fourth phase the increase of the Hotelling rent is compensated by the decrease of the (positive) shadow cost of the carbon ceiling, the energy price is constant and equal to the marginal cost of the renewable energy. On complete exhaustion of the fossil fuel, the shadow cost of carbon becomes inconsequential as the renewable resource now supplies the entire demand. With availability of the low-cost substitute, there is no longer any pure Hotelling phase.

The third feature corresponds to the capture and storage of some fraction of the carbon emissions. The sequestration of carbon could allow for continued intensive use of fossil fuels because sequestration would alleviate the environmental consequences of carbon combustion that are implicated in climate change. Such a mitigation option would likely lead to less stringent Kyoto-type constraints on greenhouse gas emissions, even if, as discussed below, the optimal policy may not be consistent with the type of policy laid down by the Protocol. However, the alleviation option is not free of charge. Additional costs are incurred by carbon sequestration, depending on the characteristics of the sink, especially its size. Chakravorty et al. (2006b) suggest a generic abatement scheme whereby the carbon sinks are of unlimited capacity. This scheme implies, however, that an emission processing cost has to be considered in determining the resource price. More interesting is the case where the capacity of the carbon sink is limited. The marginal cost of consuming one unit of fossil fuel compatible with some environmental preservation objective is thus fourfold: it includes the monetary cost of exploiting the resource, the cost of carbon processing, the scarcity rent of the resource, and the rent associated with the limited capacity of the carbon sink, both rents being endogenous. This overall cost needs to be compared with the supply cost of the renewable energy, this cost being generally higher than the exploitation cost of the fossil fuel alone. If the cost of the renewable energy is higher than the sum of the cost of fossil resource exploitation and the cost of carbon sequestration, then it is better to exploit the depletable resource before the renewable one. We need to show how the application of a capture option at the source of pollution emission leads to the competitiveness of the clean substitute discussed earlier. Carbon sequestration, of course, relaxes the constraints on fossil fuel consumption. As can be immediate inferred, if the nonrenewable resource is exhausted earlier, the renewable one kicks in earlier. So the optimal path here consists of five phases. As long as the ceiling is not reached, only the fossil resource is used and the pollution stock continues to increase. Carbon sequestration takes place until the sink is completely filled, which is the ceiling point. The next phase occurs at the ceiling without sequestration. The two following phases are identical to the three last phases that occur in the case of a pollution stock ceiling without any sequestration.

Let us assume that the clean renewable substitute is scarce, that at a price equal to its marginal cost, the market demand is greater than the available flow. In this case, absent a pollution constraint, the resources are no longer exploited sequentially. Once the fossil fuel price (i.e., the sum of the extraction cost and the Hotelling rent) is equal to the marginal cost of the renewable substitute, renewable energy becomes competitive with it and has to be exploited. But if the available flow cannot satisfy all the demand, in order to clear the market, the residual demand must be supplied by the fossil resource. In this phase of simultaneous use of both resources, prices will vary according to the same rule as in the first phase because the scarcity rent of the fossil is still growing at the interest rate. Thus the discrepancy between the energy price and the marginal cost of the renewable resource increases at a rate higher than the interest rate. This is a consequence of the nonstorability of the resource, excluding any intertemporal arbitrage, so the rent of the renewable resource is allowed to grow faster than the rise in interest rate. The share of consumption supplied by the fossil fuel decreases continuously until complete exhaustion of the resource. The backstop finally must supply all the demand. When a cap is set on carbon accumulation, the use of the clean renewable substitute begins before the pollution ceiling is reached provided that the energy price at that time is higher than the marginal cost of the substitute. By the same token, use of the substitute will begin after pollution reaches the ceiling if the energy price is lower than the marginal cost of the substitute. Furthermore, if there is an opportunity for sequestration, whether it is applied before or after using the renewable resource depends upon their respective costs.

14.1 The Model

14.1.1 Assumptions and Notations

We consider an economy in which the instantaneous gross surplus or utility, measured in monetary units and generated by an instantaneous energy consumption q_t,[6] is given by the following standard function u:

ASSUMPTION 1 $u \cdot \text{IR}_{++} \to \text{IR}_+$ is a function of class C^2, which is strictly increasing, strictly concave and which satisfies the Inada condition: $\lim_{q \downarrow 0} u'(q) = +\infty$, where $u'(q) = du/dq$.

We sometimes use p to denote the marginal surplus u' as well as (by a slight misuse of formal notation) the marginal surplus function: $p(q) = u'(q)$. The direct demand function $d(p)$ is merely the reciprocal of $p(q)$, as usually defined.

Energy needs may be supplied by two resources: a dirty nonrenewable resource, such as coal, or a clean renewable resource, such as solar energy. If X^0 represents the initial coal stock of the society, X_t the stock of coal available at time t ($X_0 = X^0$) and x_t the instantaneous coal consumption, we can write

$$\frac{dX_t}{dt} = -x_t. \tag{14.1}$$

We assume that the average cost of transforming coal into energy that is directly usable at delivery to the users, is constant and equal to c_x. Hence, c_x is also the constant marginal cost. This cost should be treated as the sum of the extraction cost sensu stricto, the cost of industrial processing of the extraction output and the cost of transportation, all of which must be borne so the energy supply can match the demand by end users. Let $\tilde{x}$ denote the flow of nonrenewable resource to be consumed, assuming an infinite available stock of the nonrenewable resource X^0 so that no rent is charged. Thus $\tilde{x}$ is the solution of $u'(x) = c_x$ that is, $\tilde{x} = d(c_x)$.

Using coal potentially generates a pollutant flow. Let ζ be the unitary carbon content of coal so that, absent any abatement policy, the instantaneous carbon flow released into the atmosphere is equal to ζx_t. However, let us assume that some carbon sequestration device is available. Let s_t be the part of the potential carbon emission flow that is sequestered, so that the effective flow, denoted by z_t, amounts to

$$z_t = \zeta x_t - s_t$$

with $s_t \geq 0$ and $\zeta x_t - s_t \geq 0$. (14.2)

We assume that the unit sequestration cost is constant (hence also the marginal cost) and equal to c_s so that the total monetary cost of sequestration is given by $c_s s_t$. $\bar{S}$ denotes the capacity of the so-called carbon sink, S^0 the initial stock of carbon contained in the sink (we postulate that $S^0 = 0$) and S_t the stock at time t ($S_0 = S^0$), we can write

$$\frac{dS_t}{dt} = s_t \quad \text{and} \quad \bar{S} - S_t \geq 0. \tag{14.3}$$

Let Z^0 be the stock of carbon in the atmosphere at the beginning of the planning period, Z_t the stock at time t ($Z_0 = Z^0$) and α, $\alpha > 0$, the instantaneous proportional rate of natural regeneration, assumed to be constant for sake of simplicity (see Kolstad and Krautkraemer, 1993)[7] so that

$$\frac{dZ_t}{dt} = z_t - \alpha Z_t. \tag{14.4}$$

Self-regeneration is merely a scheme for natural sequestration of carbon emissions into a sink of sufficiently large capacity. By that we mean that whatever the quantity of carbon to be sequestered, it can still be buried in the so-called sink. We assume that this stock of carbon cannot be larger than some threshold $\bar{Z}$:

$$\bar{Z} - Z_t \geq 0 \quad \text{and} \quad \bar{Z} - Z^0 \geq 0. \tag{14.5}$$

This constraint should be considered as some kind of damage function. The damage generated at each point of time by the stock of atmospheric carbon is equal to 0, provided that $Z < \bar{Z}$, but jumps to infinity when $Z = \bar{Z}$.[8] In the following, $\bar{x}$ denotes the flow of nonrenewable resource that could be used at the ceiling, without any sequestration scheme, that is, the solution of $dZ_t/dt = \zeta x_t - s_t - \alpha Z_t$ for $s_t = 0$ and $Z_t = \bar{Z}$; hence $\bar{x} = \alpha \bar{Z}/\zeta$. We use $\bar{p}_x$ to denote the corresponding price, $\bar{p}_x = u'(\bar{x})$. Clearly, if $\bar{p}_x$ were lower than c_x, there would be no ceiling problem. This is because even if the resource rent becomes negligible, the optimal consumption of the polluting resource will remain below $\bar{x}$. Thus we assume the following:

ASSUMPTION 2 $\bar{p}_x > c_x$, which is equivalent to $\tilde{x} > \bar{x}$.

However, if sequestration must be used, we need to consider an even stronger assumption: the total marginal cost of a clean consumption of coal, $c_x + c_s \zeta$, must be lower than $\bar{p}_x$. If not, it would always be better to stay constrained at $\bar{x}$ than relax the constraint by sequestering some part of the emission flow.

ASSUMPTION 3 $\bar{p}_x > c_x + c_s \zeta$.

As discussed in section 14.3.1, assumption 3 is a necessary but not a sufficient condition.

The other resource is a renewable resource that can be made available to the end users at a constant average cost c_y (hence the same constant marginal cost). The cost of the renewable resource is the total cost of supplying the good to the final users, so that the nonrenewable and the renewable resources are perfect substitutes for the users. Let us assume that $\bar{y}$ is the constant instantaneous flow of renewable resource available at each point of time, and that this resource is nonstorable in the long term, except at a prohibitive storage cost. Let y_t be the part of the available flow consumed at time t, so the part $\bar{y} - y_t$ of the flow that is not immediately consumed is definitely lost.

For the monetary costs alone, namely costs other than scarcity rents, we assume that the cost of the nonrenewable resource is lower than the cost of the renewable resource. In the present case this corresponds to the main renewable energies[9] and the main nonrenewable energies.

ASSUMPTION 4 $c_x < c_y$.

Let $\tilde{y}$ be the flow of renewable resource society would have to consume once the nonrenewable resource is exhausted, provided that $\bar{y}$ is sufficiently large. $\tilde{y}$ is the solution of $u'(y) = c_y : \tilde{y} = d(c_y)$. Chakravorty et al. (2006a) showed that, for $\bar{y} < \tilde{y}$ and without any sequestration opportunity, there would be many different optimal paths to choose from because rent would have to be charged for the use of the renewable resource before the nonrenewable resource is exhausted. For simplicity, we first assume (as we do later in sections 14.2 and 14.3) that the renewable resource is abundant. By abundant, we mean that at the marginal cost c_y, the quantity to be supplied is, at the very most, equal to $\bar{y}$. We also assume that $c_y > \bar{p}_x$ so that, when coal consumption is constrained by the ceiling, the renewable resource is not competitive.

ASSUMPTION 5 $\bar{y} > \tilde{y}$ and $c_y > \bar{p}_x$.

As indicated by assumptions 4 and 5 the active sequestration phase always precedes the renewable resource phase (as described later in section 14.3). For the use of renewable energy to be introduced during an active sequestration phase, we must have not only $\bar{y} < \tilde{y}$, but also $c_y < \bar{p}_{xy}$, where $\bar{p}_{xy} = u'(\bar{x} + \bar{y}) < \bar{p}_x$. Hence the moment the stock of pollution hits the ceiling level, the use of renewable energy becomes competitive as $c_y < \bar{p}_{xy}$, and moreover the two types of resources have to be combined. In this case the nonrenewable resource is limited at $\bar{x}$ by the pollution stock constraint and the renewable resource at $\bar{y}$ by the available supply. Thus, using $\bar{p}_y$ to denote the marginal gross surplus at $\bar{y}$, we can write alternatively:

ASSUMPTION 6 $\bar{y} < \tilde{y}$ and $\bar{y} < \bar{x}$ or equivalently $\bar{p}_{xy} > c_y$ and $\bar{p}_y > \bar{p}_{xy}$. Furthermore $\bar{p}_{xy} > c_x + c_s \zeta$.

By assumption 6, we should add another constraint: that the renewable energy consumption cannot be higher than $\bar{y}$,

$$\bar{y} - y_t \geq 0. \tag{14.6}$$

Let us assume that the instantaneous social rate of discount, $\rho > 0$, is constant. The objective of the social planner is to choose the resource

Chapter 14 Optimal Sequestration Policy with Ceiling on Carbon

and abatement trajectories that maximize the sum of the discounted instantaneous net surplus.

14.1.2 Problem Formulation

The social planner problem can be expressed as follows (**P**):

$$\max_{\{(s_t, x_t, y_t), t \geq 0\}} \int_0^\infty [u(x_t + y_t) - c_s s_t - c_x x_t - c_y y_t] e^{-\rho t} \, dt \tag{P}$$

subject to (14.1) to (14.6), $X_0 = X^0$, $Z_0 = Z^0 < \bar{Z}$, $S^0 = 0$, $s_t \geq 0$, $x_t \geq 0$ and $y_t \geq 0$. Let L be the current value Lagrangian for the problem (**P**):

$$L = u(x_t - y_t) - c_s s_t - c_x x_t - c_y y_t - \lambda_t x_t + \eta_t s_t + \mu_t[\zeta x_t - s_t - \alpha Z_t]$$
$$+ v_{St}[\bar{S} - S_t] + v_{Zt}[\bar{Z} - Z_t] + \bar{\gamma}_{st}[\zeta x_t - s_t] + \gamma_{st} s_t + \gamma_{xt} x_t + \bar{\gamma}_{yt}[\bar{y} - y_t]$$
$$+ \gamma_{yt} y_t.$$

The first-order conditions (FOCs) and complementary slackness conditions are

$$\partial L/\partial s_t = 0 \Leftrightarrow c_s = \eta_t - \mu_t - \bar{\gamma}_{st} + \gamma_{st}, \tag{14.7}$$

$$\partial L/\partial x_t = 0 \Leftrightarrow u'(x_t + y_t) = c_x + \lambda_t - \mu_t \zeta - \bar{\gamma}_{st} \zeta - \gamma_{xt}, \tag{14.8}$$

$$\partial L/\partial y_t = 0 \Leftrightarrow u'(x_t + y_t) = c_y + \bar{\gamma}_{yt} - \gamma_{yt}, \tag{14.9}$$

$$\bar{\gamma}_{st} \geq 0 \quad \text{and} \quad \bar{\gamma}_{st}[\zeta x_t - s_t] = 0, \tag{14.10}$$

$$\gamma_{st} \geq 0 \quad \text{and} \quad \gamma_{st} s_t = 0, \tag{14.11}$$

$$\gamma_{xt} \geq 0 \quad \text{and} \quad \gamma_{xt} x_t = 0, \tag{14.12}$$

$$\bar{\gamma}_{yt} \geq 0 \quad \text{and} \quad \bar{\gamma}_{yt}[\bar{y} - y_t] = 0, \tag{14.13}$$

$$\gamma_{yt} \geq 0 \quad \text{and} \quad \gamma_{yt} y_t = 0. \tag{14.14}$$

By assumptions 1, 4 and 5, condition (14.9) implies that $x_t + y_t$ must be at least equal to $\tilde{y}$ and that y_t must be at most equal to $\tilde{y} < \bar{y}$.[10] Thus $\bar{\gamma}_{yt}$ must be equal to 0, $t \geq 0$. The dynamics of the corresponding state variables must satisfy

$$\frac{d\lambda_t}{dt} = \rho\lambda_t - \frac{\partial \mathbf{L}}{\partial X} \Leftrightarrow \frac{d\lambda_t}{dt} = \rho\lambda_t, \tag{14.15}$$

$$\frac{d\eta_t}{dt} = \rho\eta_t - \frac{\partial \mathbf{L}}{\partial S} \Leftrightarrow \frac{d\eta_t}{dt} = \rho\eta_t + \nu_{St}, \tag{14.16}$$

$$\frac{d\mu_t}{dt} = \rho\mu_t - \frac{\partial \mathbf{L}}{\partial Z} \Leftrightarrow \frac{d\mu_t}{dt} = (\alpha + \rho)\mu_t + \nu_{Zt}, \tag{14.17}$$

with the following associated complementary slackness conditions:

$$\nu_{St} \geq 0 \quad \text{and} \quad \nu_{St}[\bar{S} - S_t] = 0, \tag{14.18}$$

$$\nu_{Zt} \geq 0 \quad \text{and} \quad \nu_{Zt}[\bar{Z} - Z_t] = 0. \tag{14.19}$$

Note that (14.15) implies that $\lambda_t = \lambda_0 e^{\rho t}$. Hence the transversality condition for X_t takes the following form:

$$\lim_{t \uparrow \infty} e^{-\rho t} \lambda_t X_t = \lambda_0 \lim_{t \uparrow \infty} X_t = 0. \tag{14.20}$$

The other transversality conditions are

$$\lim_{t \uparrow \infty} e^{-\rho t} \eta_t S_t = 0, \tag{14.21}$$

$$\lim_{t \uparrow \infty} e^{-\rho t} \mu_t Z_t = 0. \tag{14.22}$$

Clearly, the costate variables η_t and μ_t are not positive. Furthermore, given that S_t is nondecreasing and starting from $S_0 = S^0 = 0$, there must exist some time interval $[0, \bar{t})$ during which $S_t < \bar{S}$. Hence $\nu_{St} = 0$, and after integrating (14.16), we get

$$\eta_t = \eta_0 e^{\rho t}, \quad t \in [0, \bar{t}). \tag{14.23}$$

By the same argument, for any time interval $[t_0, t_1)$ during which $Z_t < \bar{Z}$, we obtain

$$\mu_t = \mu_{t_0} e^{(\alpha+\rho)(t-t_0)}, \quad t \in [t_0, t_1). \tag{14.24}$$

Since $Z^0 < \bar{Z}$, there must exist some initial interval with $t_0 = 0$ and $t_1 > 0$, so that $\mu_t = \mu_0 e^{(\alpha+\rho)t}$, $t \in [0, t_1)$. Also, since X^0 is finite, there must be some time t_2 from which $Z_t < \bar{Z}$, $t \geq t_2$, so that $\mu_t = 0$.

14.2 Hotelling and Optimal Paths with No Carbon Sink

14.2.1 Pure Hotelling Paths

Without a ceiling constraint and by assumption 5, the FOCs (14.8) and (14.9) would be

$$u'(x_t + y_t) = c_x + \lambda_0 e^{pt} - \gamma_{xt}, \tag{14.25}$$

$$u'(x_t + y_t) = c_y - \gamma_{yt}, \tag{14.26}$$

together with the complementary slackness conditions (14.12) through (14.14). Thus, if both x_t and y_t were strictly positive over some non-degenerating time interval, we would have $u'(x_t + y_t) = c_x + \lambda_0 e^{pt} = c_y$ over the interval, which is clearly not possible. Hence the resources have to be exploited sequentially: first the less costly resource (i.e., the nonrenewable resource) and next the more costly resource (i.e., the renewable resource). Moreover the initial value of the coal rent λ_0, is at most equal to $c_y - c_x$.

For any $\lambda_0 \in (0, c_y - c_x)$, let $t^H(\lambda_0)$ be that time at which $c_x + \lambda_0 e^{pt} = c_y$, and let $d_t^H(\lambda_0) = d(c_x + \lambda_0 e^{pt})$, $t \in [0, t^H(\lambda_0))$. The optimal value of λ_0, λ_0^H, is given as the unique solution of the cumulative demand-initial endowment balance equation $\int_0^{t^H(\lambda_0)} d_t(\lambda_0)\, dt = X^0$. The optimal consumption would then be the standard Hotelling solution:

$$x_t = \begin{cases} d_t^H(\lambda_0^H), & t < t^H(\lambda_0^H) \\ 0, & t^H(\lambda_0^H) \le t \end{cases} \text{ and } y_t = \begin{cases} 0, & t < t^H(\lambda_0^H) \\ \tilde{y}, & t^H(\lambda_0^H) \le t \end{cases}. \tag{14.27}$$

All the optimality conditions are satisfied by the following values of γ_{xt} and γ_{yt}:

$$\gamma_{xt} = \begin{cases} 0, & t < t^H(\lambda_0^H) \\ c_x + \lambda_0 e^{pt} - c_y, & t^H(\lambda_0^H) \le t \end{cases} \text{ and }$$

$$\gamma_{yt} = \begin{cases} c_y - c_x - \lambda_0 e^{pt}, & t < t^H(\lambda_0^H) \\ 0, & t^H(\lambda_0^H) \le t \end{cases}.$$

As a function of X^0, λ_0^H strictly decreases with[11] $\lim_{X^0 \downarrow 0} \lambda_0^H = c_y - c_x$ and $\lim_{X^0 \uparrow \infty} \lambda_0^H = 0$. Let $Z_t^H(\lambda_0)$ be the trajectory of the carbon stock generated by $d_t^H(\lambda_0)$. $Z_t^H(\lambda_0)$ is the solution of $dZ_t/dt = \zeta d_t^H(\lambda_0) - \alpha Z_t$,

$t \in [0, t^H(\lambda_0))$, together with the initial condition $Z_0 = Z^0$. Here we define $Z_m^H(\lambda_0)$ as the maximum quantity of atmospheric carbon over $[0, t^H(\lambda_0))$[12]: $Z_m^H(\lambda_0) = \sup\{Z_t^H(\lambda_0), t \in [0, t^H(\lambda_0))\}$.

By assumption 2, when λ_0 is sufficiently low (i.e., X^0 is sufficiently high), then $Z_m^H(\lambda_0) > \bar{Z}$.[13] Let $\bar{X}_0$ be the value of X_0 for which $Z_m^H(\lambda_0(X^0)) = \bar{Z}$. In terms of the ceiling constraint, it can be easily seen that, for X^0 lower than $\bar{X}_0$, the constraint would never be rigid, so the optimal consumption path could be above the standard Hotelling path as given by (14.27). In the following assumption, we treat the ceiling constraint having violated along the pure Hotelling path.

ASSUMPTION 7 $X^0 > \bar{X}^0$.

14.2.2 Optimal Paths with No Opportunity for Abatement

Let us assume that there is no opportunity for pollution abatement apart from the natural regeneration process. It has been shown by Chakravorty et al. (2006a) that by assumptions 1, 2, 4, and 5, the optimal consumption path is a four-phase path as illustrated in figure 14.1. In the first phase $[0, t_1)$, the constraint is slack, and only coal is used: $q_t = x_t = d(c_x + \lambda_0 e^{\rho t} - \mu_0 e^{(\alpha+\rho)t}\zeta)$ with λ_0 and $|\mu_0|$ sufficiently low so that $x_t > \bar{x}$. Since $x_t > \bar{x}$ and $Z_t < \bar{Z}$, then Z_t is increasing because $\zeta x_t > \alpha Z_t$: the emission rate is higher than the natural regeneration flow. At t_1, the carbon ceiling is reached and the full marginal cost of coal, $c_x + \lambda_0 e^{\rho t} - \mu_0 e^{(\alpha+\rho)t}\zeta$, is equal to $\bar{p}_x$.

The second phase $[t_1, t_2)$ occurs at the ceiling, when the coal consumption—the only energy being used—is constrained to $\bar{x}$. Thus the energy price is constant and equal to $\bar{p}_x$. Since $p_t = \bar{p}_x = c_x + \lambda_0 e^{\rho t} - \mu_0 e^{(\alpha+\rho)t}\zeta$, then $|\mu_t|$ must be decreasing during this phase. At t_2, $\mu_t = 0$, and the ceiling constraint will no longer be active from t_2 onward.

The third phase $[t_2, t_3)$ is a pure Hotelling phase during which only coal is used: $p_t = c_x + \lambda_0 e^{\rho t}$ and $q_t = x_t = d(p_t)$. Thus coal consumption decreases, as seen during the first phase, and the stock is exhausted at the end of that phase. Subsequently the price must become equal to the marginal cost of the renewable resource c_y.

In the last phase $[t_3, \infty)$, only the renewable resource is used, $q_t = y_t = \tilde{y}$ and the price is constant and equal to c_y. Both price and quantity paths are illustrated in figure 14.1. The hatched surface under the x_t curve must be equal to X^0.

Chapter 14 Optimal Sequestration Policy with Ceiling on Carbon

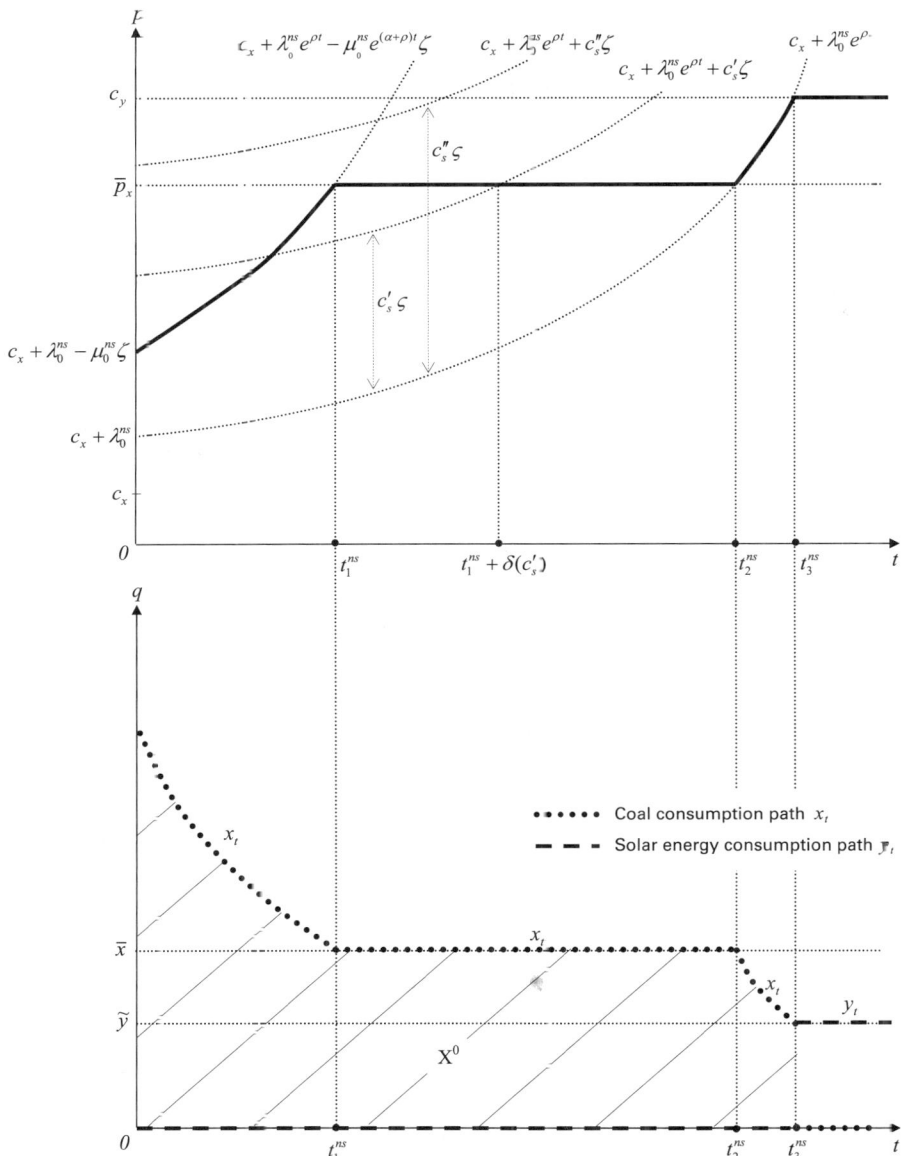

Figure 14.1
Optimal paths without the sequestration opportunity

We need to determine the optimal values λ_0^{ns}, μ_0^{ns}, t_1^{ns}, t_2^{ns}, and t_3^{ns} (ns stands here for no sequestration) of the five fundamental variables λ_0, μ_0, t_1, t_2, and t_3 to solve a system of five equations.

The first equation is the cumulative coal consumption-initial stock balance equation:

$$\int_0^{t_1} d(c_x + \lambda_0 e^{\rho t} - \mu_0 e^{(\alpha+\rho)t}\zeta)\,dt + [t_2 - t_1]\bar{x} + \int_{t_2}^{t_3} d(c_x + \lambda_0 e^{\rho t})\,dt = X^0.$$

The second equation is the price continuity equation at t_1:

$$c_x + \lambda_0 e^{\rho t_1} - \mu_0 e^{(\alpha+\rho)t_1}\zeta = \bar{p}_x.$$

The third equation is the pollution stock continuity equation at t_1:

$$Z_{t_1}(\lambda_0, \mu_0) = \bar{Z},$$

with $Z_t(\lambda_0, \mu_0)$ the solution of $dZ_t/dt = \zeta d(c_x + \lambda_0 e^{\rho t} - \mu_0 e^{(\alpha+\rho)t}\zeta) - \alpha Z_t$, $Z_0 = Z^0$.

The fourth and fifth equations are the price continuity equations at t_2 and t_3:

$$c_x + \lambda_0 e^{\rho t_2} = \bar{p}_x \quad \text{and} \quad c_x + \lambda_0 e^{\rho t_3} = c_y.$$

Chakravorty et al. (2006a) demonstrated that solving this system of equations for λ_0^{ns}, μ_0^{ns}, t_1^{ns}, t_2^{ns}, and t_3^{ns} provides values of the other multipliers that satisfy all the optimality conditions.

14.3 The Case of an Abundant Renewable Substitute

Although the coal consumption is constrained over a certain time interval $[t_1, t_2)$ under assumption 6, as shown in section 14.2.2 and illustrated in figure 14.1, it is not clear a priori whether it is worth relaxing this constraint by sequestering the carbon because sequestration is costly. In this section we start with a very simple test of the optimality of sequestration for relaxing the ceiling constraint. Assuming that it is optimal to sequester, we next determine the optimal policy according to whether the sink or reservoir capacity $\bar{S}$ is large (in section 14.3.2) or small (in section 14.3.3). Large and small capacities are endogenous characteristics of the sink that depend on all the other fundamentals of the model.

14.3.1 Testing the Optimality of the Sequestration Opportunity

Let us consider the optimal paths determined in section 14.2.2 under a forced condition of no sequestration, and assume two hypothetical values c'_s and c''_s (where $c'_s < c''_s$) of the marginal cost of sequestration. We assume that c'_s is so low that, at t_1^{ns}, $c_x + \lambda_0^{ns} e^{\rho t} + c'_s \zeta$ is lower than $\bar{p}_x$, while c''_s is so high that, at t_1^{ns}, $c_x + \lambda_0^{ns} e^{\rho t} + c''_s \zeta$ is higher than $\bar{p}_x$, as illustrated in figure 14.1.

In the first case where $c_x + \lambda_0^{ns} e^{\rho t_1^{ns}} + c'_s \zeta < \bar{p}_x$, there is a certain time interval $[t_1^{ns}, t_1^{ns} - \delta(c'_s))$, $\delta(c'_s) > 0$, during which $c_x + \lambda_0^{ns} e^{\rho t} + c'_s \zeta$ is lower than $\bar{p}_x$ while $Z_t = \bar{Z}$. Thus, over this interval, the instantaneous marginal gross surplus generated by $\bar{x}$, i.e. $\bar{p}_x$, is higher than the full marginal cost of supplying a "clean" coal to the final users (i.e. $c_x + \lambda_0^{ns} e^{\rho t} + c'_s \zeta$) provided that the shadow cost charged to the use of the sink is negligible. Whatever the capacity of the sink, this capacity will not be saturated if the rate of sequestration is sufficiently low. Hence slightly augmenting the coal consumption within the interval would still allow the net social welfare to increase even if coal consumption must be reduced later. For example, increasing the coal consumption by $dx_t > 0$ at time t within the interval and decreasing it by the same amount at some date t' within the interval (t_2^{ns}, t_3^{ns}) results, in value at time 0, in a net benefit equal to $[(\bar{p}_x - (c_x + c'_s \zeta))e^{-\rho t} - \lambda_0^{ns}] dx_t > 0$.

In the second case, the marginal cost of clean coal consumption, $c_x + \lambda_0^{ns} e^{\rho t} + c''_s \zeta$, is always higher than the marginal gross surplus of the energy consumption. Thus a sequestration scheme cannot increase the optimized value of the objective function of problem (**P**). Clearly, there exists some critical value of the sequestration marginal cost, denoted by $\bar{c}_s$, below which relaxation of the ceiling constraint must be used, and above which it must be abandoned. This threshold value is the solution of $c_x + \lambda_0^{ns} e^{\rho t_1^{ns}} + c_s \zeta = \bar{p}_x$, that is $\bar{c}_s = [\bar{p}_x - (c_x + \lambda_0^{ns} e^{\rho t_1^{ns}})]/\zeta$.

ASSUMPTION 8 $c_s < \bar{c}_s$.

In the following discussion we assume that assumption 8 applies.

14.3.2 The Large Reservoir Case

In the case where no rent is charged for the use of the sink capacity, the reservoir capacity constraint $\bar{S} - S_t$ is not active. Hence $\eta_t = 0$, $t \geq 0$.

Note that because $\eta_t = 0$ and $s_t > 0$ imply that $\gamma_{st} = 0$, the optimality condition (14.7) becomes $-\mu_t = c_s + \bar{\gamma}_{st}$. Next, substituting this value of $-\mu_t$ into the optimality condition (14.8), given that $s_t > 0$ implies $x_t > 0$, we have $\gamma_{xt} = 0$. We thus obtain

$$u'(x_t) = c_x + c_s \zeta + \lambda_0 e^{\rho t}, \tag{14.28}$$

and this in turn implies that $x_t = d(c_x + c_s\zeta + \lambda_0 e^{\rho t})$. However, only part of the emission flow, represented by $\zeta[d(c_x + c_s\zeta + \lambda_0 e^{\rho t}) - \bar{x}]$, has to be sequestrated. When the ceiling constraint is binding, the instantaneous marginal full cost of a clean unit of coal is denoted by c_m:

$$c_m = \begin{cases} c_x + \lambda_0 e^{\rho t}, & x_t < \bar{x} \\ c_x + \lambda_0 e^{\rho t} + c_s \zeta, & x_t > \bar{x} \end{cases}. \tag{14.29}$$

For $x < \bar{x}$, the regeneration rate $\alpha \bar{Z}$ is higher than the emission rate, whereas the opposite holds for $x > \bar{x}$. Then society has to sequestrate emission at the margin in the large reservoir case. The determination of x_t during the sequestration phase is illustrated in figure 14.2. As the figure shows, with the passage of time, $c_x + \lambda_0 e^{\rho t}$ and $c_x + c_s\zeta + \lambda_0 e^{\rho t}$ are shifted vertically and upward. Hence the sequestration phase is necessarily followed by a phase during which $c_x + \lambda_0 e^{\rho t} < u'(\bar{x}) < c_x + c_s\zeta + \lambda_0 e^{\rho t}$. During this phase it becomes optimal to consume $\bar{x}$. Although a ceiling is constraining on coal consumption remains in effect, it is no longer optimal to sequester carbon emissions. This phase is fol-

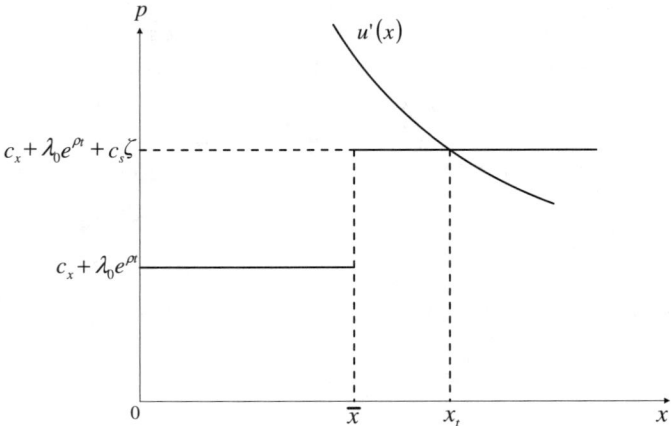

Figure 14.2
Determination of the optimal consumption rate of coal during the sequestration phase

Chapter 14 Optimal Sequestration Policy with Ceiling on Carbon

lowed by a pure Hotelling phase during which $x_t < \bar{x}$, once t is sufficiently high so that the $u'(x)$ curve intersects the horizontal line $c_x + \lambda_0 e^{\rho t}$ before $\bar{x}$. The optimal path therefore consists of five phases as illustrated in Figure 14.3.

During the first phase $[0, t_1)$, which takes place below the ceiling, $p_t = c_x + \lambda_0 e^{\rho t} - \mu_0 e^{(\alpha+\rho)t}\zeta < c_x + \lambda_0 e^{\rho t} + c_s \zeta < \bar{p}_x$ and $q_t = x_t = d(p_t) > \bar{x}$, so the pollution stock is increasing. At the end of the phase, $p_t = c_x + \lambda_0 e^{\rho t} + c_s\zeta$ and the ceiling is attained. The second phase $[t_1, t_2)$ is a phase at the ceiling: $p_t = c_x + \lambda_0 e^{\rho t} + c_s\zeta < \bar{p}_x$ and $q_t = x_t = d(p_t) > \bar{x}$. A part $\zeta[d(p_t) - \bar{x}]$ of the potential emission flow is sequestered so that the pollution flow is equal to $\zeta\bar{x}$. At the end of the phase, $p_t = \bar{p}_x$. The third phase $[t_2, t_3)$ is at the ceiling, but without sequestration: $p_t = \bar{p}_x$ and $q_t = x_t = \bar{x}$, during which $|\mu_t|$ is decreasing. At the end of the phase, $\mu_t = 0$. The fourth phase $[t_3, t_4)$ is a pure Hotelling phase: $p_t = c_x + \lambda_0 e^{\rho t}$ and $q_t = x_t = d(p_t)$. At the end of this phase the price of energy is just equal to the marginal cost of the renewable resource c_y and the coal is exhausted. The last phase $[t_4, \infty)$ is a phase during which the only renewable resource is used.

We now need to determine the values of the six variables $\lambda_0, \mu_0, t_1, t_2, t_3$, and t_4. The values are obtained by solving a system of six equations.

The first equation is the cumulative demand-supply balance equation, written here as

$$\int_0^{t_1} d(c_x + \lambda_0 e^{\rho t} - \mu_0 e^{(\alpha+\rho)t}\zeta)\,dt + \int_{t_1}^{t_2} d(c_x + \lambda_0 e^{\rho t} + c_s\zeta)\,dt + [t_3 - t_2]\bar{x}$$

$$+ \int_{t_3}^{t_4} d(c_x + \lambda_0 e^{\rho t})\,dt = X^0.$$

The second equation is the price continuity equation at t_1:

$$c_x + \lambda_0 e^{\rho t_1} - \mu_0 e^{(\alpha+\rho)t_1}\zeta = c_x + \lambda_0 e^{\rho t_1} + c_s\zeta.$$

The third equation is the pollution stock continuity equation at t_1:

$$Z_{t_1}(\lambda_0, \mu_0) = \bar{Z}.$$

The fourth through sixth equations are the price continuity equations at t_2, t_3, and t_4:

$$c_x + \lambda_0 e^{\rho t_2} + c_s\zeta = \bar{p}_x, \quad c_x + \lambda_0 e^{\rho t_3} = \bar{p}_x, \quad \text{and} \quad c_x + \lambda_0 e^{\rho t_4} = c_y.$$

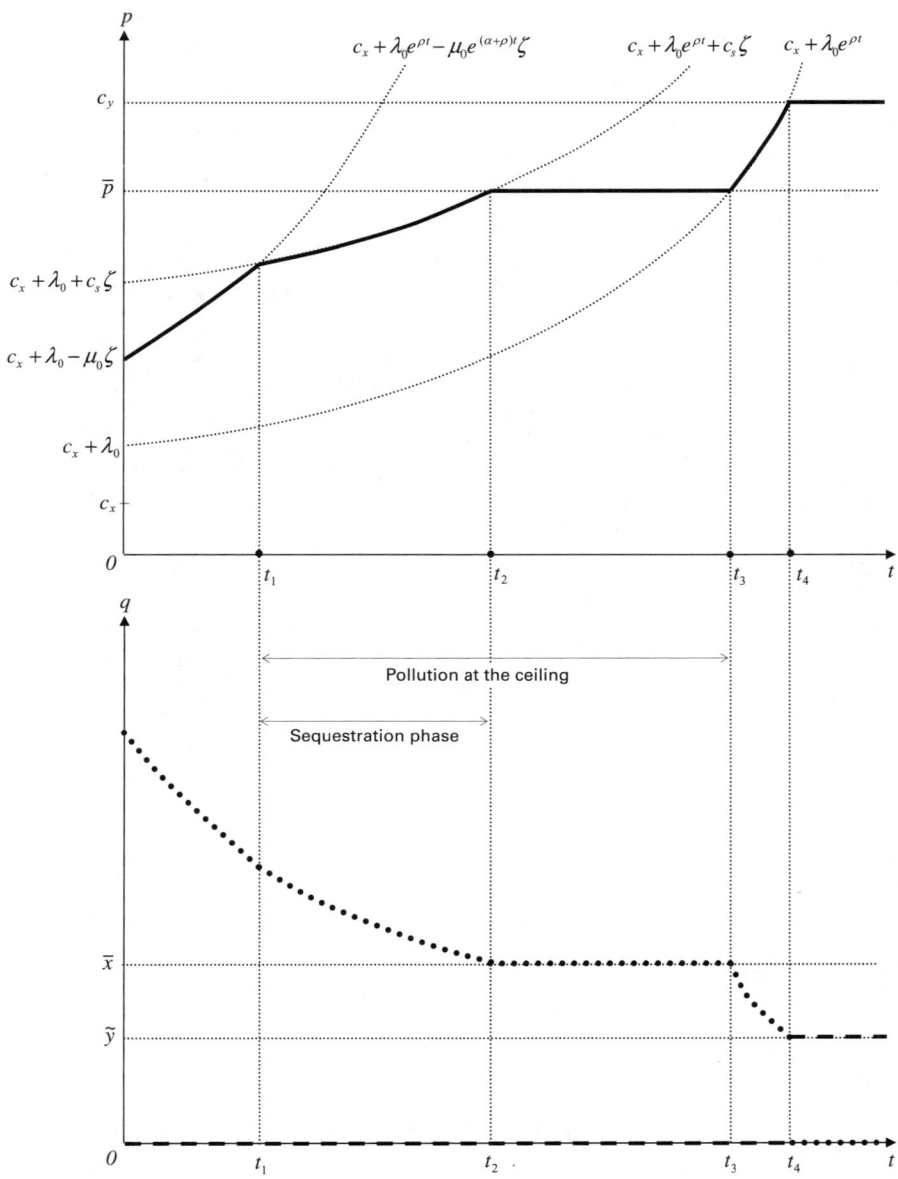

Figure 14.3
Optimal paths: Large reservoir case

Let λ_0^{lr}, μ_0^{lr}, and t_1^{lr} to t_4^{lr} (where *lr* stands for the large reservoir) represent the values obtained by solving the system of equations given above. In the appendix we show that for these values of λ_0, μ_0, t_1 to t_4, the other multipliers have values satisfying all the optimality conditions. We can now give a precise definition of a large reservoir: a reservoir or a sink is said to be large if it allows for the carbon to be effectively sequestered as prescribed by the policy above, that is,

$$\bar{S} \geq \int_{t_1^{lr}}^{t_2^{lr}} [d(c_x + \lambda_0^{lr} e^{\rho t} + c_s \zeta) - \bar{x}] \, dt.$$

The reservoir will be considered small if such a carbon mass cannot be sequestered.

14.2.3 The Small Reservoir Case

If the reservoir is small, its shadow cost η_t cannot be regarded as negligible. We know that as long as the reservoir is not saturated, the absolute value of η_t is increasing at the social rate of discount[14]: $\eta_t = \eta_0 e^{\rho t}$. Thus the full marginal cost of clean coal is given by $c_x + \lambda_0 e^{\rho t} + (c_s - \eta_0 e^{\rho t})\zeta$.

As in the case of a large reservoir, the optimal path consists of five phases. The only difference is that $p_t = c_x - \lambda_0 e^{\rho t} + (c_s - \eta_0 e^{\rho t})\zeta$ during the second phase $[t_1, t_2)$, when it is optimal to sequester part of the potential emission flow represented by $\zeta[d(p_t) - \bar{x}]$. Also at t_2, the carbon reservoir capacity $\bar{S}$ must be saturated, that is, $S_{t_2} = \bar{S}$ and $S_t < \bar{S}$, $t < t_2$. We now have to determine the values of the seven variables λ_0, μ_0, η_0, and t_1 to t_4. These variables are obtained by solving a system of seven equations, as shown below.

First is the cumulative demand-supply balance equation:

$$\int_0^{t_1} d(c_x + \lambda_0 e^{\rho t} - \mu_0 e^{(\alpha+\rho)t}\zeta) \, dt + \int_{t_1}^{t_2} d(c_x + \lambda_0 e^{\rho t} + (c_s - \eta_0 e^{\rho t})\zeta) \, dt$$

$$+ [t_3 - t_2]\bar{x} + \int_{t_3}^{t_4} d(c_x + \lambda_0 e^{\rho t}) \, dt = X^0.$$

Second is the price continuity equation at t_1

$$c_x + \lambda_0 e^{\rho t_1} - \mu_0 e^{(\alpha+\rho)t_1}\zeta = c_x + \lambda_0 e^{\rho t_1} + (c_s - \eta_0 e^{\rho t_1})\zeta.$$

Third is the pollution stock continuity equation at t_1, which is the same as in the previous case of a large reservoir:

$$Z_{t_1}(\lambda_0, \mu_0) = \bar{Z}.$$

Fourth is the price continuity equation at t_2:

$$c_x + \lambda_0 e^{\rho t_2} + (c_s - \eta_0 e^{\rho t_2})\zeta = \bar{p}_x.$$

Fifth is the reservoir capacity saturation equation at t_2:

$$\zeta \int_{t_1}^{t_2} [d(c_x + \lambda_0 e^{\rho t} + (c_s - \eta_0 e^{\rho t})\zeta) - \bar{x}] \, dt = \bar{S}.$$

Sixth and seventh are the price continuity equations at t_3 and t_4, which are the same as those used for the large reservoir:

$$c_x + \lambda_0 e^{\rho t_3} = \bar{p}_x \quad \text{and} \quad c_x + \lambda_0 e^{\rho t_4} = c_y.$$

Similarly for these values of λ_0, μ_0, η_0, and t_1 to t_4 we show that the other multipliers have values satisfying all the optimality conditions (see the appendix). The main conclusion of the analysis is that if sequestration needs to be implemented by assumption 5, it must occur before the renewable resource is used. As discussed in the next section, the optimal policy may be different where solar energy—although relatively inexpensive—is not abundant.

14.4 The Case of a Rare Renewable Substitute

Let us assume that the renewable energy is not abundant and that assumption 6 is valid. In that case, absent the opportunity to sequester, it would be optimal to use the renewable resource at the ceiling. We first show how to modify the optimality test of the sequestration option. Since the test is positive, indicating it is optimal to sequester, we next show that there are two types of optimal policy, according to whether the sequestration phase should begin before starting use of the renewable clean substitute. Otherwise, the opposite must apply.

14.4.1 Testing the Optimality of the Sequestration Opportunity

To test the optimality of the sequestration opportunity in the present case, we start by determining the optimal policy in the absence of an

Chapter 14 Optimal Sequestration Policy with Ceiling on Carbon

opportunity. In the case of a rare renewable substitute, the decision whether to sequester is endogenously determined. The first point to be noticed is that since $\bar{y} < \tilde{y}$, although the renewable resource is competitive at a price $p_e > c_y$, it cannot supply the entire market, at least for prices not too far from c_y. To determine the optimal policy, let us define $d_n(p_e)$ as that part of the energy need that must be supplied by the nonrenewable resource. The other part, should any exist, is represented by $d(p_e) - d_n(p_e)$; it has to be supplied by the renewable resource:

$$d_n(p_e) = \begin{cases} d(p_e), & p_e < c_y \\ d(p_e) - \bar{y}, & c_y \le p_e < \bar{p} \\ 0, & \bar{p}_y \le p_e \end{cases}.$$

By assumption 6, $\bar{p}_{xy} > c_y$, which means we can have two types of optimal paths according to the value of Z^0. Along the first type of path (see figure 14.4), which would appear only for sufficiently low values of Z^0, the initial price p_0 is lower than c_y. Hence we have the initial period at which $p_t = c_x + \lambda_0 e^{\rho t} - \mu_0 e^{(\alpha+\rho)t}$ and $Z_t < \bar{Z}$ are split into two phases. During this initial phase $[0, t_1)$, with $p_t < c_y$, coal as the only energy source, and $q_t = x_t = d_n(p_t) = d(p_t)$. On the other hand, in the second phase $[t_1, t_2)$, both coal and solar energy must be used, with $x_t = d_n(p_t)$ and $y_t = \bar{y}$. The phase at the ceiling $[t_2, t_3)$ begins at t_2, where $p_t = \bar{p}_{xy}$; throughout this phase, $p_t = \bar{p}_{xy}$ and $q_t = \bar{x} + \bar{y}$ while, at the end, $\mu_t = 0$. This phase is followed by a pure Hotelling phase $[t_3, t_4)$ as far as prices are concerned: $p_t = c_x + \lambda_0 e^{\rho t}$, $x_t = d_n(p_t)$ and $y_t = \bar{y}$. At t_4, $p_t = \bar{p}_y$, $x_t = 0$ and coal is exhausted. In the last phase $[t_4, \infty)$, only the renewable resource is available, so $p_t = \bar{p}_y$ and $q_t = y_t = \bar{y}$.

The values of the six endogenous variables characterizing this type of path are determined by solving a six-equation system.

One is the cumulative coal consumption–initial stock balance equation, written here as

$$\int_0^{t_2} d_n(c_x + \lambda_0 e^{\rho t} - \mu_0 e^{(\alpha+\rho)t} \zeta) \, dt + \bar{x}[t_3 - t_2] + \int_{t_3}^{t_4} d_n(c_x + \lambda_0 e^{\rho t}) \, dt = X^0.$$

Another is the pollution stock continuity equation at t_2:

$$Z_{t_2}^n(\lambda_0, \mu_0) = \bar{Z},$$

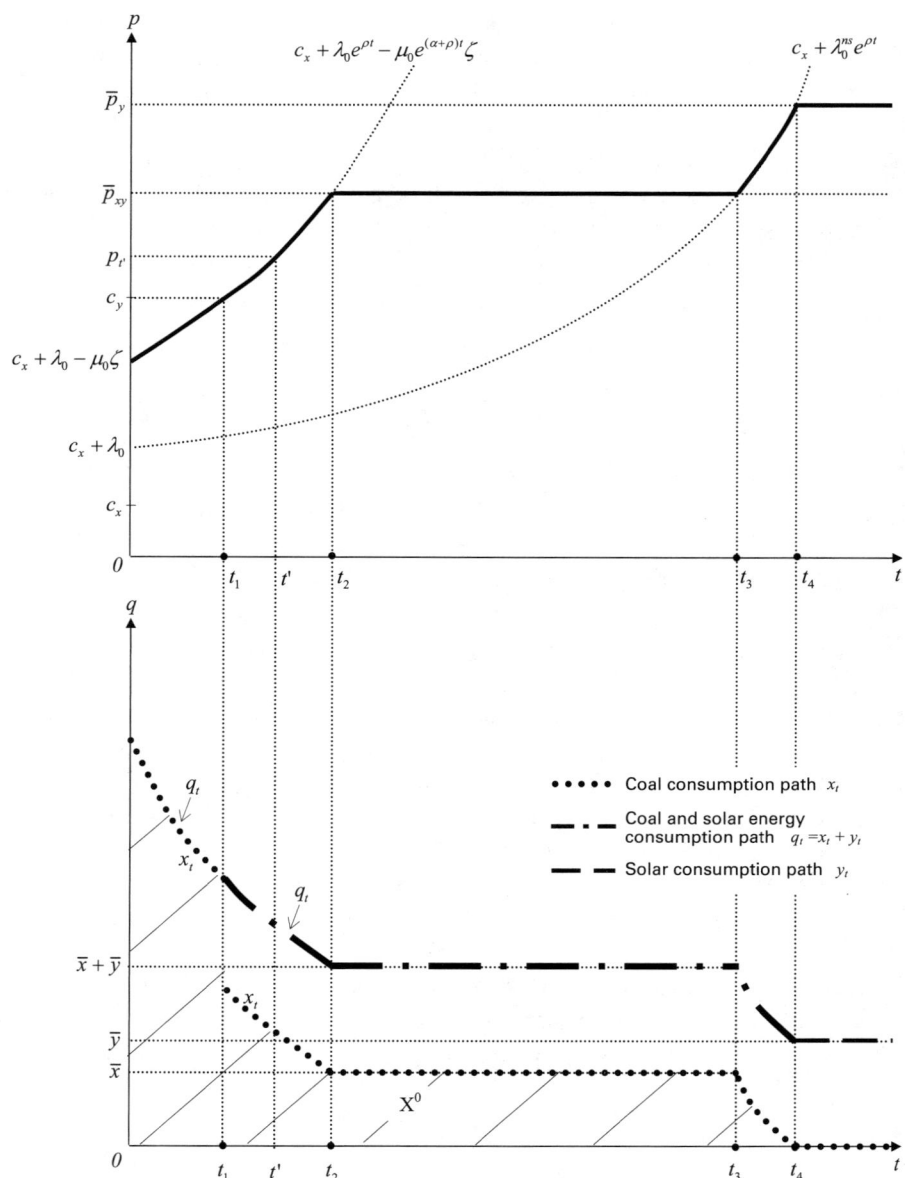

Figure 14.4
Optimal paths: No sequestration allowed, rare renewable substitute

Chapter 14 Optimal Sequestration Policy with Ceiling on Carbon

where $Z_{t_2}^n(\lambda_0, \mu_1)$ is the solution of $dZ_t/dt = \zeta d_n(c_x + \lambda_0 e^{\rho t} - \mu_0 e^{(\alpha+\rho)t}\zeta) - \alpha Z_t$,[15] $Z_0 = Z^0$.

Then there are the four price-continuity equations:

$$c_x + \lambda_0 e^{\rho t_1} - \mu_0 e^{(\alpha-\rho)t_1}\zeta = c_y,$$

$$c_x + \lambda_0 e^{\rho t_2} - \mu_0 e^{(\alpha+\rho)t_2}\zeta = c_x + \lambda_0 e^{\rho t_3} = \bar{p}_{xy},$$

$$c_x + \lambda_0 e^{\rho t_4} = \bar{p}_y.$$

The corresponding optimal paths are illustrated in figure 14.4, where the hatched surface under the curve x_t is equal to X^0.

Now, let us assume that $\{(p_t, x_t, y_t), t \geq 0\}$ is an optimal path for initial values X^0 and Z^0 of the state variables, and X_t and Z_t represent, respectively, the remaining coal stock and the pollution stock generated by x_t. We also assume a given date $t' > 0$ and problem (P) with initial conditions $X_{t'}$ and $Z_{t'}$. Let $\{(p_t', x_t', y_t'), t \geq 0\}$ be the solution of this new problem. Then this solution simply corresponds to $p_t' = p_{t'+t}$, $x_t' = x_{t'+t}$, and $y_t' = y_{t'+t}$. With this connection in mind, we see that there is a second type of optimal path starting for $p_0 > c_y$, where the renewable resource must be used from the start. These paths have only four distinct phases, since there is no longer any first phase as defined in the previous five-phase optimal path model. Clearly, the second type of path is optimal under assumption 6 when Z^0 is sufficiently high but nevertheless lower than $\bar{Z}$.

Let $\bar{t}_{xy}$ denote the time at which $p_t = \bar{p}_{xy}$. Along paths of the first type, $\bar{t}_{xy} = t_2$ at the end of the second phase, whereas along paths of the second type, $\bar{t}_{xy} = t_1$, since $\bar{p}_{xy}$ is reached at the end of the first phase $[0, t_1)$. The optimality test of the sequestration option is the same as for an abundant renewable substitute, except that $\bar{p}_{xy}$ must be taken here as the reference price instead of $\bar{p}_x$. The corresponding curve $c_x + \lambda_0 e^{\rho t} + c_s \zeta$ must be located at $t = \bar{t}_{xy}$, as illustrated in figure 14.4.

The threshold value $\bar{c}_s$ represents the average sequestration cost below which it is optimal to sequester and above which it is not. That cost is here equal to

$$\bar{c}_s = [\bar{p}_{xy} - (c_x + \lambda_0^{ns} e^{\rho \bar{t}_{xy}})]/\zeta,$$

where λ_0^{ns} is the optimal value of λ_0 under an enforced no-sequestration policy.

ASSUMPTION 9 $c_s < \bar{c}_s$.

In the following discussion, we assume that assumption 9 is valid.

14.4.2 Optimal Paths for Beginning Sequestration before Using the Renewable Substitute

By assumption 6, when sequestration is not applied, use of the renewable resource must always begin before the ceiling is reached, as shown in the preceding section. However, when sequestration has to be used, it may happen that, at the ceiling, sequestration may be the necessary scheme to first apply in order to relax the ceiling constraint. Then sequestration and the renewable substitute may be used jointly, and, finally, only the renewable resource on its own. This case is illustrated in figure 14.5, in the case of a small reservoir.

In figure 14.5 the price path, the demand functions $d(p_e)$ and $d_n(p_e)$, and the resource consumption paths are drawn in the northeast, the northwest, and the southeast quadrants, respectively. The southwest quadrant is a purely technical device to show how the quantities are derived from the price at the same time.

The optimal path consists of six phases. $[0, t_1)$ is the usual first phase of coal consumption under the ceiling: $p_t = c_x + \lambda_0 e^{\rho t} - \mu_0 e^{(\alpha+\rho)t}\zeta < c_x + \lambda_0 e^{\rho t} + (c_s - \eta_0 e^{\rho t})\zeta$. At the end of this phase, the ceiling is reached and $p_{t_1} = c_x + \lambda_0 e^{\rho t_1} + (c_s - \eta_0 e^{\rho t_1})\zeta$.

The second phase $[t_1, t_2)$ takes place at the ceiling, during which time only coal is consumed, with some part of the potential emissions being sequestered: $p_t = c_x + \lambda_0 e^{\rho t} + (c_s - \eta_0 e^{\rho t})\zeta$, $x_t = d_n(p_t) = d(p_t)$, and $s_t = [d_n(p_t) - \bar{x}]/\zeta$. At the end of this phase, $p_{t_2} = c_y$ and the renewable energy becomes competitive.

In the third phase $[t_2, t_3)$, the constraint is relaxed by the joint use of sequestration and solar energy consumption: $p_t = c_x + \lambda_0 e^{\rho t} + (c_s - \eta_0 e^{\rho t})\zeta$. Then, because some part of the energy demand is satisfied by the solar energy, the proportion of the emission flow that has to be sequestered is lower than in the previous phase: $x_t = d_n(p_t) = d(p_t) - \bar{x}$, $s_t = [d_n(p_t) - \bar{x}]/\zeta$, $y_t = \bar{y}$. At the end of the phase, $p_{t_3} = \bar{p}_{xy}$ and the capacity of the sink is saturated, $S_{t_3} = \bar{S}$, so sequestration is no longer of any help.

The fourth phase $[t_3, t_4)$ is at the ceiling with both coal and solar energy, but without sequestration: $p_t = \bar{p}_{xy}$, $x_t = \bar{x}$, and $y_t = \bar{y}$. At the end of this phase, $\mu_{t_4} = 0$.

Chapter 14 Optimal Sequestration Policy with Ceiling on Carbon

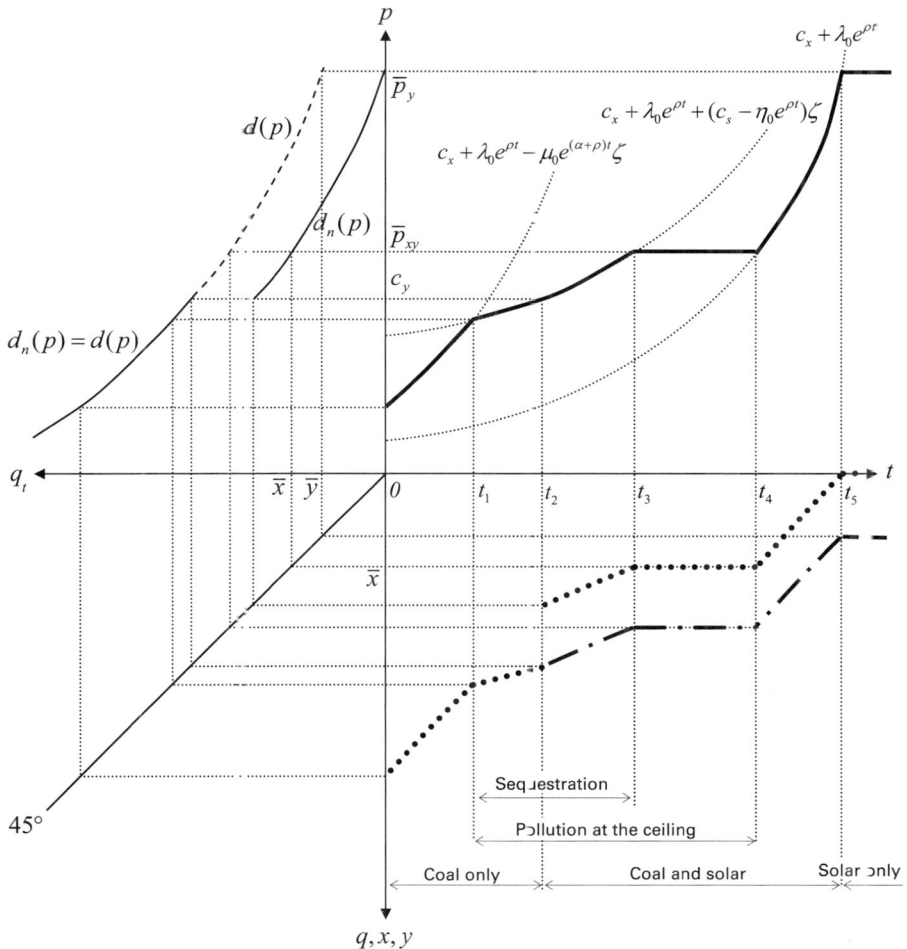

Figure 14.5
Optimal paths with a rare renewable resource: Sequestration implemented before the renewable substitute

In the fifth phase, the price path is a pure Hotelling path, $p_t = c_x + \lambda_0 e^{\rho t}$, but with part $\bar{y}$ of the energy consumption supplied by the renewable resource. The contribution from nonrenewable energy $d_n(p_t)$ decreases to zero at the end of the phase, so the price p_{t_5} is equal to $\bar{p}_y$. At this time, coal is exhausted.

The last phase $[t_5, \infty)$ is the usual phase of renewable energy consumption: $p_t = \bar{p}_y$, $x_t = 0$, and $y_t = \bar{y}$. Note that now a rent has to be borne by the user of the renewable energy, starting from 0 at $t = t_2$ and increasing up to $\bar{p}_{xy} - c_y$ at $t = t_3$, and remaining constant at this value during $[t_3, t_4)$, increasing again up to $\bar{p}_y - c_y$ during $[t_4, t_5)$, and afterward remaining constant at this level. Such a path is characterized by values of the eight variables λ_0, μ_0, η_0, and t_1 to t_5, obtained by solving a system of eight equations similar to the preceding systems.

14.4.3 Optimal Paths for Beginning Use of the Renewable Substitute before Sequestration

Figure 14.6 illustrates the case where implementation of the renewable resource must begin before resorting to sequestration, assuming that the reservoir is small. As in the preceding case, the optimal path consists of six phases.

In the first phase $[0, t_1)$, only coal is used and $p_t = c_x + \lambda_0 e^{\rho t} - \mu_0 e^{(\alpha+\rho)t}\zeta$, $q_t = x_t = d_n(p_t) = d(p_t)$. At the end of the phase, $p_t = c_y$, so the renewable substitute becomes competitive as the ceiling constraint is relaxed, $Z_{t_1}(\lambda_0, \mu_0) < \bar{Z}$. In the second phase $[t_1, t_2)$, both coal and solar energy are used, but the expression of the price remains the same because the pollution stock Z_t stays below $\bar{Z}$, though now with $x_t = d_n(p_t) < d(p_t)$ and $y_t = \bar{y}$. At the end of this phase, the pollution constraint becomes binding and $c_x + \lambda_0 e^{\rho t_2} - \mu_0 e^{(\alpha+\rho)t_2}\zeta = c_x + \lambda_0 e^{\rho t_2} + (c_s - \eta_0 e^{\rho t_2})\zeta$. The last four phases are similar to the last four of the preceding case.

It is possible for the renewable resource to be applied at the paths right at time $t = 0$. This scenario would correspond to optimal path scheme where the initial values X^0 and Z^0 of the state variables X_t and Z_t equal their values at $t' \in [t_1, t_2)$ as in the path illustrated in figure 14.6. The argument for this scenario is based on the same concept of time consistency developed in section 14.4.1.

Chapter 14 Optimal Sequestration Policy with Ceiling on Carbon 299

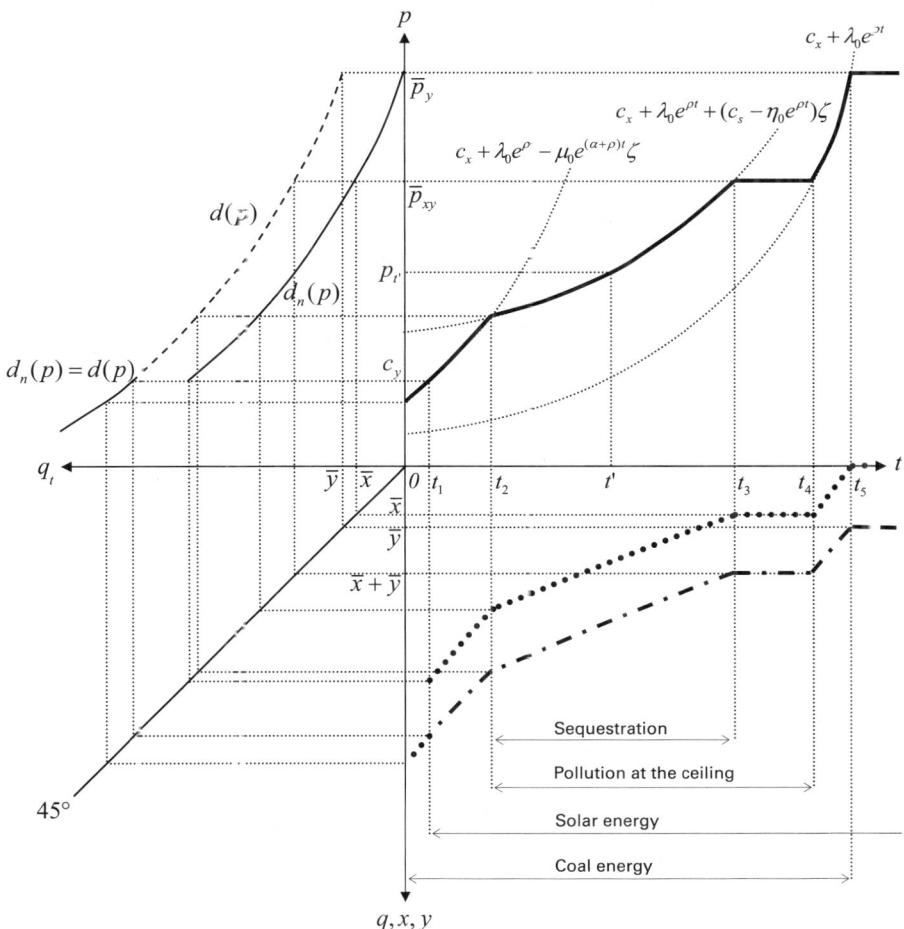

Figure 14.6
Optimal paths with a rare renewable resource: Sequestration implemented after the renewable substitute

14.5 Conclusion

We have examined the potential use of carbon sequestration in environmental policy to maintain the atmospheric carbon concentration below some threshold level. We showed that whatever the sink capacity, sequestration must be implemented once the pollution ceiling is reached. In addition our analysis suggested that the polluting fossil fuel will be exhausted in a finite time, after which the market will have to be supplied by a renewable substitute, whether or not the fossil fuel's capacity is constrained.[16] However, in contrast to other cases, where the capacity of the renewable resource flow is constrained, and is initially very affordable, the renewable resource must be used before the ceiling is reached. In that scenario the renewable resource could be regarded as a midterm option for alleviating pollution, while sequestration allows for further emission reductions in the longer term.

We showed that, more generally, the lack of sequestration before the ceiling is reached should not be seen as weakening the preventive short-term role of sequestration usually advocated as an option for such a climate change mitigation. Indeed, whether or not the renewable resource is scarce, the optimal environmental policy affects the extraction of the exhaustible resource anyway, with extraction decreasing until the pollution ceiling is reached. This reduction in consumption should be attributed to theopportunity cost of emitting one unit of carbon before the ceiling, as well as the opportunity cost of sequestering one unit once at the ceiling, these costs being added to the overall exploitation cost of the resource.

Last we discussed what we did not consider in our model: the possibility of carbon leakage. This is because geological or even oceanic sinks appear only to represent temporary storage options (see Herzog et al. 2003; Paccala 2003). The leakage phenomenon, if continuous over time, would have no short-term incidence on the optimal solution. The phase with sequestration on its own would then be extended to the entire pollution ceiling phase: once the storage capacity has been filled, sequestration would simply allow for compensating the leakage at each moment of time.

Appendix

Let $p_t^H = c_x + \lambda_0 e^{\rho t}$ be the Hotelling price path, $\hat{p}_t = p_t^H - \mu_0 e^{(\alpha+\rho)t}\zeta$ the optimal price path preceding the ceiling, and $\tilde{p}_t = p_t^H + (c_s - \eta_0 e^{\rho t})\zeta$

Chapter 14 Optimal Sequestration Policy with Ceiling on Carbon

the optimal price path followed within the sequestration phase. In the small reservoir case the complete solution of (**P**) is[17]

$$x_t = \begin{cases} d(\hat{p}_t), & t \in [0, t_1) \\ d(\tilde{p}_t), & t \in [t_1, t_2) \\ \bar{x}, & t \in [t_2, t_3) \\ d(p_t^H), & t \in [t_3, t_4) \\ 0, & t \in [t_4, \infty) \end{cases}, \quad y_t = \begin{cases} 0, & t \in [0, t_4) \\ \tilde{y}, & t \in [t_4, \infty) \end{cases},$$

$$s_t = \begin{cases} 0, & t \notin [t_1, t_2) \\ \zeta[d(\tilde{p}_t) - \bar{x}], & t \in [t_1, t_2) \end{cases}. \tag{14.30}$$

The associated Lagrange multipliers are

$$\gamma_{st} = \begin{cases} c_s + \mu_0 e^{(\alpha + \rho)t} - \eta_0 e^{\rho t}, & t \in [0, t_1) \\ 0, & t \in [t_1, t_2) \\ c_s + (p_t^H - \bar{p}_x)/\zeta, & t \in [t_2, t_3) \\ c_s, & t \in [t_3, \infty) \end{cases}, \quad \bar{\gamma}_{st} = 0, \, t \geq 0 \tag{14.31}$$

$$\gamma_{xt} = \begin{cases} 0, & t \in [0, t_4) \\ p_t^H - c_y, & t \in [t_4, \infty) \end{cases} \tag{14.32}$$

$$\gamma_{yt} = \begin{cases} c_y - \hat{p}_t, & t \in [0, t_1) \\ c_y - \tilde{p}_t, & t \in [t_1, t_2) \\ c_y - \bar{p}_x, & t \in [t_2, t_3) \\ c_y - p_t^H, & t \in [t_3, t_4) \\ 0, & t \in [t_4, \infty) \end{cases} \tag{14.33}$$

$$\eta_t = \begin{cases} \eta_0 e^{\rho t}, & t \in [0, t_2) \\ 0, & t \in [t_2, \infty) \end{cases}, \quad \nu_{St} = 0, \, t \geq 0 \tag{14.34}$$

$$\mu_t = \begin{cases} \mu_0 e^{(\alpha + \rho)t}, & t \in [0, t_1) \\ \eta_0 e^{\rho t} - c_s, & t \in [t_1, t_2) \\ (p_t^H - \bar{p}_x)/\zeta, & t \in [t_2, t_3) \\ 0, & t \in [t_3, \infty) \end{cases} \tag{14.35}$$

$$\nu_{Zt} = \begin{cases} 0, & t \in [0, t_1) \\ (\alpha + \rho)c_s - \alpha \eta_0 e^{\rho t}, & t \in [t_1, t_2) \\ [dp_t^H/dt - (\alpha + \rho)(p_t^H - \bar{p}_x)]/\zeta, & t \in [t_2, t_3) \\ 0, & t \in [t_3, \infty) \end{cases} \tag{14.36}$$

Given (14.30), it is easy to check that if λ_0, μ_0, η_0, t_1, t_2, t_3, and t_4 satisfy the system of seven equations described in subsection 14.3.3, then the Lagrange multipliers defined by (14.31)–(14.36) are such that conditions (14.7)–(14.19) hold. In other respects, since the non-renewable resource stock X_t is exhausted at t_4, the transversality condition (14.20) is satisfied. In the same way, since $\eta_t = 0$ and $S_t = \bar{S}$ for $t \geq t_2$ on the one hand, $\mu_t = 0$ and $Z_t < \bar{Z}$ for $t \geq t_4$ on the other hand, then conditions (14.21) and (14.22) are true.

Notes

The authors wish to thank an anonymous referee for providing valuable comments.

1. See Chakravorty et al. (2008) for more on the substitutions between fossil and bioenergies.

2. A carbon sink is a form of biological sequestration, since the carbon is captured by photosynthesis of trees or plants, but this is not the focus of the present chapter.

3. Carbon sequestration in partially or completely depleted oil deposits has been carried out in the North Sea since 1996 by the Norwegian company Statoil. The enhanced oil recovery consists in injecting gas into the oil well, thus increasing pressure and improving the extraction productivity.

4. Oceanic storage, despite its enormous potential as a sink, raises several issues concerning sequestration permanence and acceptability as the acidification of water that would result is toxic to marine ecosystems (see Herzog et al. 2003).

5. For the much more complex case of nonstationary demand functions, see Chakravorty et al. (2008).

6. Strictly speaking, q_t is a power, so assuming that q_t is differentiable, the energy consumed over a time interval $[t, t+dt]$ is equal to $(q_t + dq_t/dt)\, dt$.

7. It is essential for the results that the natural regeneration flow should be some increasing function f of the pollution stock. The specification $f(Z) = \alpha Z$ is assumed for the sake of analytical tractability. For a discussion of the problems raised by nonincreasing functions, see Tahvonen and Salo (1996), Tahvonen and Withagen (1996), and Toman and Withagen (1996).

8. For standard results on optimal mining with a smooth damage function, see Tahvonen (1997).

9. An important exception is hydroelectricity.

10. If $y_t > 0$ then $\gamma_{yt} = 0$, thus $u'(x_t + y_t) = c_y + \bar{\gamma}_{yt}$. Since u' is decreasing, the highest value of y_t solution of (14.9) is obtained for $x_t = 0$ and for $\bar{\gamma}_{yt} = 0$, which is possible under the abundance assumption 5. Thus in this case y_t is precisely equal to $\bar{y}$.

11. First, $t^H(\lambda_0)$ is a strictly decreasing function of λ_0 with $\lim_{\lambda_0 \downarrow 0} t^H = \infty$ and $\lim_{\lambda_0 \uparrow (c_y - c_x)} t^H = 0$, and second, for any $t \geq 0$, $\lim_{\lambda_0 \downarrow 0} d_t^H = \tilde{x}$ and $\lim_{\lambda_0 \uparrow (c_y - c_x)} d_t^H = 0$.

12. After $t^H(\lambda_0)$, use of coal is inconsequential; hence Z_t is decreasing.

13. For $\lambda_0 = 0$, $d_t^H(0) = \tilde{x}$, $t \geq 0$. Hence $dZ_t/dt = \zeta\tilde{x} - \alpha Z_t$, $Z_0 = Z^0$, yielding the solution $Z_t = \zeta\tilde{x}/\alpha + (Z^0 - \zeta\tilde{x}/\alpha)e^{-\alpha t}$, so that $\lim_{t\uparrow\infty} Z_t \equiv \tilde{Z} = \zeta\tilde{x}/\alpha$. By assumption 2, $\tilde{x} > \bar{x}$ hence $\tilde{Z} > \bar{Z}$.

14. Remember that $\eta_t < 0$.

15. Since $d_n(p_e)$ is discontinuous at $p_e = c_y$, then technically Z_t^n is obtained by solving first: $dZ_t/dt = \zeta d(c_x + \lambda_0 e^{\rho t} - \mu_0 e^{(\alpha+\rho)t}\zeta) - \alpha Z_t$, $Z_0 = Z^0$. Let $Z_t^{(1)}$ be the solution of this equation and $Z_{t_1}^{(1)}$ its value at time t_1 at which $c_x + \lambda_0 e^{\rho t} - \mu_0 e^{(\alpha+\rho)t}\zeta = c_y$. Next, by solving $dZ_t/dt = \zeta[d(c_x + \lambda_0 e^{\rho t} - \mu_0 e^{(\alpha+\rho)t}\zeta) - \bar{x}] - \alpha Z_t$, $Z_{t_1} = Z_t^{(1)}$, we obtain $Z_{t_1}^{(2)}$ as the solution of this equation. Then we can write

$$Z_t^n = \begin{cases} Z_{t_1}^{(1)}, & 0 \leq t \leq t_1 \\ Z_{t_1}^{(2)}, & t_1 \leq t \leq t_2 \end{cases}.$$

16. This resource exploitation and sequestration scheme, obtained with constant average costs, is robust to alternative specifications for the cost functions. For the sequestration or extraction costs (see Lafforgue et al. 2008, who also studied the case of multiple sequestration sinks), these specifications may depend on the cumulative sequestered carbon or the cumulative extracted fossil resource, respectively, as in Heal (1976).

17. For the large reservoir case, just set $\eta_t = 0$.

References

Chakravorty, U., B. Magné, and M. Moreaux. 2006a. Plafond de concentration atmosphérique en carbone et substitutions entre ressources énergétiques. *Annales d'Economie et de Statistique*, 81: 141–68.

Chakravorty, U., B. Magné, and M. Moreaux. 2006b A Hotelling model with a ceiling on the stock of pollution *Journal of Economic Dynamics and Control*, 30(12): 2875–2904.

Chakravorty, U., B. Magné, and M. Moreaux. 2008. A dynamic model of food and clean energy. *Journal of Economic Dynamics and Control* 32(4): 1181–1203.

Edmonds, J., J. Clarke, J. Dooley, S. H. Kim, and S. J. Smith. 2004. Stabilization of CO_2 in a B2 world: Insights on the roles of carbon capture and disposal, hydrogen, and transportation technologies. *Energy Economics* 26(4): 517–37.

Gitz, V., F. Ambrosi, B. Magné, and P. Ciais. 2005. *Is There an Optimal Timing for Sequestration to Stabilize Future Climate?* Washington, DC: American Geophysical Union.

Ha Duong, M., and D. W. Keith. 2003. Carbon storage: the economic deficiency of storing CO_2 in leaky reservoirs. *Clean Technologies and Environmental Policy* 5: 181–89.

Heal, G. 1976. The relationship between price and extraction cost for a resource with a backstop technology. *Bell Journal of Economics* 7(2): 371–78.

Herfindahl, O. L. 1967. Depletion and economic theory. In M. Gaffiney, ed., *Extractive Resources and Taxation*. Madison: University of Wisconsin Press, pp. 63–90.

Herzog, H., K. Caldeira, and J. M. Reilly. 2003. An issue of permanence: Assessing the effectiveness of temporary carbon storage. *Climatic Change* 59: 293–310.

Hotelling, H. 1931. The economics of exhaustible resources. *Journal of Political Economy* 39: 137–75.

Kolstad, C. D., and J. A. Krautkraemer. 1993. Natural resource use and the environment. In A. V. Kneese and J. L. Sweeney, eds., *Handbook of Natural Resource and Energy Economics*, vol. 3. Amsterdam: Elsevier Science, ch. 26.

Kurosawa, A. 2004. Carbon concentration target and technological choice. *Energy Economics* 26(4): 675–84.

Lafforgue, G., B. Magné, and M. Moreaux. 2008. Energy substitutions, climate change and carbon sinks. *Ecological Economics*, forthcoming.

McFarland, J. R., H. J. Herzog, and J. M. Reilly. 2003. Economic modeling of the global adoption of carbon capture and sequestration technologies. In J. Gale and Y. Kaya, eds., Greenhouse Gas Control Technologies: *Proceedings of the Sixth International Conference on Greenhouse Gas Control Technologies*. Amsterdam: Elsevier Science.

Pacala, S. W. 2003. Global constraints on reservoir leakage. In J. Gale and Y. Kaya, eds., Greenhouse Gas Control Technologies: *Proceedings of the Sixth International Conference on Greenhouse Gas Control Technologies*. Amsterdam: Elsevier Science.

Riahi, K., E. S. Rubin, M. R. Taylor, L. Schrattenholzer, and D. Hounshell. 2004. Technological learning for carbon capture and sequestration technologies. *Energy Economics* 26(4): 539–64.

Tahvonen, O. 1997. Fossil fuels, stock externalities and backstop technology. *Canadian Journal of Economics* 30(4): 855–74.

Tahvonen, O., and S. Salo. 1996. Nonconvexities in optimal pollution accumulation. *Journal of Environmental Economics and Management* 31(2): 160–77.

Tahvonen, O., and C. Withagen. 1996. Optimality of irreversible pollution accumulation. *Journal of Economic Dynamics and Control* 20(9): 1775–95.

Toman, M. A., and C. Withagen. 1996. Accumulative pollution, "clean technology" and policy design. *Resource and Energy Economics* 22(4): 367–84.

IV Models and Policies

15

Mind the Rate! Why the Rate of Global Climate Change Matters, and How Much

Philippe Ambrosi

Given the difficulties in modeling climate change damages,[1] the first attempts to assess climate policies have consisted in approximating climate change risks through critical thresholds beyond which climate change would be most dangerous. The threshold initially selected was the simplest and most readily assessable measure in the upstream causal chain linking greenhouse gases (GHGs) emissions to damages: an atmospheric concentration ceiling.

15.1 Global Mean Temperature Rise as a Tentative Metrics to Capture Climate Risks

Concentration targets, however, provide a very crude measure that fails to capture the diversity of climate change risk. The targets are not tangible, as they bypass many links from atmospheric chemistry to ultimate damages (besides the propagation of uncertainty), and refer only to long-term risk of climate change. By contrast, just one step down in the GHGs chain, we have the global mean temperature. Temperature is a better and more tangible proxy of climate change risk for a number of reasons:

• As a synthetic index of the ongoing climate change, the global mean temperature parameter incorporates uncertainty about climate dynamics.

• As a more palpable metric of climate risk, the global mean temperature parameter can be applied in regional assessments of climate change impacts, making it easier for stakeholders to connect a given magnitude of global climate change with a set of results.

• As a major determinant of vulnerability, both for ecosystems and socio-economic systems, the global mean temperature parameter allows for climate change to be accounted for by its global rate. Indeed residual damages can reach a much higher level in a situation where the climate changes so fast as to overwhelm our capacity to adapt (also socioeconomically) than in a situation where the climate changes gradually and the impacts spread more evenly, thus enabling the global community to make timely adaptation strategies.

I thus proceed in this chapter to propose a climate policy assessment within a cost-efficiency framework, using constraints referring to global mean temperature rise (its magnitude and rate). I focus on short-term policy, up to 2050, in being mindful of the prospect that the transition of energy systems toward low-carbon societies will at least last fifty years. I address three issues:

• Does the uncertainty regarding climate dynamics and the definition of climate risks lead to very stringent recommendations? In other words, does an explicit reference to the precautionary principle[2] imply significant abatement efforts as long as our knowledge has not yet progressed?

• Has learning a critical impact on short-term decision? In other words, can we wait to know more before we decide to act, and until when?

• Can we sort out these uncertainties, especially with regard to short-term decision-making?

15.2 Integrated Assessment in the Context of Climate Stabilization

In the last few years, with notable acceleration, the scientific community has concentrated on the assessment of climate policies in the context of climate stabilization. The studies have mostly accepted a +2°C (with respect to preindustrial times) dangerous climate change threshold, in line with the long-term climate policy goal stated by the European Union in 1996, and today still advocated as was re-emphasized more recently: "the global average [temperature] should not exceed the preindustrial level by more than 2°C" (Council of the European Union, 2004).[3] The International Climate Change Taskforce (2005) has also taken this view.

Beyond the +2°C threshold it is agreed that climate change entails *dangerous* impacts.[4] On a global scale, for instance, a drift of the climate system could create a shutdown of the thermohaline circulation in the North Atlantic, and regionally, there are the well-known threats of extinction to coral reefs from the warming of the ocean and increases in storm surges for small island states. Cost-efficiency studies of climate change policy and numerous impact studies have provided some important insights for policy makers on the connection between the global mean temperature and climate change risk. For instance, the Global Fast Track Assessment (Parry et al. 2001) of a +2°C dangerous impact threshold predicts a sharp increase in the number of people at risk of water shortage as the global mean temperature rise gets close to +2°C. A comprehensive review of impact studies (ECF 2004) draws a similar conclusion about the multiple impacts of a global warming that ventures beyond +2°C or +3°C.

However, remarkably fewer studies have considered the rate of climate change, and risks contingent on the pace of global warming in the deployment of impacts. Although very little information on this topic is available, global disruptions of the climate system (e.g., the thermohaline circulation) are known to be sensitive to the rate of climate change as well as to the absolute magnitude of climate change. The risks that are characterized in the literature at best relate to ecosystems. Leemans and Eickhout (2004) argue that with the ongoing rate of climate change (between +0.1°C and +0.2°C each decade) continuing over the coming decades, the decline in biodiversity within many ecosystems will soon accelerate. Leemans and van Vliet (2005) call for a low 0.05°C target in the coming decades, but this is not very safe given the projected rate of warming over the coming decades.

Other contributions, as I review next have mainly examined the component of uncertainty in climate sensitivity[5] within an allowable (short-term) GHGs emissions chart and the corresponding rigidity of climatic constraints. They can be broadly categorized by ways in which they treat the uncertainty component, by the complexity and interdisciplinary nature of the underlying models, and by the priority given to policy recommendations:

- *Probabilistic integrated assessment.* The risk of overshooting is assessed for a given climate target (absolute magnitude of global mean temperature rise or rate of climate change) within a number of emissions scenarios. Alternative probabilistic climate change projections are made that quantify the likelihood of climate change. Typical results consist of probability distributions of overshooting a given climate stabilization goal, probabilistic scenarios of climate change, investigations of how a delayed or anticipated global action alters the risk of overshooting or the likelihood of future climate outcomes (den Elzen and Meinshausen 2005a; Hare and Meinshausen 2004; Knutti et al. 2003;

Mastrandrea and Schneider 2004; Meehl et al. 2005; Meinshausen 2005; O'Neil and Oppeinheimer 2004; Wigley 2005).

- *Inverse approach.* Both safe landing (Alcamo and Kreileman 1996; Swart et al. 1998) and the tolerable windows (Kriegler and Bruckner 2004; Toth et al. 2003a, b) analyses aim at defining a corridor of allowable emissions given sets of constraints, mainly referred to as unacceptable impacts (e.g., rates of global mean temperature rise and sea-level rise) and intolerable mitigation costs but not to the exclusion of other constraints (e.g., maximal yearly decarbonization rate). Tolerable windows analysis differs from safe landing analysis in that it goes not to the global scale but to a detailed regional integrated model whereby constraints are specified by certain sectoral or regional objectives (e.g., preserving two-thirds of natural vegetation in nonfarming areas) or mitigation costs (e.g., setting an upper bound on the relative distribution of these costs between regions). Both approaches provide insights on the relative effects of certain constraints on short-term decision-making. Although these analyses do not consider emissions pathways, they guide decision-makers in their choice of criteria by delineating the admissible emissions trajectory in an allowable emissions corridor.

- *Cost-efficiency analysis.* A least-cost GHGs trajectory is defined that complies with a given climate target. As in the inverse type of analysis, uncertainty is computed by a sensitivity study (Böhringer et al. 2005; Caldeira et al. 2003; den Elzen and Meinshausen 2005b; Richels et al. 2004). However, unlike both the inverse and probabalistic approaches, cost-efficiency studies allow for the interaction between uncertainty and decision-making to be taken into account, with the objective of reduction of uncertainty in a future period. There are further two forms of cost efficiency studies that have been undertaken whether or not decision-makers require (subjective) probabilities. One is the standard decision-making model under uncertainty that aims at an optimal strategy given a decision criterion (here least cost) across a set of likely future states of the world (Manne and Richels 2005; this study). The other aims at strategies that are robust (i.e., largely insensitive) to many uncertainties in order to get round the difficulties of probability elicitations in situations of deep uncertainty or of high controversy (Hammit et al. 1992; Lempert 2002; Lempert et al. 1994, 2001; Yohe et al. 2004).

Nevertheless, whichever the approach these studies reach similar conclusions as to the importance of dealing with the uncertainties in climate sensitivity. For instance, Caldeira et al. (2003) state that the climate sensitivity index should be 4.5°C: "one should almost totally reduce emissions by 2050; by the turn of the century, almost 75 percent of the energy supply should be carbon free whatever the value of climate sensitivity." Den Elzen and Meinshausen (2005b) conclude that "[f]or achieving the 2°C target with a probability of more than 60 percent, greenhouse gas concentrations need to be stabilized at 450 ppm CO_2-equivalent or below, if the 90 percent uncertainty range for climate sensitivity is believed to be 1.5°C to 4.5°C." Kriegler and Bruckner (2004) come to similar conclusions: the lower the warming threshold and the higher the climate sensitivity (both implying stringent concentrations targets), the narrower the global carbon budget (see also Lempert et al. 1994; Hammitt et al. 1992). These studies also emphasize the consequences of a delayed global action: for instance, "the next 5 to 15 years might determine whether the risk of overshooting 2°C can be limited to a reasonable range" (Meinshausen 2005). Mastrandrea and Schneider (2005) compare the probability distributions of temperature change in specific overshooting and not overshooting scenarios that stabilize at the 500 ppm CO_2 equivalent, based on published probability distributions on climate sensitivity. They find that from 2000 to 2200 the overshooting of the 500 ppm target increases by 70 percent the probability of temporary or sustained exceedence of a 2°C above the preindustrial level. Hare and Meinshausen (2004) calculate that with each ten-year delay in action on emissions there is at least a further 0.2°C to 0.3°C warming over a 100- to 400-year time horizon. Yohe et al. (2004) add a pitch of urgency: "uncertainty [about climate sensitivity] is the reason for acting in the near term and uncertainty cannot be used as a justification for doing nothing."

The sequential decision-making with learning framework used in this chapter is close in scope to that of Manne and Richels (2005) but with two distinct differences: first, I include a rate constraint; second, I compute the value of information to rank the uncertainties in the model and to assess the influence of the date of learning (which, to my knowledge, is new in the context of climate stabilization).

15.3 RESPONSE Θ, A Cost-Efficiency Optimal Control Integrated Assessment Model

RESPONSE Θ belongs to the RESPONSE model family,[6] a generic set of stochastic climate policy optimization integrated assessment models.

It includes a simple description of climate policy costs (baseline scenario and abatement cost function; see section 3.1) and of the chain-linking emissions to climate change through reduced forms of carbon cycle and climate dynamics (see section 3.2). RESPONSE Θ seeks to minimize the discounted sum of abatement costs (15.1) subject to two climate constraints, one constraint on the magnitude of global warming since 1990, (15.2) with $\Delta\theta_{max}$ set at +2°C (in the central case), and one constraint on the decadal rate of global warming (15.3), with $\Delta\theta_{RYT}$ ranging from 0.3°C a decade to 0.1°C a decade:

$$\min_{Ab_t} \sum_{t=1990}^{2300} \frac{f(Ab_t, Ab_{t-1}, t)}{(1+\rho)^{(t-1990)}} \qquad (15.1)$$

subject to

$$(\theta_t^{At} - \theta_{1990}^{At}) \leq \Delta\theta_{max}, \qquad (15.2)$$

$$(\theta_{t+10}^{At} - \theta_t^{At}) \leq \theta_{RYT}, \qquad (15.3)$$

where $f(Ab_t, Ab_{t-1}, t)$ is the total cost of mitigation measures at time t (trillion US$), Ab_t is the abatement rate at time t (percentage of baseline emissions), ρ is the discount rate (3 or 5 percent a year), and θ_t^{At} is the global mean atmospheric temperature rise with respect to preindustrial temperatures (in Celsius). The components of the model are discussed below.

15.3.1 The Objective Function

I use the following abatement cost function (15.4), from STARTS (Lecocq 2000), which has been re-calibrated against IPCC TAR estimates for a 550 ppm target (Metz et al. 2001, ch. 8):

$$f(Ab_t, Ab_{t-1}, t) = \frac{1}{3} BK \cdot PT_t \cdot \gamma(Ab_t, Ab_{t-1}) \cdot em_t \cdot (Ab_t)^3, \qquad (15.4)$$

where BK is the initial marginal cost of backstop technology (1.2 thousand US$ tC^{-1}), PT_t is a technical change factor, $\gamma(Ab_t, Ab_{t-1})$ is a socio-economic inertia factor, and em_t is the baseline CO_2 emissions at time t (GtC).

By these specifications, abatement costs are represented as a backstop technology with convex (quadratic) marginal costs. The specifications incorporate an autonomous technical change factor, PT_t, with costs decreasing at a yearly constant rate of 1 percent down to 25 percent of their initial value. They also capture socioeconomic inertia as a cost-multiplier, through a multiplicative index, $\gamma(Ab_t, Ab_{t-1})$; see (15.5). $\gamma(.)$ is equal to unity (no additional costs) if the abatement rate between two consecutive periods increases at a rate lower than $\delta\tau$, namely the annual turnover of capital below which mitigation policies do not lead to premature retirement of productive units times the time step of the model ($\delta = 10$ years). Otherwise, $\gamma(.)$ increases linearly with the rate of increase of abatement rate between two consecutive periods. That is,

$$\gamma(Ab_t, Ab_{t-1}) = \begin{cases} 1 & \text{if } \frac{Ab_t - Ab_{t-1}}{\delta\tau} \leq 1 \\ \frac{Ab_t - Ab_{t-1}}{\delta\tau} & \text{otherwise} \end{cases} \quad (15.5)$$

Finally, em_t, baseline emissions (amount of CO_2 emissions, both from fossil fuel use and land use) come from the marker scenario for the A1 SRES family, A1B (Nakicenovic 2000): the emissions double, rising from today's level of 7 to 8 GtC each year to 16 GtC yearly by midcentury and declining after 2060 because of energy efficiency improvements and adoption of less CO_2 emitting technologies in the energy sector. Cumulative emissions sum up to 2,077 GtC (more than 2.7 times the atmospheric carbon content in 1980).

15.3.2 Carbon and Climate Dynamics

I use the carbon cycle from DICE and RICE (Nordhaus and Boyer 2000), a linear three-reservoir model (atmosphere, biosphere, plus surface ocean and deep ocean). GHGs emissions (CO_2 solely) accumulate in the atmosphere, and they are slowly removed by the biospheric and oceanic sinks. However, some of these emissions irreversibly accumulate into the atmosphere. The dynamics of carbon flows is given by (15.6):

$$\begin{pmatrix} A_{t+1} \\ B_{t+1} \\ O_{t+1} \end{pmatrix} = C_{trans} \cdot \begin{pmatrix} A_t \\ B_t \\ O_t \end{pmatrix} + \delta(1 - Ab_t)em_t \cdot u, \quad (15.6)$$

where A_t is the carbon contents of atmosphere at time t (GtC), B_t is the carbon content of the upper ocean and biosphere at time t (GtC), O_t is the carbon content of the deep ocean at time t (GtC), C_{trans} is the net transfer coefficient matrix, and u is a column vector $(1,0,0)$. Nordhaus's calibration using existing carbon-cycle models gives the following results (for a decade-long time step):

$$C_{trans} = \begin{pmatrix} 0.66616 & 0.27607 & 0 \\ 0.33384 & 0.60897 & 0.00422 \\ 0 & 0.11496 & 0.99578 \end{pmatrix}$$

with initial conditions (GtC) being

$$C_{1990} = \begin{pmatrix} 758 \\ 793 \\ 19230 \end{pmatrix}.$$

The evolution of global mean temperature is described using a two-equation perturbation model (à la Schneider and Thompson 1981) parameterized using a warming scenario from a general circulation model[7] forced by a 1 percent yearly atmospheric CO_2 increase. CO_2 is the only radiative gas considered. Since the main issue is the timing of abatement over the short run, I prioritize the description of the interaction between the atmosphere and the surface ocean and neglect the interactions within the deep ocean. So this model describes the modification of the thermal equilibrium between atmosphere and surface ocean (15.8) in response to anthropogenic greenhouse effect (15.7). Long-term climate dynamics is led by climate sensitivity.

$$F(t) = F_{2x} \frac{\log(A_t/A_{PI})}{\log 2}, \tag{15.7}$$

$$\left\{ \begin{bmatrix} \theta^{At}_{t+1} \\ \theta^{Oc}_{t+1} \end{bmatrix} = \begin{bmatrix} 1 - \sigma_1(F_{2x}/T_{2x} + \sigma_2) & \sigma_1\sigma_2 \\ \sigma_3 & 1-\sigma_3 \end{bmatrix} \begin{bmatrix} \theta^{At}_t \\ \theta^{Oc}_t \end{bmatrix} + \sigma_1 \begin{bmatrix} F(t) \\ 0 \end{bmatrix}, \right. \tag{15.8}$$

where $F(t)$ is the radiative forcing at time t ($W \cdot m^{-2}$), F_{2x} is the instantaneous radiative forcing for twice A_{PI}, set at 3.71 $W \cdot m^{-2}$, A_{PI} is the CO_2 atmospheric concentration at preindustrial times, set at 280 ppm, θ^{Oc}_t is the global mean oceanic temperature rise in preindustrial times (°C), T_{2x} is the climate sensitivity (°C), σ_1 is the transfer coefficient (set at 0.479°C $W^{-1} \cdot m^2$), σ_2 is the transfer coefficient (set at 0.109°C^{-1} $W \cdot m^{-2}$), and σ_3 is a transfer coefficient (set at 0.131).

15.3.3 Climate Sensitivity

Uncertainty regarding climate sensitivity is large, more than 3°C, and persists since the second IPCC report: "The equilibrium climate sensitivity ... was estimated to be between +1.5°C and +4.5°C in the SAR [Second Assessment Report]. This range still encompasses the estimates from the current models in active use" (see Houghton et al. 2001, p. 561). Around the same time, Wigley and Raper (2001) had proposed an ad hoc lognormal distribution, with a 90 percent confidence range from 1.5°C to 4.5°C. More recent studies have led to better characterizations of climate sensitivity and have quantified the accompanying uncertainty,[8] but this parameter remains difficult to assess from observations, given fragile historical radiative forcing and ocean heat uptake data (Andronova and Schlesinger 2001; Forest et al. 2002; Frame et al. 2005; Gregory et al. 2002; Knutti et al. 2002, 2003) or from atmosphere-ocean global circulation models, given the parametrizations of such key processes as cloud effects need improving (Murphy et al. 2004; Stainforth et al. 2005). These recent studies have produced new estimates that remain concentrated over the +1.5°C to +4.5°C range with a mean close to +3.5°C, but much higher values, admittedly with low probabilities, cannot be excluded.

To account for this uncertainty, I explore three values for T_{2X}, centered around the mean estimate $\{+2.5°C, +3.5°C, +4.5°C\}$ with the probabilities $\{1/6, 2/3, 1/6\}$ whose distribution is close to that obtained by Murphy et al (2004). To convey an idea of the consequences of this distribution for decision-making, it means that to achieve a +2°C target with at least 80 percent confidence CO_2 concentration must be stabilized at 450 ppm or below. To account for the uncertainty about climate sensitivity (i.e., the tail of the distribution for high values) and also for the various attitudes toward risk people will adopt in view of such fragile information and their degree of risk aversion, I will test alternative distributions with different weights for bad news (i.e., high climate sensitivity).

In the following, I examine RESPONSE Θ recommendations for a warming threshold set at +2°C (with no rate constraint for the moment). The uncertainty about climate sensitivity leads to very different optimal emissions trajectories, as shown by the sensitivity study in figure 15.1. It is, no doubt, a crucial uncertainty for decision-making for the decades to come: in 2010 mitigation effort may amount to 2, 9, or 17 percent of baseline emissions depending on whether climate

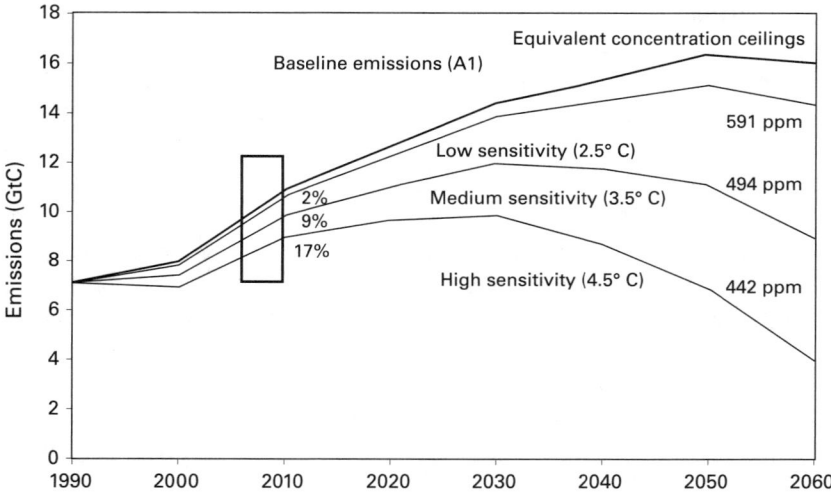

Figure 15.1
Alternative mitigation efforts for different values of climate sensitivity: Results from RESPONSE Θ

sensitivity is low (+2.5°C), medium (+3.5°C), or high (+4.5°C). Opting for a +2°C temperature ceiling means that in the long term the CO_2 atmospheric concentration would be stabilized at a stringent level as climate sensitivity is particularly high: respectively, 591 ppm, 494 ppm, and 442 ppm.

I now turn to analyze the relative influence of the two climate constraints that approximate climate risks. To this aim, I have a look at baseline temperature trajectories for different values of climate sensitivity so as to detect at what time these constraints can bite and how their respective importance varies (figure 15.2). The figure is to be read as follows: since global mean temperature is always increasing, one moves with time along each curve from left to right; some dates are indicated for ease of reading. An early kink occurs because of the sharp increase in baseline emissions in 2000. The increase in global mean temperature is related to the magnitude and duration of forcing, that is to say, to the atmospheric stock of CO_2. Since this stock is increasing up to 2150, temperature is increasing as well. The rate, however (indicated by the y-axis), depends on the increase in radiative forcing between two periods of time. This is, of course, directly related to the increment of the atmospheric stock of CO_2, that is to say, apart from

Chapter 15 Mind the Rate! Why Global Climate Change Matters

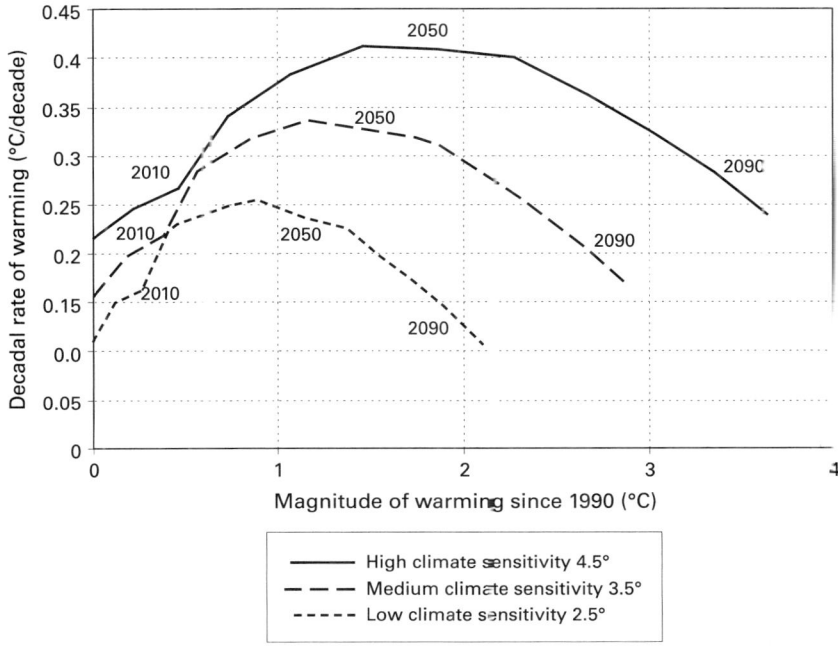

Figure 15.2
Magnitude of warming (since 1990) plotted against decadal rate of warming in the baseline scenario for three values of climate sensitivity

carbon cycle, to the time profile of emissions. As for the position of the curves, that depends on the value of climate sensitivity.

Beyond 2050 global mean temperature is increasing at a much smaller rate than in the first half of the century because GHGs emissions begin to taper off soon after 2050 in the baseline scenario. Hence from 2010 to 2050 the rate constraint makes the sharpest bite as climate sensitivity is particularly high. To slow down the acceleration, more mitigation must occur during the early half of the century. Results from the sensitivity analysis for short-term abatement are given in table 15.1. This conclusion is close to that of Metz et al. (2001, ch. 10): controlling the rate of climate change sets a significant constraint on GHGs emissions in the first half century, especially as this constraint is set at 0.1°C or 0.2°C for each decade.

Sensitivity analyses can provide useful information, especially, for extreme situations, and may guide policy makers provided the decision criterion is maximin or minimax regret, as these criteria do not

Table 15.1
Relative influence of the climate constraints on the abatement rate from 2000 to 2050

	$\Delta\theta_{max}$		
Δ_{RYT}	1°C	2°C	3°C
0.1°C per decade	>	>	>
0.2°C per decade	<	>	>
0.3°C per decade	<	<	>
0.4°C per decade	<	<	<

Note: ">" means that the influence of the rate constraint dominates the influence of the magnitude constraint for short-term decision (up to 2050) and "<" that the opposite occurs.

use probability distributions on the occurrence of future states of the world. However, sensitivity analyses do not ponder precisely the likelihood of the occurrence of future states of the world and thus may not fully guide decision. For instance, our former results, which show the short-term decision to be optimal given that the uncertainty about climate sensitivity, suggest adjusting emissions abatement from 2 to 17 percent in 2010. Furthermore sensitivity analyses do not take into account learning through sequential decision-making. Introducing learning could provide for flexibility in climate policies. These important contributing factors to sensitivity analyses are addressed next.

15.4 Optimal Climate Policy under Uncertainty on Climate Sensitivity, with Learning

To capture decision-making under uncertainty about climate sensitivity with learning, I altered RESPONSE Θ in the following manner. Uncertainty about climate sensitivity is considered discrete: there are three possible states of the world (s) to consider in terms of climate sensitivity $\{+2.5°C, +3.5°C, +4.5°C\}$; the corresponding ex ante subjective probabilities or priors are (p_S) $\{1/6, 2/3, 1/6\}$. Learning is exogenous, and information arrives at a fixed point in time (t_{info}) at the beginning of each decade in the twenty-first century. The two polar cases are perfect information ($t_{\text{info}} = 1990$) and complete uncertainty ($t_{\text{info}} = 2300$, the horizon of RESPONSE Θ):

$$\min_{Ab_t^s} \sum_s p_s \sum_{t=1990}^{2300} \frac{f(Ab_t^s, Ab_{t-1}^s, t)}{(1+\rho)^{(t-1990)}} \quad (15.9a)$$

$$\forall t \leq t_{\text{info}}, \forall (s, s') \in S, Ab_t^s = Ab_t^{s'}, \tag{15.9b}$$

subject to

$$(\theta_t^{At,s} - \theta_{1990}^{At,s}) \leq \Delta\theta_{\max}, \tag{15.10}$$

$$(\theta_{t+10}^{At,s} - \theta_t^{At,s}) \leq \bar{\sigma}_{RYT}. \tag{15.12}$$

The decision-maker adopts the best[9] sequence of actions given that information on the "true" state of the world will be available at t_{info}. Beyond this date, the decision-maker can adapt the optimal abatement profile to that information (three states of the world so three possible abatement trajectories); before this date, the decision-maker has to base the decision on subjective probabilities about the likelihood of the future states of the world and the date of disclosure of information. In numerical terms, the objective function (15.9a) is re-specified as the minimization of expected costs of abatement trajectories across the three states of the world, subject to one constraint on the magnitude of warming (15.10) and one constraint on the decadal rate of warming (15.11). Sequential decision-making is captured through an additional constraint (15.9b) that, before the disclosure of information, the decision variables have to be the same across all states of the world.

15.4.1 Complete Uncertainty and the Significant Economic Consequences of the Worst-Case Hypothesis

First to be examined are the RESPONSE Θ recommendations in the case of no learning ($t_{\text{info}} = 2300$).[10] As shown in figure 15.3, the optimal emissions path entirely adheres to the worst-case hypothesis (climate sensitivity equal to +4.5°C), generating significant economic regrets (i.e., investments in abatement technologies finally not necessary) if ultimately climate sensitivity turns out to have a lower value. In other words, results are similar to those obtained with a maximin decision criterion that focuses only on the worst case. In a cost-effectiveness framework the environmental constraints must be satisfied whatever the cost. So the optimal emissions path totally adheres to the worst-case hypothesis as it corresponds to the lowest concentration ceiling (442 ppm).

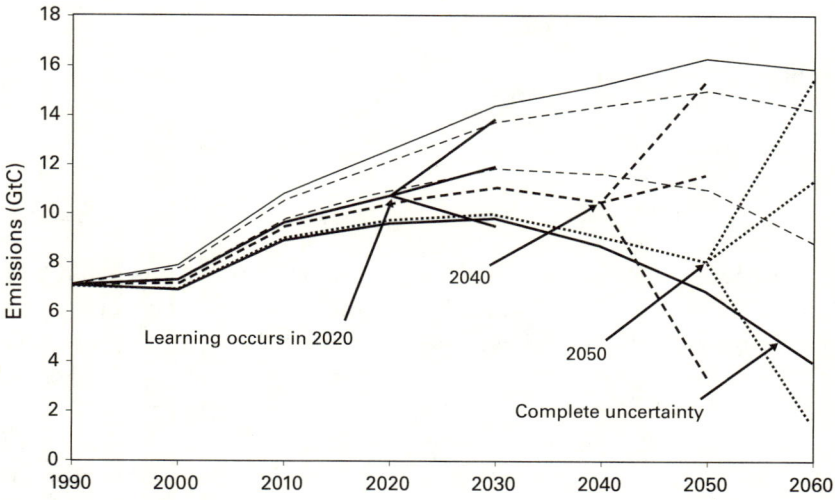

Figure 15.3
Learning and short-term flexibility

15.4.2 Key Role of Learning: Flexibility in Short-term Abatement Efforts

Now, if there is the possibility of learning in the future, short-term mitigation efforts can be relaxed, and so decrease the economic regrets. If learning occurs early, this effect is all the more pronounced. For instance, note that in figure 15.3 the abatement efforts in 2010 amount to 17 percent of baseline emissions in the no learning case; they gradually decrease to 16, 12, and 11 percent, which is very close to the central case where climate sensitivity is equal to +3.5°C, as the information becomes available respectively in 2050, 2040, and 2020. These results are similar to the conclusions reached by Ha Duong et al. (1997) in the context of a stabilization of GHGs atmospheric concentrations. There is indeed a near equivalence between aiming at a temperature threshold in the presence of uncertainty on climate sensitivity and stabilizing GHGs atmospheric concentrations at ceilings not yet known.

Such information might not be soon available (Kelly et al. 2000; Leach 2004). It could take at least fifty years to acquire from observations a reliable estimate of climate sensitivity. As a result we are forced to follow the stringent emissions pathway implied by the bottom curve in figure 15.3, making achieving a +2°C objective costly in the coming decades.

Chapter 15 Mind the Rate! Why Global Climate Change Matters   321

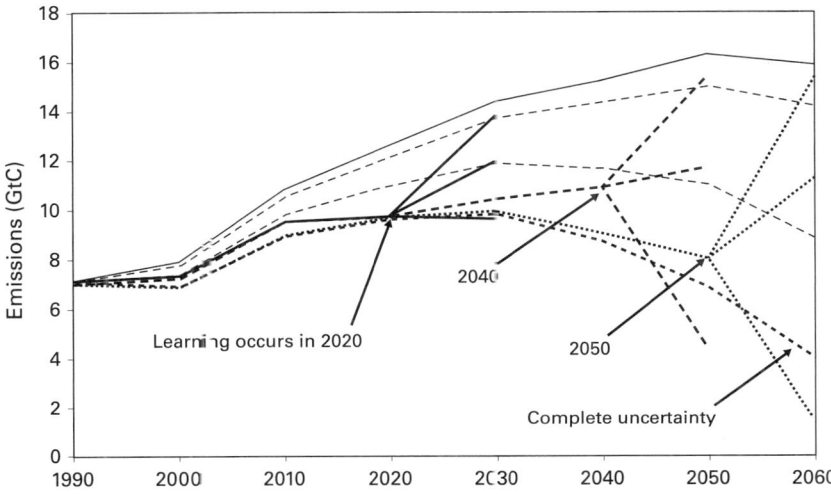

Figure 15.4
Significant short-term regrets with the rate constraint

15.4.3 Significant Short-term Costs with the Rate Constraint

The rate constraint almost neutralizes short-term benefits from learning because of the significant mitigation efforts. So any flexibility allowed for future learning is lost (see the short-term policy recommendations in figure 15.4 for the rate constraint set at 0.3°C per decade). In other words, in the presence of uncertainty about climate sensitivity, the influence of the rate constraint on short-term decision-making is considerably enlarged as there has to be taken into account the eventuality that climate sensitivity may equal +4.5°C, and thus result in more intense and faster warming.

These are weighty conclusions for the decision-maker to consider. Besides a *central* belief, I have tested a *neutral* belief as an equiprobable distribution whose mean is equal to +3.5°C with {1/3, 1/3, 1/3}, an *optimistic* belief whose left-skewed distribution has a mean equal to +3°C with {2/3, 1/6, 1/6}, and last, a *pessimistic* belief whose right-skewed distribution has a mean equal to +4°C with {1/6, 1/6, 2/3}. Whatever the prior of the decision-maker, the optimal abatement rate in 2020 (learning occurring in 2020) amounts to about 22.0 percent of the baseline emissions, and this is very close to the worst-case hypothesis under certainty for the same time period (23.6 percent) and almost

Table 15.2
Influence of the discount rate on the optimal decision for the certainty case and in the presence of uncertainty about climate sensitivity with learning

Discount rate (% per year)	Optimal abatement rate in 2020 (% of baseline emissions)	
	Certainty case (climate sensitivity: 3.5°C)	Learning in 2020
1	27.3	28.0
3	18.1	22.1
5	12.8	21.9
10	6.0	21.6

twice the optimal abatement effort in the central case for the same time period (12.8 percent).

All results are robust to discounting, which is, in the certainty case, the most decisive socioeconomic parameter for decision-making with the effect on passing on emissions reduction efforts to future generations (see table 15.2, left column). In the certainty case, the optimal abatement rate differs widely depending on the value of the discount rate (from 6 to 27 percent as the discount rate decreases from 10 to 1 percent a year): in the cost-efficiency framework, a high discount rate reduces the discounted value of future costs, so the bulk of the mitigation is best postponed for several decades. In the learning case (see table 15.2, right column), the optimal abatement rate is systematically greater than the optimal rate in the certainty case. In general, the influence of the discount rate is much less important than before. Delayed action is still desirable (the mitigation effort decreases as the discount rate increases), but for 3, 5, and 10 percent a year the optimal abatement rates are almost comparable. The optimal response that stands out is for only a 1 percent yearly discount rate. Whereas in the certainty case sharp controversies can arise over the choice of a correct value for the discount rate (since the optimal abatement rate may accordingly vary by a factor of 4), conflicting views concerning this parameter are not anymore as decisive when following a sequential decision approach. This is especially the case if the debate focuses around values such as 3 to 5 percent a year (a plausible range given the baseline scenario growth rate and standard assumptions regarding the economic agent preferences), which unanimously leads to a mitigation rate of 22 percent.

Just as the constraint on the magnitude of warming, the rate constraint is still unknown. Some have argued for a very tight constraint, while others have advocated much looser constraint, be they quite optimistic about our future adaptive capacities or less concerned by endangered ecosystems. It will no doubt take time before a social consensus is reached or major breakthroughs in climate or impacts science help in defining unambiguously what constitutes a "dangerous interference with the climate system" or help in better quantifying climate sensitivity. In order to rank the relative influences on short-term decision-making of the three uncertainties described above, I next compute the value of information associated with each of the three parameters.

15.5 Ranking Uncertainties Using Information Value

The measure of information value I used is the expected value of perfect information (EVPI). Here EVPI measures the opportunity value of possessing a piece of information when making a decision. In a dynamic perspective, EVPI is computed as the maximal willingness to pay to obtain today this piece of information rather than waiting for it to emerge at a later time. EVPI is classically defined as the difference between the expected value of the objective function in the "act–then learn" case (a policy must be adopted before the disclosure of information) and in the "learn–then act" case (the value of the parameter is known from the outset and a policy is adopted accordingly). In RESPONSE Θ notation it takes the form

$EVPI(t_{info})$

$$= \underbrace{(\sum_s p_s \sum_{t=1990}^{2300} f(Ab^s_{ATL(t_{info}),t}, Ab^s_{AT\_(t_{info}),t-1}, t)/(1+\rho)^{(t-1990)})}_{\text{Act–then learn policy}}$$

$$- \underbrace{(\sum_s p_s \sum_{t=1990}^{2300} f(Ab^s_{LTA,t}, Ab^s_{LTA,t-1}, t)/(1+\rho)^{(t-1990)})}_{\text{Learn–then act policy}}$$

where $Ab^s_{ATL(t\_info),t}$ is the optimal abatement rate, with the state of the world s being disclosed at time t_{info}, and $Ab^s_{LTA,t}$ is the optimal abatement rate in the certainty case, with the state of the world being s. EVPI allows today's decision-maker to rank the relative importance of a set of uncertain parameters. For instance, EVPI *ad infinitum*, EVPI(2300), reveals the magnitude of the regret at never possessing definitive information to incorporate in the decision process. EVPI(2300) measures the costs of complete uncertainty and the amount one is willing to pay today to obtain immediately (as opposed to never) the needed knowledge in order to fine-tune precautionary climate policies. Moreover the time profile of EVPI shows the opportunity to accelerate the reduction of uncertainties: if EVPI increases sharply between two points in time, it is then vital for today's decision-maker to get the information at the beginning of this period.

I explore the following sets for climate sensitivity $\{+2.5°C, +3.5°C, +4.5°C\}$, the magnitude constraint $\{+1°C, +2°C, +3°C\}$, and the rate constraint $\{0.1°C \text{ per decade}, 0.2°C \text{ per decade}, 0.3°C \text{ per decade}\}$. The accompanying subjective probability distribution is $\{1/6, 2/3, 1/6\}$.

Once again, the prominent influence of the rate constraint is confirmed (figure 15.5). Unquestionably, EVPI for this parameter shows the fastest increase (in 2020, more than 60 percent of its value *ad infini*-

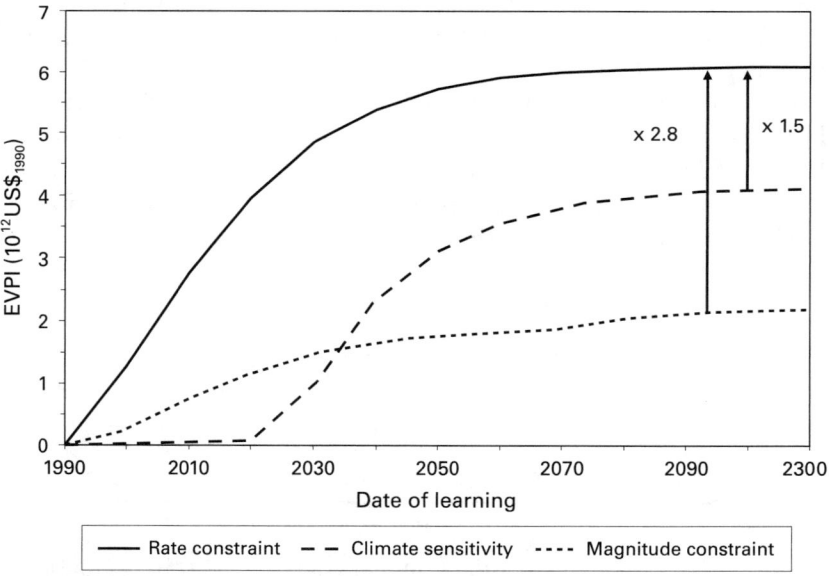

Figure 15.5
Value of information for climate sensitivity, rate, and magnitude constraints as a function of date of learning.

tum; almost 90 percent in 2040, 20 years later) and is the highest value *ad infinitum* (2.8 times greater than EVPI for climate sensitivity and 1.5 times greater than EVPI for the magnitude constraint). The respective dominances of the other two parameters vary with time: EVPI for climate sensitivity is greater than EVPI for the magnitude constraint until 2035, and the situation then reverses (in 2100, EVPI for the magnitude constraint is greater than EVPI for climate sensitivity by a factor of 1.8). The *ad infinitum* value of information attached to climate sensitivity increases faster (in 2020, more than 50 percent of its value *ad infinitum*, almost 70 percent ten years later, in 2030), whereas the value of information for the magnitude constraint increases sharply from 2020 onward (in forty years, from 2020 to 2060, it grows by more than 80 percent).

There exists thus a high opportunity cost of not knowing before 2040 the scientifically objective or socially acceptable values for the rate constraint, climate sensitivity, and the magnitude constraint, and on a closer horizon (between now and 2020) of not knowing the first two parameters. Beyond 2040 there still exists, of course, a benefit to discovering the values for these parameters. However, with regard to short-term flexibility in the abatement effort, the late acquisition of information is not that decisive (since the optimal emissions pathway then adheres to the worst-case hypothesis).

These results qualitatively hold for alternative values of the discount rate (figure 15.6), that is, for the prominence of the rate constraint (highest value and fastest increase at least for the next forty years), for climate sensitivity (second highest value for the next thirty years), and for the magnitude constraint (sharp increase from 2020 onward). Interestingly, use of the EVPI measure makes clear the balance between short term and long term as induced by the discount rate. In a world with high discounting, mitigation efforts tend to be postponed. The uncertainty regarding rate constraint or climate sensitivity (when combined with the rate constraint in the short term) is more critical for the short term because significant abatement rates must be faced that might preferably be delayed. Conversely, in a world with low discounting, mitigation efforts are spread more evenly across time as the long-term horizon is given more weight. This way the uncertainties that become more significant over the long term can be given more consideration today: the value of information associated with the magnitude constraint—related to the atmospheric stock of carbon and to

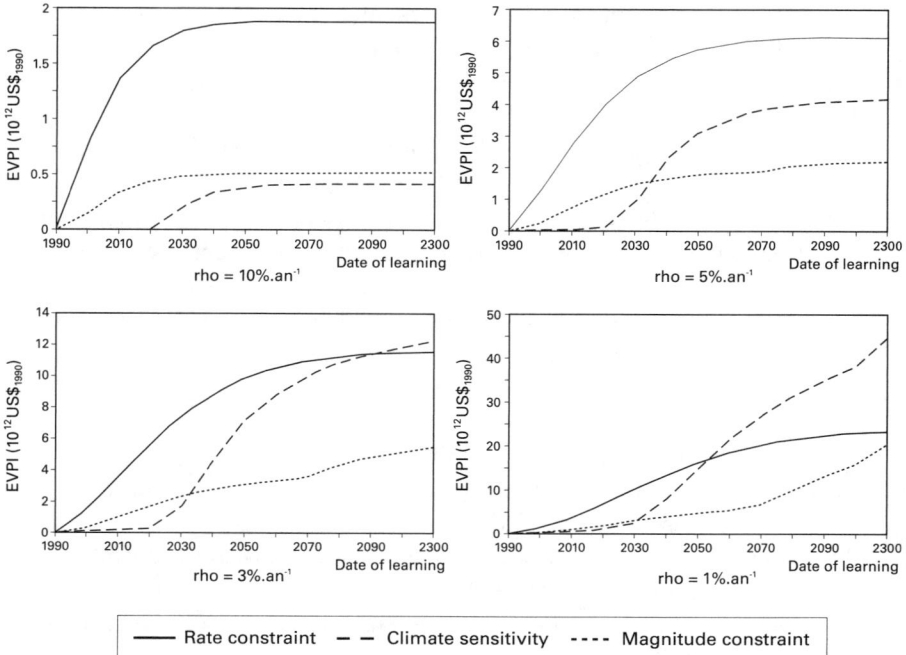

Figure 15.6
Value of information for climate sensitivity, rate, and magnitude constraints as a function of the discount rate

climate sensitivity—increases gradually over time and in long term attains the highest EVPI.

As figure 15.7 shows, the relative importance of the three uncertainties remains unaltered for various beliefs of the decision-maker, but the scale of the EVPIs decreases as the decision-maker becomes more pessimistic. In the best case indeed, the decision-maker must accept significant abatement efforts as long as the uncertainties are not reduced even if the worst case is given a low weight. Any information is therefore valuable because it makes it possible to avoid entrapment by the (less likely) worst case.

As for defining proxies of climate risks based on the global mean temperature, the most critical information relates first, to the socially tolerable rate of climate change, which is a transient characteristic of risks, and, second, to the critical magnitude of climate change, which is a long-term constraint.

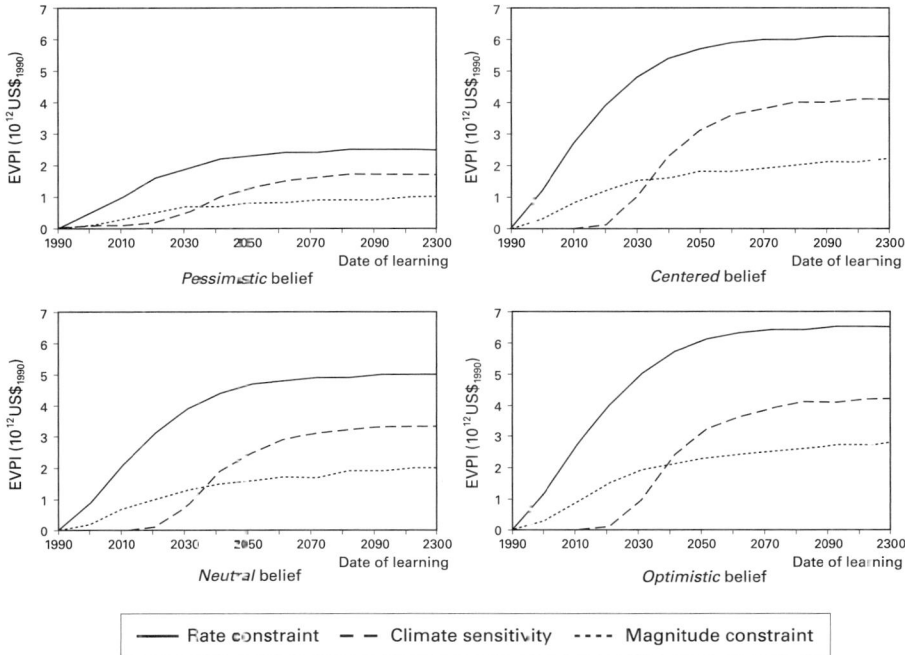

Figure 15.7
Value of information for climate sensitivity, rate, and magnitude constraints as a function of the priors of the decision-maker

15.6 Conclusion

As demonstrated in this chapter, there is a high opportunity cost of not knowing, before 2030, the value of climate sensitivity. This information is useful in order to mitigate any significant economic missteps by a precautionary policy that addresses such uncertainty. Because information on the extent of climate sensitivity might not be available in the next decades, decision-makers need to follow relatively stringent emissions pathways. In this context, a +2°C objective might be therefore considered unacceptable.

Furthermore I find that in stepping from an abstract definition of climate risks based on desirable concentration ceilings to a more palpable definition based on long-term global protection (magnitude of climate change) and transient protection (rate of climate change), any precautionary climate policy should start with a short-term abatement policy because of the rate constraint, even if the long-term objectives, concentration ceiling, or warming threshold are equivalent. Moreover tougher short-term abatement efforts may be required in the presence of uncertainty about climate sensitivity. It also becomes clear that the uncertainty about the rate constraint is more important for short-term decision-making than the uncertainty about climate sensitivity or magnitude of warming. Because the critical rate of climate change (i.e. the transience of climate risks) matters much more than the long-term objective of climate policy (i.e., the critical magnitude of climate change), more research is needed on characterizing climate change risks so that decision-makers can agree on a safe guardrail to slow the rate of global warming.

Notes

1. Among the main difficulties in assessing and modeling damages is our limited knowledge of the complex dynamics over the long term (i.e., the shape of the damage function capable of handling vulnerability thresholds or irreversible effects) and their relation to the socioeconomic development pathways (i.e., adaptation strategies in an uncertain context). For a review of these many shortcomings, see, for instance, Ambrosi (2004) and Hitz et Smith (2004). The findings and opinions expressed in this chapter are the sole responsibility of the author. They do not necessarily reflect the views of the World Bank.

2. "The parties should take precautionary measures to anticipate, prevent, or minimize the causes of climate change and mitigate its adverse effects. Where there are threats of serious or irreversible damages, lack of scientific certainty should not be used as a reason for postponing such measures . . . ." (UNFCCC 1992, Art. 3, para. 3).

3. The explicit reference to preindustrial times suggests a tighter constraint (1.5°C since 1990) given the observed global warming over the past centuries.

4. This does not mean that a +2°C target can be regarded as safe. Warren's review (2005) suggests that local impacts are not entirely benign, even for low warming; setting a target involves value judgments about different categories of impacts, our concerns for vulnerable regions and for our descendants, as well as our attitude toward risk.

5. Climate sensitivity is defined as the global mean temperature rise at equilibrium for a constant atmospheric forcing, set at twice the preindustrial level (i.e., 2×280 ppm = 560 ppm). It is a very uncertain parameter: were CO_2 atmospheric concentration stabilized at 550 ppm, the global mean temperature would rise by approximately +1.5°C to +4.5°C (its preindustrial value) suggesting a very wide and uncertain range of impacts. The uncertainty about climate sensitivity is the major contributor to the uncertainty in global warming projections for a given concentration pathway; other sources of uncertainty relate to, in order of importance, radiative forcing (esp. aerosols), ocean heat uptake, and carbon cycle dynamics.

6. See Ambrosi et al. (2003) for a more detailed presentation of the model.

7. Available at http://www.lmd.jussieu.fr/Climat/couplage/ipsl_ccm2/index.html, data kindly provided by Friedlingstein and Le Treut.

8. For a review of these studies, the methodologies followed, their limits, and their results, see the National Academies (2003) or IPCC WGI (2004).

9. Meaning here "the least expensive abatement trajectory which satisfies the environmental constraints."

10. Or, similarly, a situation where the decision-making neglects the eventuality of future learning.

References

Alcamo, J., and E. Kreileman. 1996. Emissions scenarios and global climate protection *Global Environmental Change* 6(4): 305–34.

Ambrosi, Ph. 2004. Amplitude et calendrier des politiques de réduction des émissions en réponse aux risques climatiques: leçons des modèles intégrés. Thèse de doctorat. *Economie de l'environnement*. Ecole des Hautes Etudes en Sciences Sociales (EHESS): Paris.

Ambrosi, Ph., J. C. Hourcade, S. Hallegatte, F. Lecocq, P. Dumas, and M. Ha Duong. 2003. Optimal control models and elicitation of attitudes towards climate damages. *Environmental Modeling and Assessment* 8(3): 133–47.

Andronova, N. G., and M. E. Schlesinger. 2001. Objective estimation of the probability density function for climate sensitivity. *Journal of Geophysical Research* 106(D19): 22605–11.

Böhringer, C., A. Löschel, and T. F. Rutherford. 2005. Efficiency gains from "what"—flexibility in Climate Policy—An integrated CGE assessment. *Energy Journal*, forthcoming.

Caldeira, K., A. K. Jain, and M. I. Hoffert. 2003. Climate sensitivity uncertainty and the need for energy without CO_2 emission. *Science* 28(5515): 2052–54.

Council of the European Union. 2004. Press Release: 2632nd Council Meeting. Council of the European Union: Brussels (Belgium).

ECF. 2004. What is dangerous climate change? Initial results of a symposium on Key Vulnerable Regions, Climate Change and Article 2 of the UNFCCC. ECF and PIK.

den Elzen, M. G. J., and M. Meinshausen. 2005a. Emission implications of long-term climate targets. In H. J. Schellnhuber et al., eds., *Avoiding Dangerous Climate Change*. Cambridge: Cambridge University Press, pp. 299–310.

den Elzen, M. G. J., and M. Meinshausen. 2005b. Meeting the EU 2°C climate target: Global and regional emission implications. 44. Netherlands Environemental Assessment Agency (MNP associated the RIVM), Bilthoven.

Frame, D. J., B. B. B. Booth, J. A. Kettleborough, D. A. Stainforth, J. M. Gregory, M. Collins, and M. R. Allen. 2005. Constraining climate forecasts: The role of prior assumptions. *Geophysical Research Letters* 32:L09702, pp. doi:10.1029/2004GL022241.

Forest, C. E., P. H. Stone, A. P. Sokolov, M. R. Allen, and M. D. Webster. 2002. Quantifying uncertainties in climate system properties with the use of recent climate observations. *Science* 295(5552): 113–17.

Gregory, J. M., R. J. Stouffer, S. C. B. Raper, P. A. Stott, and N. A. Rayner. 2002. An observationally based estimate of the climate sensitivity. *Journal of Climate* 15(22): 3117–21.

Ha Duong, M., M. Grubb, and J. C. Hourcade. 1997. Influence of socioeconomic inertia and uncertainty on optimal CO_2-emission abatement. *Nature* 390: 270–74.

Hammitt, J. K., R. J. Lempert, and M. E. Schlesinger. 1992. A sequential-decision strategy for abating climate change. *Nature* 357: 315–18.

Hare, B., and M. Meinshausen. 2004. How much warming are we committed to and how much can be avoided? 45. Postdam Institute for Climate Impact Research (PIK), Potsdam, Germany.

Hitz, S., and J. Smith. 2004. Estimating global impacts from climate change. In J. Corfee-Morlot and S. Agrawala, eds., *The benefits of climate change policies: analytical and framework issues*. Paris: OCDE/OECD, pp. 31–82.

Houghton, J. T., Y. Ding, D. J. Griggs, P. J. Noguer, P. J. van der Linden, X. Dai, K. Maskell, and C. A. Johnson, eds. 2001. *Climate Change 2001: The Scientific Basis. Contribution of Working Group I to the Third Assessment Report of the Intergovernmental Panel on Climate Change*. Cambridge: Cambridge University Press.

International Climate Change Taskforce. 2005. Meeting the climate challenge: recommandations of the International Climate Change Taskforce.

IPCC WGI. 2004. "Workshop on Climate Sensitivity." 177. IPCC, Paris.

Kelly, D., C. D. Kolstad, M. E. Schlesinger, and N. G. Andronova. 2000. Learning about climate sensitivity from the instrumental temperature record.

Knutti, R., T. F. Stocker, F. Joos, and G.-K. Plattner. 2003. Probabilistic climate change projections using neural networks. *Climate Dynamics* 21: 257–72.

Knutti, R., T. F. Stocker, F. Joos, and G.-K. Plattner. 2002. Constraints on radiative forcing and future climate change from observations and climate model ensembles. *Nature* 416(6882): 719–23.

Krause, F., W. Bach, and J. Koomey. 1989. Energy policy in the greenhouse. Final report. Vol. 1: From warming fate to warming limit: benchmarks for a global climate convention. International Project for Sustainable Energy Paths (IPSEP), El Cerrito, CA.

Kriegler, E., and T. Bruckner. 2004. Sensitivity analysis of emissions corridors for the 21st century. *Climatic Change* 66(3): 345–87.

Leach, A. 2005. The Climate Change Learning Curve. 33. HEC Montreal, Montreal.

Leemans, R., and A. van Vliet. 2005. Responses of species to changes in climate determine climate protection targets. In H. J. Schellnhuber et al., eds., *Avoiding Dangerous Climate Change*. Cambridge: Cambridge University Press, pp. 135–41.

Leemans, R., and B. Eickhout. 2004. Another reason for concern: regional and global impacts on ecosystems for different levels of climate change. *Global Environmental Change* 14: 219–28.

Lecocq, F. 2000. Distribution spatiale et temporelle des coûts des politiques publiques sous incertitudes: théorie et pratique dans le cas de l'effet de serre. *Sciences de l'environnement*. Paris: Ecole Nationale du Génie Rural, des Eaux et Forêts (ENGREF).

Lempert, R. J. 2002. A new decision sciences for complex systems. *Proceedings of the National Academy of Sciences (PNAS)* 99(suppl. 3): 7309–13.

Lempert, R. J., and M. E. Schlesinger. 2001. Climate-change strategy needs to be robust. *Nature* 412(26 July): 375.

Lempert, R. J., M. E. Schlesinger, and J. K. Hammitt. 1994. The impact of potential abrupt climate changes on near-term policy choices. *Climatic Change* 24: 351–76.

Manne, A. S., and R. Richels. 2005. Global climate decisions under uncertainty. International Energy Workshop 2005 (IEW 2005), Kyoto.

Mastrandea, M. D., and S. H. Schneider. 2004. Probabilistic integrated assessment of "dangerous" climate change. *Science* 304: 571–75.

Meehl, G. A., M. A. Washington, W. D. Collins, J. M. Arblaster, A. Hu, L. E. Buja, W. G. Strand, and H. Teng. 2005. How much more global warming and sea level rise? *Science* 307(5716): 1769–72.

Meinshausen, M. 2005. On the risk of overshooting +2°C, in H. J. Schellnhuber et al., eds., *Avoiding Dangerous Climate Change*. Cambridge: Cambridge University Press, pp. 265–79.

Metz, B., D. Ogunlade, R. Swart, and J. Pan, eds. 2001. *Climate Change 2001: Mitigation. Contribution of Working Group III to the Third Assessment Report of the Intergovernmental Panel on Climate Change*. Cambridge: Cambridge University Press.

Murphy, J. M., D. M. H. Sexton, D. N. Barnett, G. S. Jones, M. J. Webb, M. Collins, and D. A. Stainforth. 2004. Quantification of modelling uncertainties in a large ensemble of climate change simulations. *Nature* 430(7001): 768–72.

Nakicenovic, N., ed. 2000. *Special Report on Emissions Scenarios: A special Report of Working Group III of the Intergovernmental Panel on Climate Change*. Cambridge: Cambridge University Press.

National Academies. 2003. Estimating climate sensitivity: Report of a workshop. 41. Steering committee on Probabilistic Estimates of Climate Sensitivity, Board on Atmospheric Sciences and Climate, Division of Earth and Life Studies, the National Academies, Washington, DC.

Nordhaus, W., and J. G. Boyer. 2000. *Warming the World: Economics Models of Climate Change*. Cambridge: MIT Press.

O'Neill, B. C., and M. Oppenheimer. 2004. Climate change impacts are sensitivie to the concentration stabilisation path. *PNAS* 1001(47): 16411–17.

Parry, M., N. Arnell, T. McMichael, R. Nicholls, P. Martens, S. Kovats, M. Livermore, C Rosenzweig, A. Iglesias, and G. Fischer. 2001. Millons at risk: Defining critical climate change threats and targets. *Global Environmental Change* 11: 181–83.

Richels, R., A. S. Manne, and T. M. L. Wigley. 2004. Moving beyond concentrations: The challenge of limiting temperature change. AEI-Brookings joint Center for Regulatory Studies, Washington, DC.

Schneider, S. H., and S. L. Thompson. 1981. Atmospheric CO_2 and climate: importance of the transient response. *Journal of Geophysical Research* 86: 3135–47.

Stainforth, D. A., T. Aina, C. Christensen, M. Collins, N. Faull, D. J. Frame, J. A. Kettleborough, S. Knight, A. Martin, J. M. Murphy, C. Piani, D. Sexton, L. A. Smith, R. A Spicer, A. J. Thorpe, and M. R. Allen. 2005. Uncertainty in predictions of the climate response to rising levels of greenhouse gases. *Nature* 433: 7024.

Swart, R., M. Berk, M. Janssen, E. Kreileman, and R. Leemans. 1998. The safe landing approach: Risks and trade-offs in climate change. In J. Alcamo, R. Leemans, and E. Kreileman, eds., *Global change scenarios of the 21st Century: Results from the IMAGE 2.1 Model.* Oxford: Pergamon/Elsevier Science, pp. 193–218.

Toth, F. L., T. Bruckner, H.-M. Füssel, M. Leimbach, and G. Petschel-Held. 2003a. Integrated assessment of long-term climate policies. Part 1: Model presentation. *Climatic Change* 56(1–2): 37–56.

Toth, F. L., T. Bruckner, H.-M. Füssel, M. Leimbach, and G. Petschel-Held. 2003b. Integrated assessment of long-term climate policies. Part 2: Model results and uncertainty analysis. *Climatic Change* 56(1–2): 37–56.

UNFCCC. 1992. United Nations Framework Convention on Climate Change (UNFCCC).

Warren, R. 2005. "Impacts of Global Climate Change at Different Annual Mean Global Temperature Increases.", in H. J. Schellnhuber (Chief Ed.), *Avoiding Dangerous Climate Change*, Cambridge University Press, pp. 93–131.

Wigley, T. M. L. 2005. The climate change commitment. *Science* 307(5716): 1766–69.

Wigley, T. M. L., and S. C. B. Raper. 2001. Interpretation of high projection for global-mean warming. *Science* 293(5529): 451–54.

Yohe, G., N. G. Andronova, and M. E. Schlesinger. 2004. To hedge or not against an uncertain climate future? *Science* 306: 416–17.

16

Leakage from Climate Policies and Border-Tax Adjustment: Lessons from a Geographic Model of the Cement Industry

Damien Demailly and Philippe Quirion

A recurrent concern raised by industry against climate policies is the fear of competitive distortions, industrial relocations, and carbon leakage in case of asymmetric constraints. The recent entry into force of the Kyoto Protocol is unlikely to reduce these concerns since the developed countries that have ratified it only account for 35 percent of world energy-related CO_2 emissions (Enerdata 2005, p. 9).

Intuition suggests that some industrial sectors may be affected by strong carbon asymmetric constraints. However, it should also be noted that many carbon-intensive sectors are typically weakly exposed to international competition. International trade of cement for example accounts for less than 7 percent of the world consumption, mostly because of the existence of significant transportation costs.

Current representations of international trade such as the well-known Armington (1969) specification and similar functional forms make it difficult to assess both the magnitude and the determinant of carbon leakage. The Armington specification assumes that products are differentiated by their place of production. For example, the chemicals produced by different countries are not considered perfect substitutes. In applied models this tool is used in such a way that it takes all the grounds for imperfect substitution as one—heterogeneity of products throughout the world, of national preferences, and of transportation costs. This takes the form of the Armington substitution elasticity, or a parameter with an equivalent meaning, which is either econometrically calibrated—what is difficult[1]—or merely guesstimated.

Our intuition is that, even though the use of the Armington specification is probably the best compromise for most sectors, especially aggregated ones, progress can be made through an alternative approach for the sectors dealing with relatively homogeneous products whose trade is not much affected by national preferences, and where transportation costs and capacity constraints are central to explain international trade patterns. Many GHG-intensive sectors fit with these characteristics.

Such an alternative must have three objectives. First, it has to satisfactorily represent transportation costs, notably by not treating markets as dimensionless points. Second, it must explicitly take into account capacity shortages. Third, investment decisions in new production capacities have to be modeled realistically.

We developed a spatial international trade model, GEO, that serves these objectives in a number of ways:

• Drops the imperfect substitution assumption among goods produced in different places.
• Makes explicit the transportation costs, for both road and sea transportation, utilizing a spatial representation of the world including 15,500 consuming "areas."
• Represents the competition among producers in every consumption area, taking into account their differentiated marginal production costs, transportation costs and capacity shortages by assuming that producers subjected to such a constraint deliver their production in the most profitable areas.
• Justifies investment decisions by explicitly representing the producer's expectation of which amount of product can be sold and where.

We applied this model to cement for three reasons. First, the characteristics of the cement sector are particularly suited to the use of GEO. Second, this sector is an important greenhouse gas emitter: it accounts for around 5 percent of global anthropogenic CO_2 emissions (IEA 1999). Third, it is potentially one of the most impacted by a climate policy: among twelve EU15 industry sectors, nonmetallic minerals—mostly cement—have the second direct CO_2 emission/turnover ratio (Quirion and Hourcade, 2004).[2] Cement manufacturers thus claim that an ambitious climate policy would impose an additional burden that may jeopardize their competitiveness and induce carbon leakage (e.g., British Cement Association 2004).

To represent the cement industry, we use a modified version of CEMSIM, a recursive bottom-up model built by the IPTS team (see Szabo et al. 2003, 2006). GEO and CEMSIM are integrated, allowing us to build a business-as-usual scenario until 2030 and three climate policy scenarios.

A CO_2 tax at 15 euros per ton, which is equivalent to auctioned emission allowances with the same price, in the Kyoto Protocol annex-B countries that have ratified it (hereafter referred to as "the annex B-R") entails significant emissions reductions in these countries. However, an important carbon leakage occurs.

The same policy with border-tax adjustments (BTA), namely a rebate on cement exports and a taxation of imported cement in annex B-R countries, is simulated. Two BTAs are tested. In the first scenario, all exported production is exempted from the climate policy and imports of cement from the rest of the world are taxed in accordance with the CO_2 intensity of the cement production in the exporting country. In the second BTA scenario, exports benefit from a rebate corresponding only to the least CO_2-intensive technology available at a large scale, and all imports are taxed to the same level. Such a system is proposed by Ismer and Neuhoff (2004) who argue that it is compatible with the WTO rules, contrary to the first one we test. In the two BTA scenarios the carbon leakage decreases. It is even replaced by a slight spillover in the first one. However, in both cases, the cement price in annex B-R countries increases more than without BTA, further impacting the cement consumers in these countries.

16.1 The GEO-CEMSIM Model

16.1.1 GEO

In GEO, the world is modeled as an ensemble of cement consumption "areas," on the one hand, and producing countries, on the other hand.[5] An area is characterized by its geographical position on the globe. The areas used are the $1° \times 1°$ squares defined in the EDGAR database (RIVM 2001), but we subdivided the squares with a high population and dropped the squares where the population is negligible (figure 16.1). Because of the computation constraints, we assume that the totality of the market of a given area is taken by the producers of only one country which, given that we have 15,500 areas, is an acceptable approximation, at least in a first step.

A producing country is first characterized by its variable production cost, its production capacity, and the intensity of the competition among its domestic producers. We assume that a Cournot oligopoly competition takes place among producers of the same country, since it is well known that the cement market has far from pure competition (Johnson and Parkman 1983). A country is also characterized by

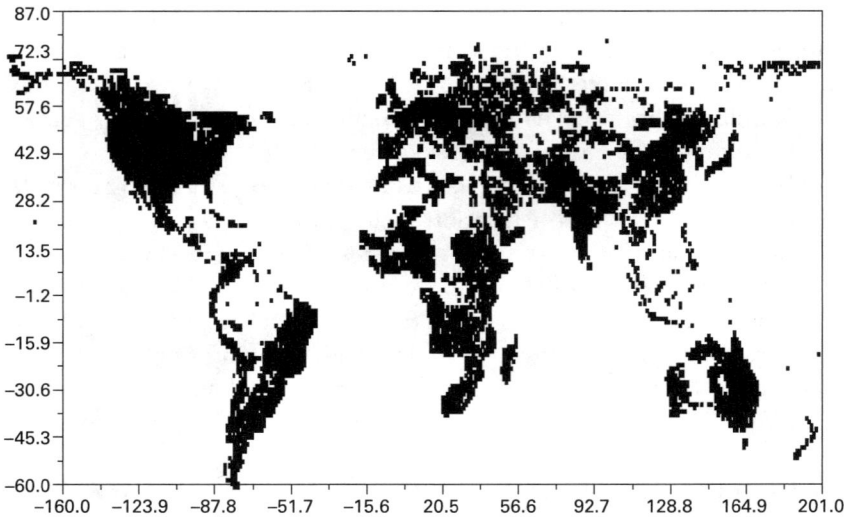

Figure 16.1
Areas of GEO

harbors able to trade cement. There are 1,600 such sea harbors in the world, according to Lloyd's list (2004). 7,500 border posts, which we label "land harbors," are defined every 25 km on land borders in order to allow for modeling land trade.

GEO then calculates the minimum transportation cost from every producing country to every area, using road national transportation costs and international sea transportation costs. A fixed and a variable transportation cost are distinguished for each transportation mode.

We assume that a producing country is ready to sell its production in an area at any price bigger than the sum of its variable cost and the transportation cost to this area, subject to a capacity constraint. When the latter is binding, a producing country sells its production in the most profitable areas. Of course, the set of "most profitable areas" depends on other producing countries' behavior, hence the need for an adequate algorithm to determine simultaneously what supplier takes each area. The cement price a firm applies in an area is limited by a double competition pressure: international competition from the other producing countries (Bertrand competition), and national competition pressure from firms of the same producing country (Cournot competition). The number of firms in the Cournot model is calibrated to match the price/cost margin in the calibration year (1997), and assumed identical thereafter.

> **GEO in Game Theory**
>
> For GEO equilibrium, every country c must maximize, subject to a capacity constraint, its profit realized on all areas of the world. It is defined by: $\Pi_c = \sum_{area} \Pi_{c,area}$, where
>
> $\Pi_{c,area} = \max[0, (P_{c,area} - VC_c - TC_{c \to area}) \cdot conso_{area}]$.
>
> VC_c is the variable cost of country c, $TC_{c \to area}$ its transportation cost to the area considered, and $conso_{area}$ the consumption in the latter. $P_{c,area}$ is the price country c would apply in the area if it would take the market. It is defined by
>
> $P_{c,area} = \min[P_{c,area}^{Cournot}, P_{c,area}^{Bertrand}]$,
>
> $P_{c,area}^{Cournot} = (VC_c + TC_{c \to area}) \cdot \left(1 + \dfrac{1}{N_c \varepsilon - 1}\right)$,
>
> $P_{c,area}^{Bertrand} = \min_{i \neq c}[VC_i + TC_{i \to area} + \lambda_i]$,
>
> where ε is the price elasticity of the demand Countries are subjected to a capacity constraint and λ_i is the Lagrange multiplier of this constraint for country i.
>
> GEO leads to the Nash equilibrium of the game found by a *tâtonnement* process. During the game Lagrange multipliers increase from 0 until all capacity constraints are saturated.

In every area the cement supplier thus applies a profit margin[4] that is the minimum between the profit margin defined by the international competition and the profit margin defined by the national oligopolistic competition. Using the variable cost and capacity constraint of every country as well as the minimum transportation cost between every producing country and every consuming area, GEO gives for every area the cement price and where it comes from. At the country level it gives the production and the average cement price (which is the weighted sum of the prices in the areas of this country).

Every country has a capacity constraint, which is not fixed but may be relaxed every year by investing in new capacities. In GEO a country builds new capacities for the market of a given area if it expects not only to sell its new production there but also to cover its fixed construction cost, despite the competition of the existing and future capacities of the others.

In order not to mislead the reader of our quantitative results, it is useful to place here two caveats. First, we model no inertia in trade, whereas in the real world, for a cement manufacturer, exporting to a new market requires time, notably to develop a distribution network. Second, the assumption of Bertrand competition among producers of different countries seems too harsh because there is some oligopolistic behavior among them. However, it is the best compromise we found to date between modeling constraints and realism. As a consequence real world changes are likely to be smoother and less intense than modeled.

16.1.2 CEMSIM

An inverted U-shape curve of "intensity of use" relates the evolution of cement consumption to the per capita GDP. As in the original CEMSIM IPTS model, the demand curve for cement is assumed isoelastic, with a price elasticity of 0.2, a value close to that estimated by La Cour and Mollgaard (2002, cited and used by IEA 2004).

CEMSIM pays particular attention to fuel and technology dynamics. Seven technologies are included, characterized by energy, material, and labor consumptions; an investment cost; and a set of retrofitting options. We modified the original CEMSIM model to introduce more flexibility in the content of clinker, the carbon-intensive intermediary product, in cement and in the choice of nonprimary fuels, following discussions with French cement industrials.

We stress that the quantification of some technical flexibility (clinker ratio, retrofitting, and fuel choice) is very difficult, so our quantitative results should be taken with some caution. In addition CEMSIM does not include any breakthrough technology such as alternatives to limestone-based cement, which could dramatically reduce CO_2 emissions (Prebay et al. 2006), because the cost of these technologies is difficult to assess.

The main exogenous variables of CEMSIM are GDP, population, electricity, and primary fuel prices, all taken from the POLES model developed by LEPII-EPE. Primary fuel prices are higher under business-as-usual than under mitigation policies, since in POLES these policies reduce fuel demand, thus world fuel prices. Prices of other fuels (waste and wood fuels, petroleum coke) are calibrated.

We use 1998 and 1997 data on consumption, production capacity, energy demand (CEMBUREAU 1999, 2002), and cement bilateral trade (OECD series C) to calibrate the GEO-CEMSIM model, which is then recursively run with a yearly step. (For further details on the combination of GEO and CEMSIM, see the appendix; for further details on

GEO and CEMSIM, see Demailly and Quirion 2005, or Szabo et al. 2003 for the latter.)

16.2 World Cement Industry in the Business-as-Usual Scenario

The business-as-usual (BAU) scenario is a necessary preliminary step to assess the impacts of a carbon mitigation policy. Moreover it provides interesting insights. To present the results, we aggregate the 47 producing regions of our model to form 12 regions:

1. Europe: EU25, Bulgaria, Romania, and the rest of Western Europe
2. R&U: Russia and Ukraine
3. Japan
4. Canada
5. USA
6. RJAN: Rest of "Japan, Australia, and New Zealand" (mostly Australia and New Zealand)
7. TRR: Turkey, rest of the ex-USSR and rest of Central and Eastern Europe
8. LAM: Latin America
9. India
10. China
11. ROA: Rest of Asia
12. A&ME: Africa and Middle-East

The first four regions have ratified the Kyoto Protocol and will implement climate policies in the next sections.[5] We label them the annex B-R countries.

16.2.1 Increasing Share of Developing Countries in Worldwide Consumption

At the world level, cement consumption is estimated to increase from 1630 Mt in 2000 to 2900 Mt in 2030, corresponding to an annual 2 percent growth rate

At the regional level, the evolution of cement consumption is highly dependent on the inverted U-shape hypothesis for the consumption path. The model predicts a high growth in developing regions. China and R&U peak around 2020, whereas the consumption in India, TRR,

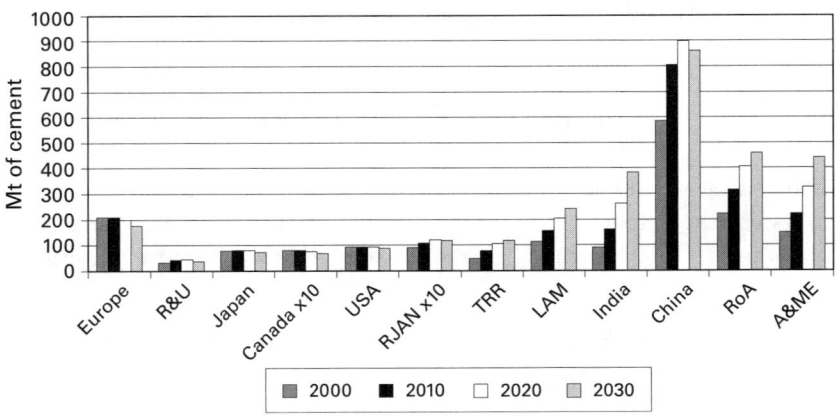

Figure 16.2
Consumption in BAU

A&ME, LAM, and ROA is still growing in 2030. On the contrary, no developed region sees its consumption growing after 2020. Whereas these regions represented 24 percent of the world consumption in 2000, they are projected to represent only 13 percent in 2030.

16.2.2 Domestic Excess Capacities and Decrease in Cement Trade Flows

A feature of international cement trade is that according to our model and as in the real world according to industry experts, very few capacities are built in order to export. Almost all the cement traded come from "domestic excess capacities." The domestic excess capacities are the capacities built for the domestic market but not fully utilized. For example, such capacities exist in countries with growing demand because its producers anticipate this growth by oversizing their new plants. The higher the growth of the consumption in a country, the higher is the amount of domestic excess capacities of its producers.

Since very few export capacities are built, some countries see their exports limited by their capacity. Had they bigger domestic excess capacities, they would export more. Why don't they build export capacities? Because the expected gains of such capacities would not be sufficient to cover their investment cost.

According to our model the intensity of international trade drops between 2002 and 2004 from 7 to 4 percent of the world production

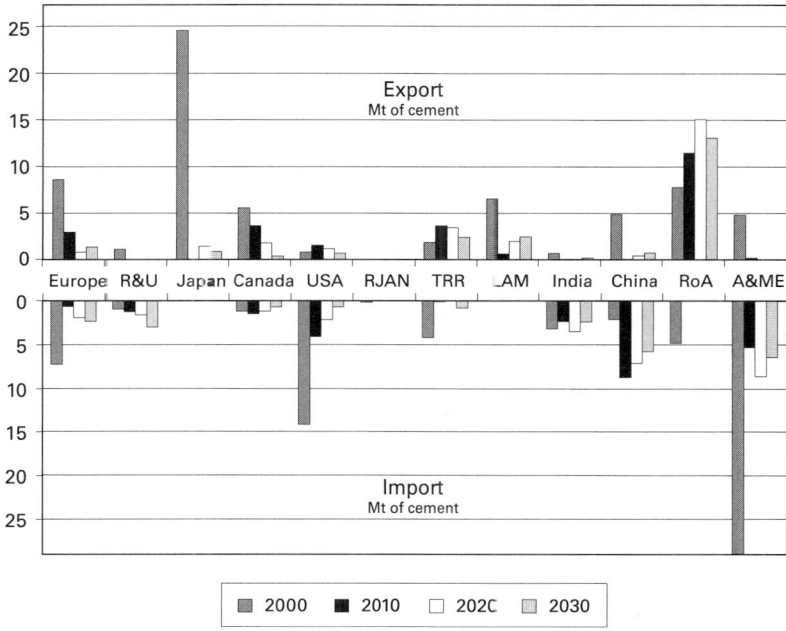

Figure 16.3
Exports and imports in BAU

because of the exogenous increase in sea transportation costs between these two time periods. This increase, observed in reality, is due to the scarcity of transport capacities, which originates in their intensive use to supply Chinese economic growth. Despite the stabilization of the sea transportation costs after 2004, the intensity of international trade keeps on decreasing until 2030 in our model. This is mostly because, after 2010, the growth of the consumption slows down in many countries. Therefore their amounts of domestic excess capacities drop, and so does their ability to export.

In a few cases cement trade is due to a lack of production capacities in the importing country (e.g., the Netherlands around 2000). But it is generally driven by differences in production costs and, as we have just seen, may be limited by capacity shortages.

Concerning the annex B-R countries, we observe that European imports drop after the increase in sea transportation costs and come mostly from TRR, A&ME, and LAM. Such a drop occurs in Canada, which only keeps on importing cement from USA. R&U imports some cement during all the simulation, from Europe, TRR, ROA, and China, whereas Japan does not.

After the increase in sea transportation costs, exports of European, Japanese, and R&U producers drop. Then European exports focus on the nearest countries. Since Canada uses mostly road to export cement to USA, its main client, Canadian exports are not so much impacted by this increase.

For most of the countries, cement trade is marginal, the ratio (export-import)/production being very rarely higher than 10 percent in absolute value. Therefore, in general, the national production almost equals the national consumption.

16.2.3 CO_2 Emission/Production Relative Decoupling

CO_2 emissions are projected to grow by 55 percent from 1320 $MtCO_2$ in 2000 to 2035 in 2030 corresponding to a 1.5 percent average annual growth rate. The relative decoupling with the consumption, which grows on average by 2 percent a year, is mostly due to the decrease in fuel consumption per ton of cement, thanks to the use of more efficient machines.

Unsurprisingly, the spatial distribution of emissions is roughly correlated with the increase in production. China's share of world emissions decreases through time but it remains the largest CO_2 emitter with more than 30 percent of the world emissions in 2030. The share of developed countries drops from 22 to 12 percent by 2030.

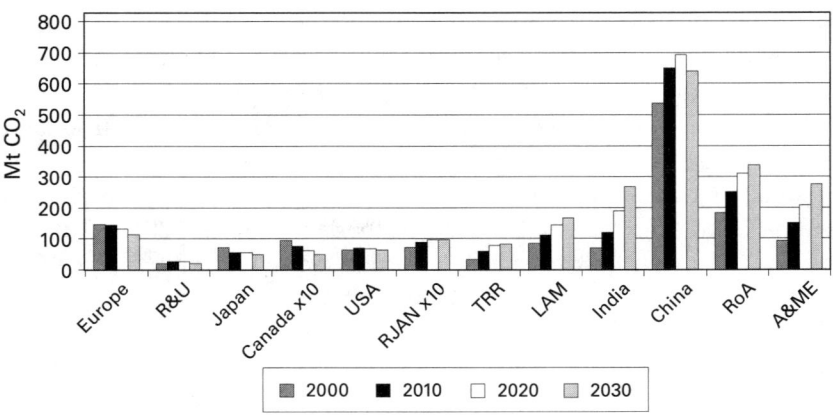

Figure 16.4
CO_2 emissions in BAU

16.3 Climate Policy without Border-Tax Adjustment Scenario ("No BTA")

16.3.1 Definition of No BTA

In the no BTA scenario, we assume that annex B-R countries implement a CO_2 tax or a CO_2 emission trading scheme[6] with auctioned allowances, without revenue recycling (thereafter: "the climate policy"). For 2008 to 2012 we rely on the estimation of the POLES model, assuming that Russia and Ukraine use their market power to raise the international CO_2 price up to 15 euros per ton (Szabo et al. 2003). We assume that this price is sustained until the end of the simulation period and that no non–annex B-R country takes on emission targets until 2030. We do not take into account the Clean Development Mechanism, since very few CDM projects are implemented in the cement sector.

This climate policy cannot be considered as the most likely outcome of the climate negotiations but has the advantage of simplicity as a benchmark for comparative analysis.

16.3.2 Technological Changes Triggered by the Carbon Value

The carbon value triggers different mechanisms in CEMSIM-GEO: reduction of cement demand due to the increase in production costs and in prices; substitution between clinker and added materials in cement composition; substitution between high- and low-carbon fuels (from coal, oil, and petroleum coke to gas, waste, and wood fuels); retrofitting of carbon-intensive technologies to low-carbon technologies; changes in technological choices for new plants.

Compared with BAU, variable production costs in annex B-R countries increase in average by 10€ per ton of cement (30 percent) from 2008 on.

Regarding the non–annex B-R countries, they generally benefit from lower production costs, thanks to the decrease in annex B-R demand for carbon-intensive fuels, and therefore in prices.

16.3.3 Significant Impacts of the Climate Policy on Trade Flows but No Building of Export Capacities

Consequently the climate policy has significant impacts on cement trade. Compared with the BAU case, Europe, Canada, and Japan stop

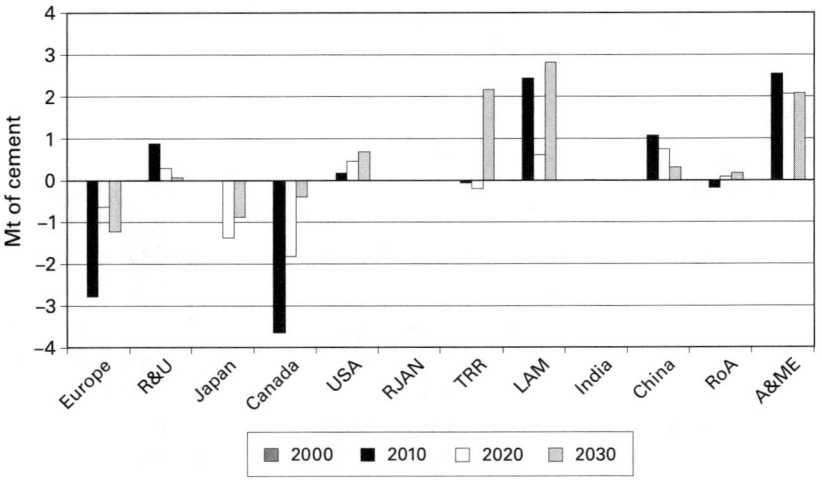

Figure 16.5
No BTA: Change in exports compared to BAU

exporting. Inversely R&U, which proves to be less impacted than European countries as we will see below, increases its exports to Europe.

Compared with BAU, Canada increases its imports from 19 to 26 percent of its consumption in 2010,[7] R&U increases from 3 to 10 percent and Europe from 1 to 4 percent. Japan, as in BAU, does not import cement: the carbon constraint is not strong enough to outweigh its cost competitiveness observed in BAU.

One interesting point is that only few capacities are built in non–annex B-R to export in annex B-R, despite the higher production cost of the latter. Exports keep on coming from domestic excess capacities, although the exports of some non–annex B-R countries are not limited by transport costs but by capacity shortages. This illustrates the fact that the rise in annex B-R production costs is not high enough to outweigh not only the transportation costs but also the investment costs.

16.3.4 Significant Drop in the Production of Annex B-R

The growth in variable production costs highly impacts the industry of annex B-R countries for two reasons: fall in domestic consumption (3 percent on average in 2010, 4 percent after) and lower market shares in the world cement market. Finally, the production of the Annex B-R countries drops in average by 7.5 percent in 2010, 2020, and 2030.

Chapter 16 Leakage from Climate Policies and Border-Tax Adjustment   345

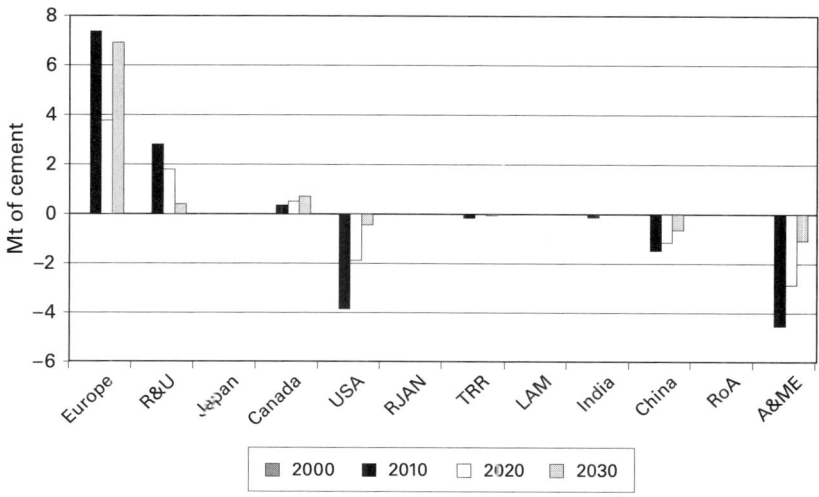

Figure 16.6
No BTA: Change in imports compared to BAU

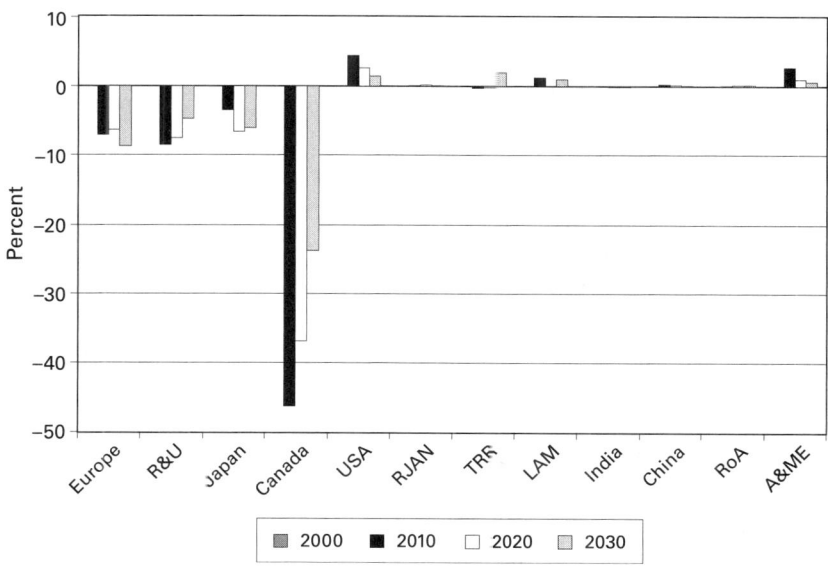

Figure 16.7
No BTA: Production compared to BAU

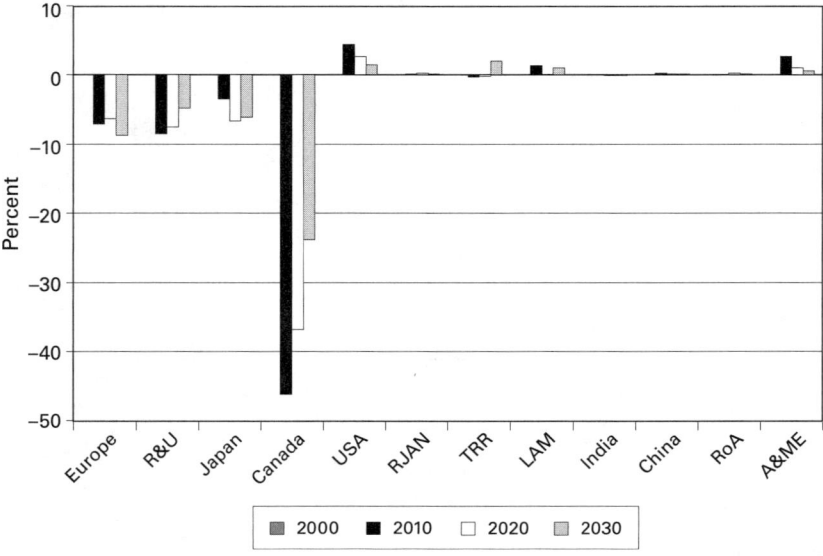

Figure 16.8
No BTA: CO_2 emissions compared to BAU

16.3.5 Significant Drop in CO_2 Emissions from Annex B-R and Significant Carbon Leakage

Emissions per ton of cement in annex B-R countries decrease with the implementation of the climate policy: from −12 percent in 2010 to −15 percent in 2030. The magnitude of this drop is roughly the same in all annex B-R countries, R&U excepted which turns out to have more technical flexibility.

Cumulated with the fall in production, this decarbonization leads to a decrease in carbon emissions of annex B-R ranging from 18 percent in 2010 up to 22 percent in 2030.

Part of these reductions is compensated by an emissions increase in non−annex B-R countries. These countries are less carbon efficient than annex B-R countries and the gap increases with the implementation of the climate policy.[8] The resulting leakage rate (emissions increase in countries outside the annex B-R divided by the emissions decrease in annex B-R countries) equals 25 percent in 2010, 13 percent in 2020, and 16 percent in 2030. These figures are close to the upper bound of the range of economywide leakage estimates of 5 to 20 percent presented in the IPCC third assessment report (see Hourcade and Shukla 2001). All in all, world emissions decrease by around 2 percent in 2010, 2020, and 2030.

16.4 Climate Policy with Border-Tax Adjustments Scenarios

16.4.1 Definition of the Two BTA Scenarios

One way of preventing carbon leakage and limiting the effects on competitiveness of a fragmented climate regime is to impose border-tax adjustments (BTA): tax exemption of GHG-intensive products and materials exported to non–annex B-R countries; border tax on the importation of these products and materials from outside annex B-R.

Using analytical models, several authors have demonstrated the rationale for BTA for dealing with international pollutions: Markusen (1975), Hoel (1996), and Maestad (1998) In particular, Hoel (1996) showed that BTA are a better response to pollution leakage than the usually applied differentiation of the tax level between the exposed and the sheltered sector. More recently Mathiesen and Mæstad (2002), with a partial equilibrium world model of the steel industry, and Majocchi and Missaglia (2001), with a general equilibrium model, quantified the impact of BTA. It turns out that BTA efficiently prevent adverse impact on the domestic industry of a carbon constraint.

We first assess this system in the "complete BTA" scenario, but its compatibility with the WTO/GATT is controversial; see Hoerner (1998) for an early discussion and Ismer and Neuhoff (2004) for an up-to-date synthesis. The latter two authors conclude that to be WTO compatible, the BTA should be set "at the level of additional costs incurred for procurement of CO_2 emission permits during production of processed materials using the best available technology." This is why, without pushing this juridical discussion further, we provide an application of the BTA proposed by Ismer and Neuhoff in the WTO BTA scenario. We take as best available technology the dry rotary kiln with pre-heater and pre-calciner fueled by natural gas.[9]

In the rest of this section we address two questions: Do the two BTA scenarios effectively prevent CO_2 leakage, and could non–annex B-R countries attack these systems on the ground that they suffer too much of them?

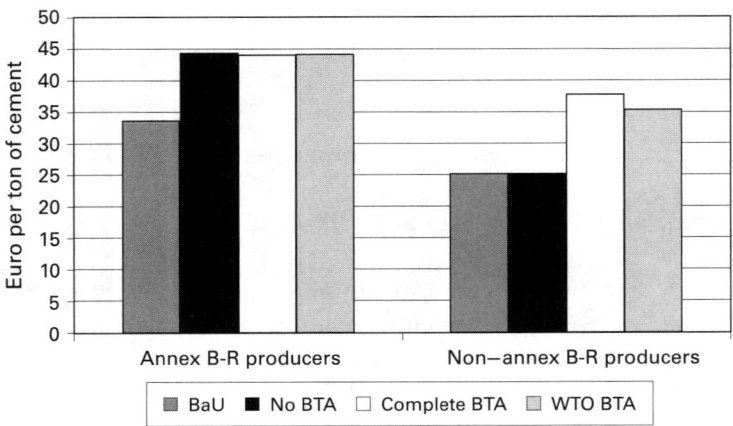

Figure 16.9
Variable production cost in 2010 on annex B-R markets

16.4.2 Differentiated Impacts on Cost Competitiveness

We now present the results of the two BTA scenarios in comparison with BAU. For simplicity sake, we present the results for annex B-R and non–annex B-R countries in aggregate, and for 2010 only. We insist on the differences of the impacts inside and outside annex B-R markets.

Annex B-R Markets
Under *complete* BTA, annex B-R variable production cost in annex B-R markets increases in average by 10.5€ per ton of cement in 2010. It increases in average by 12.5€ for the non–annex B-R countries. Thus competition terms on annex B-R markets are modified in favor of annex B-R countries, which are in general more carbon efficient than the others. Indeed, not only do they use more energy-efficient technologies and less carbon-intensive fuels already in BAU, but the climate policy also leads them to reduce their CO_2 emissions per ton of cement (especially by decreasing their clinker rates); in contrast, non–annex B-R countries do not. Therefore the complete BTA system tends to improve the cost competitiveness of annex B-R countries in their territory.

Under *WTO* BTA, the "after-tax production cost" of non–annex B-R countries in annex B-R markets is the cost of the least carbon-intensive

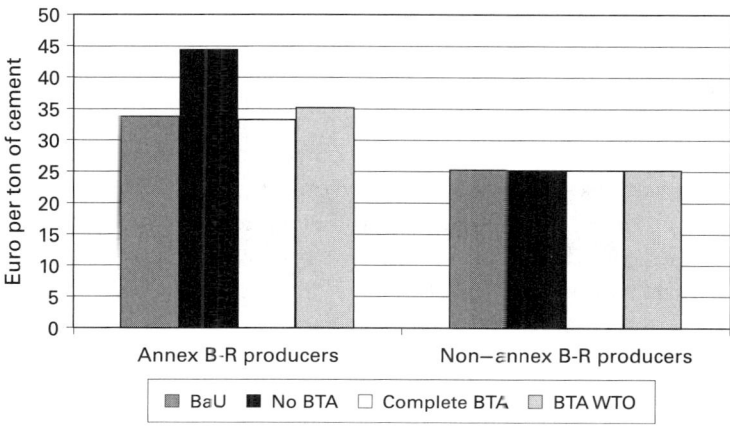

Figure 16.10
Variable production cost in 2010 on non-annex B-R markets

technology. Their variable production cost increases on average by 10€ in 2010. Therefore the BTA WTO system results in a slight degradation of the cost competitiveness of annex B-R countries in their own markets.

Non–Annex B-R Markets

On the markets outside annex B-R, the complete BTA still impacts variable production costs:

• Countries implementing the climate policy speed the retrofitting of their plants toward efficient technologies.

• In the other direction, their average fuel costs increase (in general). This is due to the higher utilization of low C fuels that are, most of the time, more expensive.

Finally, we observe that most of the annex B-R countries increase their cost competitiveness on the markets outside their territory in 2010 (−0.4€ on average). However, these gains only last a few years after the implementation of the system.

Under the WTO BTA, the cost competitiveness of annex B-R countries on the non–annex B-R markets suffers a bit, since the technologies less efficient than the best available technology are partially taxed. Their variable production cost increases, on average by 1.5€ in 2010.

16.4.3 Increase in Annex B-R Net Exports

Annex B-R Markets
In the complete BTA scenario, as we have just seen, most annex B-R countries increase their cost competitiveness inside their territory. Therefore non–annex B-R countries lose some market shares, paving the way to a possible qualification of this system as protectionist. Annex B-R countries decrease their imports not only because of the increase in their competitiveness but also because the cement price increase makes their consumption drop.

Under the WTO BTA scenario, annex B-R countries import more cement from non–annex B-R countries than in BAU, since their cost competitiveness is a bit reduced. This system is thus not protectionist vis-à-vis non–annex B-R countries.

Non–Annex B-R Markets
In the complete BTA scenario, but also to a lesser extent in the WTO BTA scenario, annex B-R countries increase their exports to non–annex B-R. In the latter case, this is due to the increase in capacities available for exports, following the drop in consumption in these countries, which compensates more than the small decrease in cost competitiveness. In the complete BTA case, this is also due to the temporary increase in their cost competitiveness.

To sum up, in the impact of BTA on trade and competitiveness the complete BTA scenario could be qualified as protectionist. Although it treats domestic and foreign producers in a similar way (they pay the same cost per ton of CO_2), it gives a competitive advantage to annex B-R producers, who use cleaner production technologies. In contrast, in the WTO BTA scenario, annex B-R countries suffer from a slightly higher cost increase than their competitors, which results in a small increase in their imports. However, as is apparent from figure 16.11, their exports rise by a larger extent despite this relative variable cost increase because some of their production capacities become available for exports. Therefore, should the WTO BTA policy be considered as distorting competition in favor of annex B-R countries? This seems highly dubious as this increase in net exports originates only in the drop in domestic consumption; the same would occur also following a macroeconomic recession, for example.

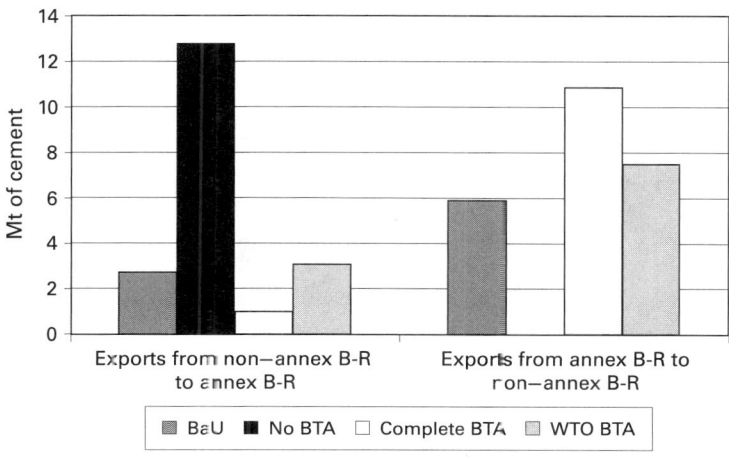

Figure 16.11
Imports and exports in 2010

16.4.4 Higher Consumption Drop but Lower Production Drop with Than without BTA

In the two BTA scenarios domestic prices increase in annex B-R countries more than without BTA: 21 percent compared with BAU in 2010 without BTA, 27 percent with complete BTA, and 26 percent with WTO BTA. Consequently consumption in annex B-R drops more significantly: on average by 4 percent in 2010, in contrast to 3 percent in the no BTA scenario.

From the total production point of view, this drop is, in both scenarios, more than halved by the gains on the international market: production in annex B-R decreases by 2 percent in 2010 under the complete BTA, by 3 percent in 2010 under the WTO BTA, instead of 7.5 percent in no BTA.

It is worth noting that under WTO BTA total production actually rises a little in non–annex B-R countries, which further reduces the rationale for attacking this scenario as distorting competition to the detriment of these countries. Indeed average cement prices in these countries tend to decrease under the higher pressure of the annex B-R, leading to the increase in their consumption. This rise offsets the increase in their net imports.

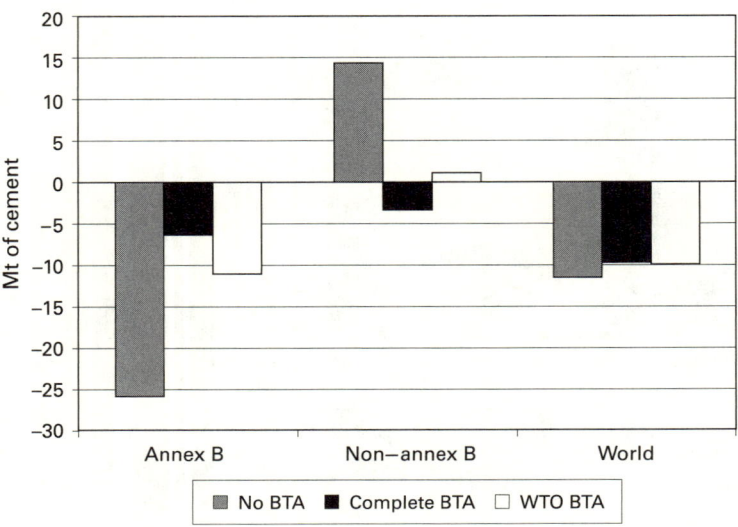

Figure 16.12
Production in 2010 compared to BAU

16.4.5 BTA: Efficient Reduction of the Carbon Leakage

Under complete BTA, annex B-R emissions decrease by around 13 percent in 2010 compared to BAU, less than under no BTA. At the same time non–annex B-R emissions also decrease, because of a little decrease in their production, although very slightly. The spillover rate (abatement in non–annex B-R/abatement in annex B-R) is 6 percent. Finally, world emissions decrease by 2 percent, a little more than under no BTA.

Under the WTO BTA, annex B-R emissions decrease compared to BAU, a little more than under complete BTA, whereas emissions from non–annex B-R increase a little. The slight spillover observed in the complete BTA is replaced by a slight leakage: around 4 percent in 2010. The reduction in world emissions is a little lower than under the complete BTA but higher than under no BTA.

16.5 Conclusions

Some of the messages delivered by our model are straightforward, for example, the fact that a CO_2 tax or auctioned allowances without reve-

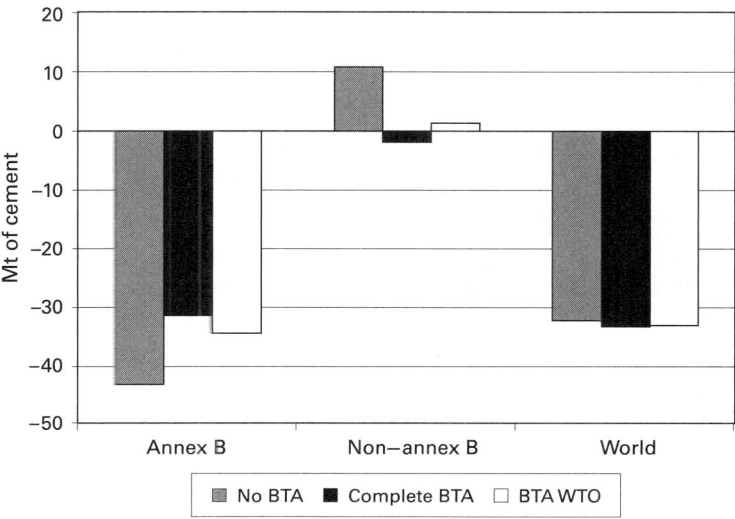

Figure 16.18
CO_2 emission in 2010 compared to BAU

nue recycling at a price of 15€ per ton of CO_2 in the annex B-R cement industry leads to a significant carbon leakage (20 percent), despite the importance of the transportation costs of this product and the capacity shortages. Even though this result does not justify the withdrawal of nonglobal climate policies (Baron 2006), since about 80 percent of the abatement in annex B-R remains, it indicates that these policies should take seriously into consideration the risks of leakage and that the tools able to tackle this issue should be further studied.

Other insights are related to the comparison between the border-tax adjustments systems tested. The complete BTA system prevents efficiently carbon leakage and even leads to a slight positive spillover for annex B-R countries. However, it could lead to WTO conflicts, since it can be accused to be unduly protectionist. The WTO BTA system, which is designed to be WTO compatible, avoids such a risk. It constitutes a less efficient hedging against carbon leakage but realizes a very acceptable environmental achievement: a 4 percent leakage instead of 20 percent without BTA. This moderate environmental loss suggests that this system should be accepted because the environmental efficiency of the complete BTA may be proved to be illusory for political reasons. Note that in both cases, the impact of the BTA systems on cement prices above the carbon constrained scenario without BTA is significant: about 5 percent; this is the price to pay by cement consumers to secure a higher environmental efficiency of the climate policy and to protect employment.

Beyond the comparison of these two types of BTA, we attempted to demonstrate that our approach helps disentangle the mechanisms at stake in carbon leakage that are merged with other issues in the Armington specification, namely the transportation costs, the capacity shortages, and the investment dynamics to expand capacities. One robust conclusion is that for a 15€/t CO_2 tax, even without BTA, there is no incentive in unconstrained countries to create significant new capacities devoted to export. Indeed the rise in production costs of constrained countries is not high enough to outweigh the transportation costs nor the investment costs. Therefore exports keep on coming from capacities built for domestic consumption and not fully used.

It should be noted that our preceding analysis has been conducted assuming a scenario in which international transport remains unaffected by the carbon constraint. Obviously such a constraint would raise the cost of transportation, hence would shelter constrained countries from international competition without taking the form of an explicit tool like BTA. Our next step will be then to scrutinize at what level such a carbon price on international transport would start to offset the impact of an asymmetric constraint.

Appendix: Combination of GEO and CEMSIM

In this appendix we provide further details on the combination of GEO and CEMSIM. For a complete presentation of these two models, see Demailly and Quirion (2005). Like CEMSIM, GEO-CEMSIM is a recursive model: every one-year period, until 2030, five modules are run.

Consumption Module (from CEMSIM)

An inverted U-shape curve of "intensity of use" relates the evolution of cement consumption to the per capita GDP. Consumption in a given country also depends on its average cement price in the previous period, delivered by GEO, following an isoelastic demand curve.

Variable Cost Module (from CEMSIM)

CEMSIM pays particular attention to fuel and technology dynamics. Seven technologies are included, characterized by energy, material, and labor consumptions. Five fuels are differentiated—oil, gas, coal, coke, and waste and wood fuels—whose mix evolves at every period depending on relative fuel prices. CEMSIM is thus able to determine the average variable cost (VC) of every country at every period.

Competition Module (GEO)

Using consumption and VC from previous modules and capacity data from modules 4 and 5 in the previous period, GEO determines trade, production and prices for every country.

Retirement and Retrofitting (from CEMSIM)

Technologies in CEMSIM are also characterized by a lifetime, an investment cost and a set of retrofitting options. Thus, at the start of every period, some plants retire and some are retrofitted.

New Investment Module (GEO)

As new plants are built, what determines the capacity constraint of countries for the following period? The amount of new capacities built is determined by a version of GEO where existing capacities compete with "capacities that have to be built": existing capacities compete with VC while capacities that have to be built compete with VC plus their investment cost. Thus a country builds new capacities for the market of a given area if it expects not only to sell its new production there but also to cover its fixed construction cost, despite the competition of the existing and future capacities of the others.

A feature of cement international trade is that almost all the cement traded comes from "domestic excess capacities." The domestic excess capacities are the capacities built for the domestic market but that are not fully used for it. For example, such capacities exist in countries with growing demand because its producers anticipate this growth by oversizing their new plants. In order to represent the existence of domestic excess capacities in our model, a country building new capacities in period t takes into account consumption expected in period $t + N$, where N is calibrated to fit real data.

Notes

The present analysis has benefited from a deep collaboration with the Institute for Prospective Technological Studies (IPTS—Joint Research Centre—European Commission). Our analysis is partly based on the world cement model CEMSIM developed by L. Szabo, I. Hidalgo, J. C. Ciscar, A. Soria, and P. Russ, from the IPTS. We thank them and the IPTS for the explanations on the model, for the free access to a world cement industry database compatible with the model structure and for having hosted one of us at the IPTS for two months.

We also thank for their useful comments two anonymous referees, Jim Cust, François Gusdorf, Jean-Charles Hourcade, Henry Tulkens, and participants at the CIRED seminar and at the David Bradford Memorial Conference on the Design of Climate Policy (CESifo Venice Summer Institute). At last we thank Françoise Le Gallo for providing data on international cement trade.

1. Standard methods are likely to underestimate this coefficient (Erkell-Rousse and Mirza 2002). For example, if an exporting country increases the quality of its products vis-à-vis its competitors (i.e., if its non–price competitiveness is improved), it will typically increase both its export level and its price. If in econometric estimations this quality effect is not controlled, there will appear wrongly to be a positive correlation between the export price and quantity exported (or at least the observed correlation will be "less negative" than if quality was taken into account). As a consequence export elasticities (i.e., the decrease in exports following an increase in export price) will be underestimated, and likewise the Armington elasticities. There are no alternative econometric methods that can lead to robust results (Erkell-Rousse and Mirza 2002).

2. Only electricity generation has a higher ratio, but this sector is largely sheltered from international competition by transmission losses.

3. Consuming areas are grouped together to form consuming countries. In GEO we have the same 47 consuming and producing countries.

4. We define profit margin as the ratio (cement price – variable production cost – transportation cost)/(variable production cost + transportation cost).

5. Unfortunately, since New Zealand is merged with Australia in our set of 47 producing areas, we have to assume that it does not implement the Kyoto Protocol although it has ratified this agreement.

6. In the European Union the ETS allowances are not auctioned but provided for free, but the quantity distributed is influenced by EU decisions. A firm closing an installation will generally stop receiving allowances, and free allowances are distributed for new installations (Schleich and Betz 2005; Demailly and Quirion 2008). As a consequence, for a given CO_2 price, the competitiveness of the ETS and its impact on emissions are much lower than that of the policy we simulate.

7. Figures for 2020 and 2030 are presented in Demailly and Quirion (2005).

8. Notice that Chinese and Indian emissions decrease and RJAN's emissions increase, marginally, although these countries are not impacted directly by the ETS (they do not

increase or decrease their imports or their exports compared to BAU). This is because their consumption levels are only indirectly affected by the changes in competition pressure.

9. A less CO_2 intensive solution is to burn waste and wood fuels instead of gas. However, because we assume that this solution cannot be generalized in the case of limited availability of these fuels, we did not retain it as "the best available technology."

References

Armington, P. 1969. A theory of demand for products distinguished by place of production. *IMF Staff papers* 16: 159–78.

Baron, R. 2006. Compétitivité et politique climatique. Notes de l'IDDRI, 11. IDDRI, Paris. Available at ⟨http://www.iddri.org/iddri/telecharge/notes/11_competitivite-climat.pdf⟩.

British Cement Association. 2004. Memorandum by the British Cement Association, UK Parliament, European Union Committee. Available at ⟨http://www.publications.parliament.uk/pa/ld200304/ldselect/ldeucom/179/4050502.htm⟩.

CEMBUREAU. 1999. *World Statistical Review* (1913–1995), no. 18; (1994–1995) nos. 19–20. Brussels.

CEMBUREAU. 2002. *World Cement Directory* (1996, 2002), nos. 21, 27. Brussels.

Demailly, D., and P. Quirion. 2005. The competitiveness impacts of CO_2 emissions reductions in the cement sector. Report for the OECD. Available at ⟨http://appli1.oecd.org/olis/2004doc.nsf/linkto/com-env-epoc-ctpa-cfa(2004)68-final⟩.

Demailly, D., and P. Quirion. 2008. European emission trading scheme and competitiveness: A case study on the iron and steel industry. *Energy Economics* 30(4): 2009–27.

Enerdata. 2005. *Bilan des émissions de CO_2-énergie dans monde et enjeux du Protocole de Kyoto en Europe*. Enerdata, Grenoble, France. Available at ⟨http://www.enerdata.fr/enerdatafr/press_release/Enerdata_bilan_mondial_GES_2003.pdf⟩.

Erkell-Rousse, H., and D. Mirza. 2002. Import price elasticities: Reconsidering the evidence. *Canadian Journal of Economics* 35(2): 282–306.

Hoel, M. 1996. Should a carbon tax be differentiated across sectors. *Journal of Public Economics* 59(1): 17–32.

Hoerner, J. A. 1998. The role of border tax adjustments in environmental taxation: Theory and U.S. experience. Presented at the International workshop on market based instruments and international trade of the Institute for environmental studies, Amsterdam.

Hourcade, J.-C., and P. Shukla. 2001. Global, regional, and national costs and ancillary benefits of mitigation. In *Climate Change 2001: IPCC Third Assessment Report*. Cambridge: Cambridge University Press, ch. 8.

International Energy Agency. 1999. The reduction of greenhouse gas emission from the cement industry. Greenhouse Gas R&D Programme. IEA, Paris.

International Energy Agency. 2004. Industrial competitiveness under the European Union emissions trading scheme. Information paper. IEA, Paris.

Ismer, R., and K. Neuhoff. 2004. Border tax adjustments: A feasible way to address nonparticipation in emission trading. Cambridge Working paper in economics 0409. Cambridge University.

Johnson, R. N. and A. Parkman. 1983. Spatial monopoly, non-zero profits and entry deterrence: The case of cement. *Review of Economics and Statistics* 65(3): 431–39.

La Cour, L., and H. P. Mollgaard. 2002. Market domination: Tests applied to the Danish cement industry. *European Journal of Law and Economics* 14: 99–127.

Lloyd's List. 2004. *Ports of the World*. London: Lloyd's.

Mæstad, O. 1998. On the efficiency of green trade policies. *Environmental and Resource Economics* 11: 1–18.

Majocchi, A., and M. Missaglia. 2001. Environmental taxes and border tax adjustment. Presented at Conference Stato o Mercato? Intervento pubblico a architettura dei mercati, Pavia Università, October 5–6.

Markusen, J. 1975. International externalities and optimal tax structures. *Journal of International Economics* 5: 15–29.

Mathiesen, L., and O. Mæstad. 2002. Climate policy and the steel industry: achieving global emission reductions by an incomplete climate agreement. Discussion paper 20/02. Norwegian School of Economics and Business Administration, Bergen.

Prebay, Y., S. Ando, E. Desarnaud, and T. Desbarbieux. 2006. Les enjeux du développement durable au sein de l'Industrie du Ciment : Réduction des émissions de CO_2. Atelier Changement Climatique, Ecole Nationale des Ponts et Chaussées. Available at ⟨http://www.enpc.fr/fr/formations/ecole_virt/trav-eleves/cc/cc0506/ciment.pdf⟩.

Quirion, P., and J.-C. Hourcade. 2004. Does the CO_2 emission trading directive threaten the competitiveness of European industry? Quantification and comparison to exchange rates fluctuations. Presented at EAERE Annual Conference, Budapest.

RIVM. 2001. Edgar database. RIVM, Bilthoven, Netherlands. Available at ⟨http://arch.rivm.nl/env/int/coredata/edgar/intro.html⟩.

Schleich, J., and R. Betz. 2005. Incentives for energy efficiency and innovation in the European emission trading system. In: *Proceedings of the ECEEE Summer Study*, Mandelieu. Available at ⟨http://www.eceee.org/library_links/proceedings/2005/abstract/7124schleich.lasso⟩.

Szabo, L., I. Hidalgo, J. C. Ciscar, A. Soria, and P. Russ. 2003. Energy consumption and CO_2 emissions from the world cement industry. DG JRC-IPTS report. Technical Report Series EUR 20769 EN.

Szabo, L., I. Hidalgo, J. C. Ciscar, and A. Soria. 2006. CO_2 emission trading within the EU and annex B-R countries: The cement industry case. *Energy Policy* 34(1): 72–87.

17

The Global Warming Potential Paradox: Implications for the Design of Climate Policy

Stéphane De Cara, Elodie Galko, and Pierre-Alain Jayet

Few concepts derived from natural sciences have made their way into international law. The global warming potential (GWP) is one of them. Article 5.3 of the Kyoto Protocol states that "the global warming potentials used to calculate the carbon dioxide equivalence of anthropogenic emissions ... of greenhouse gases ... shall be those accepted by the Intergovernmental Panel on Climate Change" (UNFCCC 1997). The wording of the Kyoto Protocol therefore passes the "legally binding" nature of the Kyoto emission targets onto the GWP concept itself.

The success of the GWP in the international negotiation arena may be seen as a failed attempt by economists to have sound economics translated into policy instruments. As soon as the early 1990s, while the GWP concept was gaining momentum in both the scientific and policy debates, economists questioned its use for greenhouse gas (GHG) comparison purposes (Eckaus 1992; Schmalensee 1993; Reilly and Richards 1993). The concept was attacked on the ground that it misleads the economically sound choice of the mitigation mix. In other words, the GWP is a wrong metric for comparing various GHGs. More than a dozen years of research later, one is left with the conclusion that the fundamental economic message contained in those criticisms was not successfully conveyed. This was summarized by Bradford (2001):

In general, natural scientists have been attracted to the GWP concept because of its purely physical quality. Although economists have argued that the trade-offs cannot be inferred from physical properties alone, but have an inherent economic and policy dimension in terms of targets, the message has been slow to be accepted in the scientific community.

The GWP has stood as a key feature in all assessment reports hitherto published by the IPCC (Houghton et al. 1990, 1995, 2001). Despite the caveats that were included in the latest IPCC Assessment Report (Ramaswamy et al. 2001), the importance of the GWP is not likely to fade away any time soon, and certainly not before the end of the first Kyoto commitment period. The concept is even commonly used by economists. Partly because of the status conferred by its inclusion in the Kyoto Protocol, and partly because the inherent economic inconsistencies it implies have been overlooked, the vast majority of economic assessments of the costs and/or the benefits of multi-gas mitigation strategies rely on the GWP concept.

Yet the result that the GWP is ill-defined is robust from an economic standpoint. It has been confirmed by a number of studies that, following up on the aforementioned pioneering works, have proposed alternative indexes (Kandlikar 1995, 1996; Hammit et al. 1996; Bradford and Keller 2000; Manne and Richels 2001; Shine et al. 2005b) or provided empirical assessments of the concept's implications (Michaelis 1999; Smith and Wigley 2000; O'Neil 2000; Tol et al. 2003; O'Neil 2003; Kurosawa 2004; Sarofim et al. 2005). See Fuglestvedt et al. (2003) for a comprehensive review. The question then is: What may explain the success of the GWP in international negotiations? Or, stated differently, what are the aspects of the GWP that proved compelling enough to outweigh the major shortcomings that critics have consistently been pointing out? The "paradox" that is referred to in the title of this chapter does not lie in the definition of the GWP per se, which provides, after all, just one means of weighing GHGs and which has its own rationale based on natural sciences. The paradox is that despite the many criticisms by economists, the GWP sets the relative prices in the GHG trade-offs and therefore affects the economic instruments that have been designed in the aftermath of the Kyoto Protocol. Alternative GHG indexes proposed by economists did not encounter the same success.

In the first part of this chapter, we address this question by reviewing the GWP-related literature. We first present and discuss the main shortcomings of the GWP concept from an economic standpoint. Alternative GHG indexes that have been proposed in the economic literature are next reviewed. In contrast to the GWP's focus on impacts on radiative forcing only, those indexes reflect the marginal economic damage and/or abatement costs related to GHG emissions. In addition they account for the differences in GHGs' atmospheric lifetimes through discounting. In the light of this literature review, we examine some possible explanations for the success of the GWP over alternative indexes. In particular, the GWP, although imperfect, is likely to provide a less vulnerable basis for the negotiation of multi-GHG mitigation targets. The "slow" recognition of the GWP's economic dimensions, as underlined by Bradford, may then be explained by the difficulty of reaching an agreement on the economic parameters that underlie the computation of alternative economic indexes.

Two strategies may be envisaged to address multigreenhouse gas issues. The continuation of frontal attacks to the GWP is one. This would involve trying to fit important economic concepts such as discounting, marginal abatement costs, and marginal damage into the definition of a GHG index in the hope that this will eventually prove more successful than it has been in the last decade. Another approach consists in analyzing the implications of GWP-based economic instruments. Given the aspects of the GWP that made it so successful as a policy concept, this approach considers that the choice of multi-GHG targets is constrained by the use of the GWP. The challenge thus consists in adapting economic instruments in order to account for the bias induced by the use of an imperfect metric. In this sense, this approach focuses on second-best instruments, insofar as the design of multi-GHG targets is constrained by a predetermined GHG equivalence rule. This latter approach is the one explored in the present chapter.

The first problem with relying on an imperfect GHG metric such as the GWP is its implication for abatement decisions. An imperfect metric would likely distort total abatement costs and result in underabatement in some gases and overabatement in others, causing GHG concentrations to deviate from their first-best trajectories. This effect would not only modify the climate response but also—as soon as nonlinearities in damage are accounted for—modify marginal damage and shadow price ratios. The use of GWP-based instruments therefore impacts both abatement costs and damage. In this chapter we distinguish between these two kinds of impacts by way of a general analytical framework. We propose a measure of the distortion induced by the use of the GWP. The bias is geometrically interpreted as depending on the parameters defining marginal abatement costs and a measure of the angle between the vector of GWPs and the vector of GHG shadow prices. Such a synthetic measure of the GWP-related distortion contrasts with previous analyses in the literature, which have addressed this issue on a gas-by-gas basis.

17.1 GWP: An Imperfect but Successful Concept

17.1.1 Shortcomings of the GWP from an Economic Standpoint

The GWP measures the time-integrated radiative forcing caused by one emission pulse of one gas relatively to that of a reference gas—typically CO_2—over a given time horizon—typically 100 years. The GWP formula (Ramaswamy et al. 2001, p. 385) is written as

$$\text{GWP}_{j,CO_2} = \frac{\int_0^{\hat{T}} f_j(t).\theta_j(t)\,dt}{\int_0^{\hat{T}} f_{CO_2}(t).\theta_{CO_2}(t)\,dt}, \qquad (17.1)$$

where $f_j(t)$ represents the time-dependent decay in abundance of the instantaneous release of one mass unit of gas j at time $t = 0$; $\theta_j(t)$ is the instantaneous radiative efficiency due to a unit increase in atmospheric abundance of gas j, and CO_2 is taken as the reference gas.

All is fine as long as the use of the GWP is restricted to a synthetic reporting of the aggregate radiative impact caused by various GHGs within a given period of time. Such an index may well be useful in identifying and prioritizing mitigation options. But there is a conceptual leap between this use of the GWP and the interpretation of the GWP as an indicator of GHG relative prices. This leap was taken in article 5.3 of the Kyoto Protocol. A direct consequence of the Kyoto multi-gas targets, expressed in tons of CO_2-equivalent and based on the use of GWP, was to set the values of abatements in methane, nitrous oxide, and other GHGs relatively to that of CO_2. A project that entails an emission reduction of one ton of methane is thus entitled to 21 times[1] as many credits as one project that entails an emission reduction of one ton of CO_2. As soon as the GWP was introduced as a substitute for relative prices, it became an easy target for economists.

First, as indicated in equation (17.1), the GWP depends on the use of a finite time horizon, $\hat{T}$, that is the same for all GHGs. Indeed, despite the sophisticated modeling required to describe radiative efficiency and atmospheric behavior of the various GHGs, when it comes to setting $\hat{T}$, the choice is arbitrary. A 100-year time horizon is the convention taken in the national emissions inventories, although the IPCC also reports 20-year and 500-year GWPs. There is no scientific argument supporting the use of 100-year instead of any other time horizon. Because of the large differences between GHGs' atmospheric lifetimes, the index is quite sensitive to the chosen time horizon (see table 17.1).

Chapter 17 The Global Warming Potential Paradox

Table 17.1
Global warming potentials of selected GHGs

Gas	Lifetime (years)	Global warming potential time horizon ($\hat{T}$)					
		20 years		100 years		500 years	
CO_2	variable	1	(1)	1	(1)*	1	(1)
CH_4	12	62	(56)	23	(21)*	7	(6.5)
N_2O	114.0	275	(280)	296	(310)*	156	(170)
HFC-134a	13.8	3,300	(3,400)	1,300	(1,300)*	400	(420)
HFC-152a	1.4	460	(410)	120	(140)*	37	(42)
SF_6	3,200	15,100	(16,300)	22,200	(23,900)*	32,400	(34,900)
CF_4	50,000	3,900	(4,400)	5,700	(6,500)*	8,900	(10,000)

Source: Adapted from Ramaswamy et al. (2001) and Houghton et al. (1995).
Note: Figures in parentheses give the 1995 IPCC Second Assessment Report estimate. The asterisk indicates the 100-year GWP from the SAR as the one used in the Kyoto Protocol.

Second, the treatment of time implied by the use of the GWP is questionable. By use of a cutoff time, $\hat{T}$, and no discounting, the GWP concept substantially diverges from the traditional economic treatment of time. Although discounting is one of the most debated issues in economics (Weitzman 1998), it is widely accepted that comparison of costs and benefits occurring at various points in time should be based on a consistent measure of welfare. This requires accounting for the possibility of intertemporal arbitrage (e.g., through saving and investment), for future growth, and for agents' impatience when comparing a change in today's consumption with a change in future consumption. If examined through the lenses of discounting theory, the GWP implies a discount rate that is flat and equal to zero for the next hundred years (if $\hat{T} = 100$) and jumps to infinity for the subsequent future (Reilly et al. 2001). This reasoning would hardly be supported by economic theory (see also Fearnside 2002; Tol 2002b). The GWP is insensitive to whether the climate impact is felt in the first year or in the ninetieth year. Climate impacts occurring after one hundred years are ignored. Two abatement profiles involving different gas mixes may well be equivalent in terms of CO_2-equivalent but significantly differ in terms of climate and welfare impacts (Fuglestvedt et al. 2000).

Third, the GWP implies equivalence factors that are constant over time, since it is based on today's atmospheric compositions in each GHG and radiative forcing.[2] Given the dynamic nature of climate and atmospheric responses, today's GWP is likely to provide an inaccurate index for comparing future emissions of GHGs.

The fourth and perhaps the most fundamental argument encompasses the three aforementioned criticisms. In a market-based economy, the price of any good is the reflection of the value attached to this particular good relative to that of a numéraire. Relative prices are thus intrinsically linked to a measure of welfare. Using the GWP as the price of any GHG relatively to CO_2 therefore implies a confusion between impact—as calculated in the GWP, expressed in terms of time-integrated radiative forcing—and damage. In short, what matters in the economic analysis of climate change is not climate change per se but how it affects welfare in the broadest sense (Hammit et al. 1996). By targeting impacts instead of damages, the GWP sends a wrong signal as soon as it is interpreted as the relative price of any two GHGs.

7.1.2 Alternative Economic GHG Indexes

Two main routes have been followed in the economics literature to deal with the GWP issue. The first approach (cost-benefit) aims at finding abatement trajectories in all GHGs that minimize the sum of abatement and damage costs. Examples of this approach are found in Kandlikar (1996) and Moslener and Requate (2007). The optimal level of emissions of each gas is associated with a shadow price reflecting the marginal impact on welfare of those particular emissions. The optimal shadow price of gas j is equal to the present value of the flow of marginal damage due to an emission of gas j. Along an optimal trajectory it is also equal to the marginal abatement cost of gas j. The shadow price ratio between any two gases j and k therefore provides an appropriate measure of the relative social value of emissions in these two gases. As such, it provides an index that can be used as the relative price between gas j and k to decentralize the optimum.

The second approach (cost-effectiveness) consists in finding the cost-minimizing trajectories to meet a given target, which is typically expressed in terms of concentration (e.g., stabilization), aggregate radiative forcing, or temperature change ceiling (Manne and Richels 2001; O'Neil 2003; Sarofim et al. 2005). In this case the shadow price ratio between any two gases is equal to the ratio of their marginal abatement costs. As no explicit specification of the damage function is required, the cost-effectiveness approach rules out major difficulties regarding the economic evaluation of climate damages and mitigation benefits.

The least-cost abatement trajectories to meet the prescribed target can be decentralized by using the shadow price ratio as the relative price between GHGs.

Unlike the GWP, the shadow price ratio obtained from either cost-benefit or cost-effectiveness models accounts for welfare dimensions in a consistent and explicit manner. Moreover the shadow price ratio distinguishes among damage and/or abatement costs occurring at different points in time through discounting. The relative shadow prices of greenhouse gases in cost-effectiveness approaches depend on the level of the target and the time frame within which this target has to be met (Manne and Richels 2001). As an illustration, if the target is expressed as a temperature ceiling not to be exceeded at the end of the century, GHGs characterized by an atmospheric lifetime in the range of the decade—such as methane—cannot play a significant role in meeting the target in the short run. Abatements of longer lived gases are thus preferable in the early periods. Indeed the social value of short-lived GHGs relative to CO_2 is low in the short run, and it starts rising as the target gets closer. Cost-benefit analyses do not show the same pattern because economic damages are explicitly and endogenously modeled.

17.1.3 Why Did GWP Succeed While Economic Indexes Failed?

Alone, any of the criticisms discussed in section 17.1.1 could have been sufficient to downplay the importance of the GWP and promote use of better-suited, alternative GHG indexes. The success of the GWP is better understood if the GWP is viewed as a policy concept (Demeritt 2001). As with many policy concepts its development resulted from compromises that were perceived as necessary for policy steps to be taken. In addition the development of the GWP built on the experience accumulated with the ozone issue in the development of the ozone depletion potential index (ODP). That concept provided the methodological basis for the first GWP estimates (O'Neil 2000).

The importance of non-CO_2 gases was a key driver for the development of a GHG index. In the early development stages of the climate issue, research was focused almost exclusively on CO_2 emissions,[3] not only because of the prime role of CO_2 in the enhanced greenhouse effect phenomenon but also because of its straightforward link with fossil fuel use and therefore with economic growth. At the time the Kyoto targets were being negotiated it became clear that the restriction of mitigation efforts to the sole energy and transport sectors would lead to unnecessarily high abatement costs (Hayhoe et al. 1999; Reilly et al. 1999). Other mitigation options were needed to lower the overall abatement burden. Expanding the "basket of gases" to include non-CO_2 gases in the agreement was one means of putting forward significant reduction targets that would not have been affordable otherwise. Multi-gas targets are also important in reaching a broader stable agreement (De Cara and Rotillon 2003). The GWP, perceived as a scientifically sound concept, could be fairly easily explained to the public and accepted by negotiating parties. At the time the Kyoto Protocol was drafted, this proved useful in setting multi-gas rather than CO_2-only targets.

Alternative indexes proposed in the economics literature were not as successful. Arguably the amount of information required to compute these indexes is much greater than for the GWP. In addition to knowledge of climate and atmospheric responses—also needed for the computation of GWPs—they require an estimate of (current and future) abatement costs for all GHGs. Until recently most of the modeling effort was oriented toward CO_2 mitigation in the energy sector, and only little was known about non-CO_2 abatement costs. In cost-benefit approaches the challenge is even greater as an explicit specification of the economic damage costs is required. Relative shadow prices obtained from either cost-effectiveness or cost-benefit approaches are, however, sensitive to the specifications regarding climate and atmospheric systems, abatement costs, damage (only in cost-benefit analyses), discount rate, and time horizon. As discussed in Fuglestvedt et al. (2003, tab. V), the resulting range of relative shadow prices estimated with various models and assumptions is admittedly wide.

Moreover uncertainties are compounded at all stages of a causality chain that proceeds from emissions to the translation of climate change impacts into economic terms (for an illustration, see Fuglestvedt et al. 2003, fig. 1). An economic GHG index inherently encompasses all the links in this causality chain and therefore adds several degrees of complexity for practical purposes. As a result the foundations of such an index appeared vulnerable to strategic manipulations by the parties. The same argument applies to the introduction of discounting in the index. Using a discounted welfare-based index forces one to explicitly state how the long run is valued against the short run. Arguably reaching an agreement among Kyoto negotiators on a common explicit discount rate would have been quite challenging.

The main GHGs—such as carbon dioxide, methane, and nitrous oxide—are sufficiently long-lived to be considered well-mixed in the atmosphere.[4] Yet climate models generally agree on the point that climatic impacts will not be uniformly distributed across regions (Watson et al. 1997), and neither will be damage costs. As soon as economic damages are entered into the computation of a GHG index, the question of accounting for spatially differentiated damages arises. Directional "exchange rates"—such as those analyzed by Førsund and Nædval (1998) in the sulfur emission problem—provide an interesting analogy in this respect. In concrete terms, the damage estimates would involve a region-differentiated or even a country-differentiated GHG index. Conceivably, given the uncertainties affecting regional damage estimates, this could again open the way to strategic manipulations and undermine the likelihood of reaching an agreement on multi-gas targets.

Where do we stand? First, with the scope of economic research about climate change expanding beyond the sole energy sector, non-CO_2 emission sources and abatement costs are now better understood (Hayhoe et al. 2000; McCarl and Schneider 2001; Reilly et al. 2003; De Cara et al. 2005; USEPA 2006). Second, more results are now available on the assessment of climate change economic costs, including spatially differentiated ones, and this field is still being intensively researched (Mendelsohn et al. 1998; Tol 2002a). All uncertainties are far from being resolved, but some progress has been made in quantifying these uncertainties (Tol 2005). The latest economic results can be helpful in designing multi-GHG instruments, even though the design of such instruments remains constrained by the use of the GWP.

17.2 Optimal Multi-GHG Abatement Path

In this section we develop an analytical framework to investigate multiple greenhouse gas issues. We begin with an examination of the analytical properties of the first-best solution to use them later as a benchmark in the analysis of GWP-based instruments.

Let n greenhouse gases be indexed by $j = 1, \ldots, n$. Next denote the n-vector[5] of atmospheric concentrations at time t by $\mathbf{z}_t = (z_{1t}, \ldots, z_{nt})$. The rate at which GHGs are removed from the atmosphere differs from one gas to the other. Most of the literature that examines multi-gas issues from an analytical perspective assumes exponential decay processes characterized by constant decay rates (e.g., Moslener and

Requate 2007). This assumption greatly simplifies the computation of the optimal control problem by restricting it to a linear differential system. However, it overlooks two important features of the atmospheric behavior of GHGs (Houghton et al. 2001). First, because of the complexity of the exchanges with different carbon reservoirs (atmosphere, ocean, terrestrial carbon pool) characterized by different transfer speeds among them, the carbon cycle can hardly be reduced to a simple, constant-rate decay process. Second, interactions among the various GHGs can significantly impact the speed at which they are removed from the atmosphere. We adopt here a more general formulation of the decay process. We represent this process by the n-vector valued function $\mathbf{f}(\mathbf{z}_t) = (f_1(\mathbf{z}_t), \ldots, f_n(\mathbf{z}_t))$. Each entry of $\mathbf{f}(\mathbf{z})$, denoted by $f_j(\mathbf{z})$, describes the decay process of gas j as a function of the full vector of concentrations.

GHG emissions are denoted by the n-vector $\mathbf{e}_t = (e_{1t}, \ldots, e_{nt})$ and are measured in mass unit of each gas (tons of CO_2, tons of methane, etc.). Net emissions are decomposed into two components: (1) business-as-usual emissions, which are assumed exogenous and are denoted by $\bar{\mathbf{e}}_t = (\bar{e}_{1t}, \ldots, \bar{e}_{nt})$, and (2) abatements, which are denoted by $\mathbf{a}_t = (a_{1t}, \ldots, a_{nt})$. The equation of motion over time of concentration is thus

$$\dot{z}_{jt} = -f_j(z_{1t}, \ldots, z_{nt}) + \bar{e}_{jt} - a_{jt} \quad \text{for } j = 1, \ldots, n, \tag{17.2}$$

or, in vector form,

$$\dot{\mathbf{z}}_t = -\mathbf{f}(\mathbf{z}_t) + \bar{\mathbf{e}}_t - \mathbf{a}_t. \tag{17.3}$$

The effect of GHGs on climate is summarized through a global measure of climate change, $\theta(\mathbf{z}_t)$, for instance, the change in global mean surface temperature. $\theta(.)$ depends on the full vector of concentrations, accommodating the possible countereffects that some gases can exert on the radiative impact of other gases. It is particularly important to account for the interactions between the radiative forcing of methane and that of nitrous oxide in the atmosphere (Ramaswamy et al. 2001).

The economic loss due to climate change, denoted by $D(\theta(\mathbf{z}_t))$, is an increasing and convex function of the change in global mean surface temperature $(D(0) = 0, (dD/d\theta)(.) > 0, (d^2D/d\theta^2)(.) \geq 0)$. Abatement costs at time t, denoted by $C(\mathbf{a}_t)$, depend on the level of abatements in all GHGs. Again, a general formulation is important to account for potential interactions among the processes governing emissions, as well as among mitigation strategies. Agriculture provides an interesting illustration of such interactions. Mitigation strategies with respect to enteric fermentation (mostly methane) can take the form of reducing livestock numbers and/or modifying the way animals are fed. Both options have impacts on emissions from manure management (methane, but also nitrous oxide) and emissions from agricultural soils (mostly nitrous oxide). Separability of abatement costs between methane and nitrous oxide is thus not justified. For simplicity, we retain a quadratic formulation[6]:

$$C(a_{1t}, \ldots, a_{nt}) = \frac{1}{2} \sum_{i=1}^{n} \sum_{k=1}^{n} c_{jk} a_{jt} a_{kt} \tag{17.4}$$

or, in matrix form,

$$C(\mathbf{a}_t) = \frac{1}{2} \mathbf{a}_t' \mathbf{C} \mathbf{a}_t. \tag{17.5}$$

Equation (17.5) assumes linear marginal abatement costs with respect to the full vector of abatements ($\mathbf{Ca}$). In order to fulfill the usual convexity requirements, $\mathbf{C}$ is an $n \times n$ symmetric and positive definite matrix. See Moslener and Requate (2007) for a discussion of the importance of nondiagonal entries in matrix $\mathbf{C}$. In addition it should be noted that equation (17.5) implies that the abatement cost function is constant[7] over time.

We now turn to the problem faced by a (risk-neutral) social planner who seeks to set optimal abatement trajectories in all gases in order to minimize the discounted sum of damage and abatement costs. ρ denotes the (constant) social discount rate. The corresponding program[8] is

$$\min_{\mathbf{a}_t} \int_0^{+\infty} [C(\mathbf{a}_t) + D(\theta(\mathbf{z}_t))] e^{-\rho t} \, dt$$

subject to

$$\dot{\mathbf{z}}_t = -\mathbf{f}(\mathbf{z}_t) + \bar{\mathbf{e}}_t - \mathbf{a}_t. \tag{17.6}$$

The first-order optimality conditions for program (17.6) are given by

$$\mathbf{a}_t^* \in \arg\min_{\mathbf{a}_t} H^C = C(\mathbf{a}_t) + D(\theta(\mathbf{z}_t^*)) + \boldsymbol{\mu}_t^{'*}(-\mathbf{f}(\mathbf{z}_t^*) + \bar{\mathbf{e}}_t - \mathbf{a}_t), \tag{17.7a}$$

$$\dot{\mu}_{jt}^* = \rho \mu_{jt}^* - \frac{\partial H^C}{\partial z_j} \quad \text{for all } j = 1, \ldots, n, \tag{17.7b}$$

where H^C denotes the current Hamiltonian of program (17.6), μ_t denotes the n-vector of adjoint variables (or shadow prices) associated with the n equations of motion. The convention regarding the sign of shadow prices in equations (17.7a) and (17.7b) implies that the public bads (concentrations of GHGs) are assigned a positive price. Optimal levels of state, control, and adjoint variables are denoted by a star.

The relationships (17.7a) and (17.7b), together with the equation of motion of z_t (17.3), initial concentrations z_0, and transversality conditions, characterize the optimal abatement trajectories. Equation (17.7a) implies that at all points in time, optimal abatement in all gases should be such that the marginal abatement cost in gas j is equal to the respective shadow price, μ_{jt}. From equations (17.5) and (17.7a) we can derive the optimal abatement supply in all gases as a function of the vector of shadow prices:

$$\mathbf{a}_t^* = \mathbf{C}^{-1} \boldsymbol{\mu}_t^*. \tag{17.8}$$

Equations (17.7b) can be rewritten in matrix form as

$$\dot{\boldsymbol{\mu}}_t^* = (\rho \mathbf{I}_n + \mathbf{J}_\mathbf{f}(\mathbf{z}_t^*)) \boldsymbol{\mu}_t^* - \frac{dD}{d\theta}(\theta(\mathbf{z}_t^*)) \mathbf{J}_\theta(\mathbf{z}_t^*), \tag{17.9}$$

where $\mathbf{I}_n$ is the $n \times n$ identity matrix, $\mathbf{J}_\mathbf{f}(\mathbf{z}_t)$ is the $n \times n$ Jacobian matrix of $\mathbf{f}(\mathbf{z}_t)$, with generic entry $(\partial f_j / \partial z_k)(\mathbf{z}_t)$, and $\mathbf{J}_\theta(\mathbf{z}_t)$ is the $n \times 1$ Jacobian matrix of $\theta(\mathbf{z}_t)$, with generic entry $(\partial \theta / \partial z_j)(\mathbf{z}_t)$. The jth row of $\mathbf{J}_\mathbf{f}(\mathbf{z}_t)$ is thus the profile of the marginal impact of a change in the atmospheric composition on gas j's concentration. Similarly the jth entry of $\mathbf{J}_\theta(\mathbf{z}_t)$ reflects the marginal impact of gas j on global temperature.

Equation (17.9) indicates that each individual shadow price changes over time in such a way that it equals the present value of damage caused by a marginal increase in emissions of the respective gas. Introducing optimal abatements from equation (17.8) into equation (17.3) yields

$$\dot{\mathbf{z}}_t^* = -\mathbf{f}(\mathbf{z}_t^*) + \bar{\mathbf{e}}_t - \mathbf{C}^{-1} \boldsymbol{\mu}_t^*. \tag{17.10}$$

If $D(.)$ is linear with respect to θ, and $\mathbf{f}(.)$ and $\theta(.)$ are both linear with respect to $\mathbf{z}_t$, then equations (17.9) and (17.10) reduce to a $2n$ linear first-order differential system with constant coefficients. The subsystem (17.9) in $\boldsymbol{\mu}_t$ can then be solved independently of $\mathbf{z}_t$.

This case is examined in the appendix, where the shadow price of any gas j is shown to be constant over time. Abatements in all gases are thus also constant over time, and so is the shadow price ratio between any two gases. Even in this simple case, the shadow price ratio between any gas j and gas 1 differs from the GWP for two reasons rooted in the very definition of the GWP: absence of discounting, and use of a finite time horizon. The GWP index is found to undervalue (overvalue) gases that are shorter lived (longer lived) than the reference gas if and only the discount rate is greater than a threshold value $\bar{\rho}$. $\bar{\rho}$ depends on the time horizon used in the GWP ($\hat{T}$) and on the decay rates of gases j and 1. (See equation 17.28 in the appendix.) As $\bar{\rho}$ is decreasing with respect to $\hat{T}$, the greater is $\hat{T}$, the more likely is the GWP to undervalue short-lived GHGs.

As soon as the nonlinearities in damage climate responses, or atmospheric concentrations are taken into account, the optimal shadow price ratio varies over time as it reflects the changes in marginal damage caused by the changes in concentrations. This illustrates one of the major sources of the bias induced by the GWP, which is by definition constant over time.

17.3 GWP-Based Abatement Targets

In this section we examine the implications of GWP-based abatement targets. Consider that a GWP-like GHG index has been agreed upon. This index converts emissions of any gas j into "equivalent" emissions of one particular reference gas. Without loss of generality, we assume that gas 1 is taken as the reference gas. For clarity of the exposition and in order to stay in line with the Kyoto Protocol's terminology, we refer to gas 1 as CO_2. Let $\gamma = (1, \gamma_2, \ldots, \gamma_n)$ be the n-vector of conversion coefficients of gas j into CO_2. By analogy with the definition of the GWP, all entries of are assumed to be constant over time. It should be noted that γ encompasses the standard definition of the GWP as a particular case but also covers any kind of constant GHG index. As an illustration, CO_2-only strategies can also be analyzed using this framework (in this case $\gamma_j = 0$ for all $j \geq 2$). Total CO_2-equivalent abatement at time t is

$$\sum_{j=1}^{n} \gamma_j a_{jt} = \gamma' \mathbf{a}_t. \tag{17.11}$$

We proceed in two steps. First, we assume that a CO_2-equivalent target for the entire planning horizon has been set by the social planner.

Agents adjust their abatement decisions in order to minimize the cost of meeting this target at all times. Second, the social planner chooses the optimal target knowing agents' responses. This second step is examined in section 17.4.

Let us thus first assume that an aggregate target, α_t based on the γ-index, has been fixed. The corresponding cost minimization program is

$$\min_{\mathbf{a}_t} C(\mathbf{a}_t)$$

subject to

$$\gamma' \mathbf{a}_t \geq \alpha_t, \tag{17.12}$$

which leads to the n following first-order conditions:

$$\mathbf{C}\mathbf{a}_t = \lambda_t \gamma. \tag{17.13}$$

The abatement profile at time t should be such that the marginal abatement cost in each gas is equal to the shadow price (λ_t) associated with the γ-aggregated target times the respective value of the GHG index. In the optimum λ_t should thus be equal to the marginal abatement cost of CO_2. If abatements are to be traded through a single emission permit market, the equilibrium price of gas j on this market should be equal to $\gamma_j \lambda_t$. This illustrates the fact that a multi-gas target sets the prices of GHGs relative to that of CO_2.

As the constraint in program (17.12) should be binding in the optimum, we can eliminate λ_t. The cost-minimizing abatement vector is denoted by a tilde and is obtained as a function of the CO_2-equivalent target and γ:

$$\tilde{\mathbf{a}}(\alpha_t) = \frac{\alpha_t}{\gamma' \mathbf{C}^{-1} \gamma} \mathbf{C}^{-1} \gamma. \tag{17.14}$$

Equation (17.14) taken together with equations (17.5) and (17.8) can be rearranged to allow a comparison of the total abatement costs under first-best and GWP-based regimes:

$$C(\tilde{\mathbf{a}}(\alpha_t)) \leq C(\mathbf{a}_t^*) \Leftrightarrow \alpha_t \leq \sqrt{\mu_t'^* \mathbf{C}^{-1} \mu_t^* \cdot \gamma' \mathbf{C}^{-1} \gamma}. \tag{17.15}$$

If α_t is not too large, the corresponding GWP-based abatement target results in lower total abatement costs. Therefore, if the inequality on the right-hand side of (17.15) holds for the entire planning horizon, the present value of the total damage under the first-best regime is necessarily lower than under the GWP-based regime (by definition of the first-best regime).

If α_t is set equal to $\alpha_t^* = \gamma' \mathbf{a}_t^*$, with $\mathbf{a}_t^*$ defined as in equation (17.8), then the GWP-based abatement is

$$\tilde{\mathbf{a}}(\alpha_t^*) = \frac{\gamma' \mathbf{C}^{-1} \boldsymbol{\mu}_t^*}{\gamma' \mathbf{C}^{-1} \gamma} \mathbf{C}^{-1} \gamma. \tag{17.16}$$

The comparison of the first-best and GWP-based abatement vectors for the same GWP-aggregated abatement at time t (α_t^*) yields

$$\tilde{\mathbf{a}}(\alpha_t^*) - \mathbf{a}_t^* = \mathbf{C}^{-1} \left(\frac{\gamma' \mathbf{C}^{-1} \boldsymbol{\mu}_t^*}{\gamma' \mathbf{C}^{-1} \gamma} \gamma - \boldsymbol{\mu}_t^* \right). \tag{17.17}$$

The only solution for the full profile of abatement $\tilde{\mathbf{a}}(\alpha_t^*)$ to be equal to the first-best full profile of abatements is such that $\gamma = k \boldsymbol{\mu}_t^*$, where k is any positive real scalar. Given the additional convention that $\gamma_1 = 1$, the only solution for $\tilde{\mathbf{a}}(\alpha_t^*) = \mathbf{a}_t^*$ to hold is that $\gamma_j = \mu_{jt}^*/\mu_{1t}^*$ for all $j = 1, \ldots, n$, and at all times. This result embeds the essence of the critical views of the GWP concept and illustrates the fundamental economic result with regard to the GHG equivalence factor. The (first-best) equivalence rule should be based on the shadow prices ratios.

For the same CO_2-equivalent target ($\alpha_t = \alpha_t^* = \gamma' \mathbf{C}^{-1} \boldsymbol{\mu}_t^*$) at any point in time, the GWP-based abatement target leads to lower abatement costs than under the first-best regime. This is readily verified by noticing that the inequality in (17.15) holds in this case as a direct application of the Cauchy-Schwarz inequality. Therefore, if the CO_2-equivalent target under the GWP-based regime is set equal to α_t^* for the whole planning horizon, abatement costs are lower and environmental damages are larger than under the first-best regime, although both regimes lead to the same GWP-aggregated abatement.

The factor $\gamma' \mathbf{C}^{-1} \boldsymbol{\mu}_t^*/\gamma' \mathbf{C}^{-1} \gamma$ appearing on the right-hand side of equation (17.16) measures the distortion induced by the GWP. This factor can be geometrically interpreted as depending on a particular measure of the angle between γ and $\boldsymbol{\mu}_t^*$. Given our assumptions on matrix $\mathbf{C}$ (symmetric and positive definite), $\mathbf{C}^{-1}$ defines a norm $\|.\|$ in $\Re^n$. Let $\cos(.,.)$ denote the generalized definition[9] of cosine based on the inner product associated with $\mathbf{C}^{-1}$. The bias can be rewritten as

$$B_t^* = \frac{\gamma' \mathbf{C}^{-1} \boldsymbol{\mu}_t^*}{\gamma' \mathbf{C}^{-1} \gamma} = \frac{\|\boldsymbol{\mu}_t^*\|}{|\gamma|} \cos(\gamma, \boldsymbol{\mu}_t^*). \tag{17.18}$$

The first factor $\|\boldsymbol{\mu}_t^*\|/\|\boldsymbol{\gamma}\|$ is a scaling factor that provides an aggregate measure of the GWP-related bias. It is large (small) when the weighted sum of the shadow prices is large (small) relative to the weighted sum of the GWPs at time t. When this factor is close to one, the weighted sum of the GWPs reflects adequately the total magnitude of the social value of emissions, although the weighted summation may hide some countervailing effects. The second factor summarizes how (in)accurately the relative social value of every individual gases at time t is reflected in γ. A value of $\cos(\gamma, \boldsymbol{\mu}_t^*)$ close to zero indicates that γ provides a poor approximation of the vector of shadow prices, whereas a value close to one signals that welfare impacts of every individual gas are well captured in γ, at least in their relative directions if not in their total absolute magnitude. This measure uses the norm defined by $\mathbf{C}^{-1}$. In other words, the measure of the distortion weights the various GHGs according to their respective contribution to marginal abatement costs. GHGs characterized by steep-sloped marginal abatement cost are assigned a low weight in this measure.

Most papers that consider GWP-related economic issues examine the difference between GWPs and the respective shadow prices relative to CO_2 on a gas-by-gas basis (e.g., Kandlikar 1996; Manne and Richels 2001; Börhinger et al. 2005). B_t^* provides a comprehensive and synthetic measure that captures in a scalar the distortion caused by the GWP at any point in time. It can be easily computed using standard inputs (GWPs, marginal abatement cost parametrization) and outputs (shadow prices) of economic models of climate change. The computation of B_t^* can rely on shadow prices from either cost-effectiveness or cost-benefit models and can be extended to more general specifications of the abatement cost function.[10] As this measure weighs each gas according to the parametrization of abatement costs, it may be useful to account appropriately for GHGs that are characterized by a substantial discrepancy between the GWP and shadow prices but play only a limited role in the optimal abatement mix. B_t^* may be used to test if (or when, as it varies in time) the GWP-related bias is large. Equation (17.18) may then be applied to distinguish between the "magnitude" and "direction" effects of this distortion.

17.4 Second-best, GWP-Based Abatement Targets

The next step in our analysis consists in finding the best possible CO_2-equivalent target. It is established that GWP-based quantity instru-

Chapter 17 The Global Warming Potential Paradox

ments lead to a distortion in the abatement mix. This section examines a class of second-best instruments insofar as the social planner's choice is assumed to be constrained by the use of the GWP as the GHG index. As a result the only available command variable is the GWP-aggregated target. The corresponding social planner's problem is

$$\min_{\alpha_t} \int_0^{+\infty} [C(\tilde{\mathbf{a}}(\alpha_t)) + D(\theta(\mathbf{z}_t))] e^{-\rho t}\, dt$$

subject to

$$\dot{\mathbf{z}}_t = -\mathbf{f}(\mathbf{z}_t) + \bar{\mathbf{e}}_t - \tilde{\mathbf{a}}(\alpha_t). \tag{17.19}$$

For any level of the CO_2-equivalent target α_t, abatements are supplied according to equation (17.14). The current Hamiltonian $\hat{H}^C$ is formed by introducing (17.14) into the objective function and into the equation of motion of $\mathbf{z}_t$ and then deriving the following first-order optimality conditions (optimal values are signaled with a hat):

$$\hat{\alpha}_t \in \arg\min_{\alpha_t} \hat{H}^C = \frac{\alpha_t^2}{2\gamma' \mathbf{C}^{-1}\gamma} + D(\theta(\hat{\mathbf{z}}_t))$$

$$+ \hat{\boldsymbol{\mu}}_t' \left(-\mathbf{f}(\hat{\mathbf{z}}_t) + \bar{\mathbf{e}}_t - \frac{\alpha_t \mathbf{C}^{-1}\gamma}{\gamma' \mathbf{C}^{-1}\gamma} \right), \tag{17.20a}$$

$$\dot{\hat{\boldsymbol{\mu}}}_t = (\rho \mathbf{I}_n + \mathbf{J}_\mathbf{f}(\hat{\mathbf{z}}_t))\hat{\boldsymbol{\mu}}_t - \frac{dD}{d\theta}(\theta(\hat{\mathbf{z}}_t))\mathbf{J}_\theta(\hat{\mathbf{z}}_t). \tag{17.20b}$$

Solving problem (17.20a) for $\hat{\alpha}_t$ yields

$$\hat{\alpha}_t = \gamma' \mathbf{C}^{-1} \hat{\boldsymbol{\mu}}_t. \tag{17.21}$$

The second-best CO_2-equivalent target depends on the full vector of shadow prices derived from program (17.19). Introducing the expression of $\hat{\alpha}_t$ into equation (17.14) gives the second-best vector of abatements:

$$\hat{\mathbf{a}}_t = \frac{\gamma' \mathbf{C}^{-1} \hat{\boldsymbol{\mu}}_t}{\gamma' \mathbf{C}^{-1} \gamma} \mathbf{C}^{-1} \boldsymbol{\gamma} \tag{17.22}$$

The interpretation of $\hat{B}_t = \gamma' \mathbf{C}^{-1} \hat{\boldsymbol{\mu}}_t / \gamma' \mathbf{C}^{-1} \gamma$ is similar to that of B_t^*. $\hat{B}_t$ provides a measure of the distortion induced by the use of the GWP relative to the modified shadow prices from program (17.19). $\hat{\alpha}_t$ is larger (smaller) than α_t^* if and only if $\hat{B}_t$ is larger (smaller) than B_t^*. Therefore the second-best GWP-based target must be adjusted upward relative to the first-best CO_2-equivalent abatement if and only if the bias between shadow prices and the GWPs is larger under the second-best regime than under the first-best regime.

We first examine the implications on abatement costs. The ranking of abatement costs under the three regimes at any time t depends on the level of the second-best GWP-based target:

$$C(\hat{\mathbf{a}}_t) \leq C(\tilde{\mathbf{a}}(\alpha_t^*)) < C(\mathbf{a}_t^*) \Leftrightarrow \hat{\alpha}_t \leq \alpha_t^*, \tag{17.23a}$$

$$C(\tilde{\mathbf{a}}(\alpha_t^*)) < C(\hat{\mathbf{a}}_t) \leq C(\mathbf{a}_t^*) \Leftrightarrow \alpha_t^* < \hat{\alpha}_t \leq \sqrt{\mu_t^{'*}\mathbf{C}^{-1}\mu_t^*.\gamma'\mathbf{C}^{-1}\gamma}, \tag{17.23b}$$

$$C(\tilde{\mathbf{a}}(\alpha_t^*)) < C(\mathbf{a}_t^*) < C(\hat{\mathbf{a}}_t) \Leftrightarrow \hat{\alpha}_t > \sqrt{\mu_t^{'*}\mathbf{C}^{-1}\mu_t^*.\gamma'\mathbf{C}^{-1}\gamma}. \tag{17.23c}$$

This is illustrated in figure 17.1, together with an interpretation of the bounds based on the definitions of B_t^* and $\hat{B}_t$. The first two cases are characterized by a second-best CO_2-equivalent target at time t that is not too large but possibly larger than the CO_2-equivalence of the first-best abatements. In these two cases, the second-best abatement path allows for some saving on abatement costs compared to the first-best regime. In the third case, the second-best GWP-based target is sufficiently large compared to the CO_2-equivalence of the first-best abatement to induce larger abatement costs.

If, for the entire planning horizon, the bias under the second-best regime is not too large relative to the first-best regime—that is, if case (17.23a) or (17.23b) prevails—then we know by definition of the first-best regime that the present value of total damage is higher under the second-best regime. How the second-best CO_2-equivalent target compares with the CO_2 equivalence of first-best abatements depends

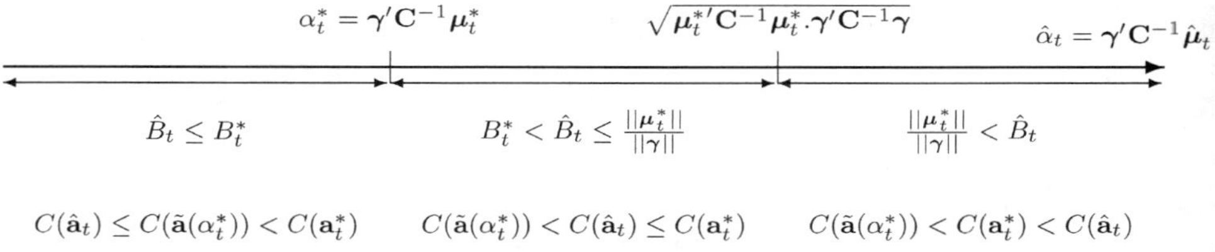

Figure 17.1
Ranking of abatement costs under GWP-based abatement target $(\tilde{\mathbf{a}}(\alpha_t^*))$, second-best GWP-based abatement target $(\hat{\mathbf{a}}_t)$, and first-best abatement target $(\mathbf{a}_t^*)$

on the gap between the shadow prices under the first-best and the second-best regimes. Solving the full system is, in general, required to determine this gap.

The difference between shadow prices under the first-best and the second-best regimes is linked to the nonlinearity in damage, atmospheric behavior, and/or temperature change. In the linear case examined in the appendix, the first-best and second-best shadow prices are equal. As a consequence, if the linear approximation is considered appropriate and if the GWP is to be used because of external reasons, the prescription is to use the CO_2-equivalence of the first-best abatements as the GWP-aggregated target for the entire planning horizon. As seen above, this will result in abatements in individual gases that differ from their first-best levels, lower total abatement costs at all time, and greater present value of total environmental damage. As soon as nonlinearities are taken into account, the second-best GWP-based target has to be adapted so as to accommodate the impact on marginal damage.

17.5 Discussion: Policy and Economic Implications

Flexibility is a key component of a cost-effective climate policy design. As it relates to the trade-offs among the GHGs, the debate over the GHG equivalence rule is straightforwardly linked to the "what"-flexibility issue (Böhringer et al. 2005). Nevertheless, it cannot be disconnected from the analysis of "where"- and "when"-flexibility. First, as the choice of any GHG index affects the trade-off between short- and long-lived GHGs (Aaheim 1999; Michaelowa 2003), it is also a determinant for the timing of the mitigation strategies. Second, the relative contribution of non-CO_2 emissions varies widely across sectors and countries. Therefore the relative weights attached to non-CO_2 gases are a crucial issue for developing countries—for which non-CO_2 sources, such as from agriculture, are important—in future climate agreements.

The multi-gas issue can theoretically be resolved by setting a GHG index that adequately reflects marginal abatement costs and the flow of future marginal damage. Such an index has to be updated on a regular basis in order to account for the changes in atmospheric concentrations and therefore in damage.

In contrast, the GWP is known to distort the trade-offs among GHGs. As the GWP implicitly assumes a linear link between radiative forcing and damage, this distortion might be viewed to some extent as resulting from a first-order approximation. Nevertheless, even in the case of linear damage, linear atmospheric behavior, and linear temperature change, the GWP is not equal to the relative social value of any two gases. Because of the absence of discounting and the use of cutoff time in the definition of the GWP, a bias remains and tends to distort abatement decisions.

Findings from cost-effectiveness studies suggest that short-lived GHGs—such as methane—are overvalued by the GWP in the short run. The contribution of short-lived GHGs to meet a given target in the distant future is indeed low relative to that of longer lived GHGs. Therefore the social value of abatements in short-lived GHGs rises only when approaching the target. The use of the GWP would thus result in overabatements in methane at least in the near term. This conclusion depends on the cost-effectiveness modeling framework. Under the simple linear case in a cost-benefit framework, whether the GWP under- or overvalues the relative prices of shorter lived GHGs depends on the time horizon used in the GWP calculation and on the discount rate. Abatements in short-lived GHGs may actually be more desirable than the abatement reflected in the GWP if the discount rate is sufficiently large.

The resilience of the GWP to criticisms suggests that it provides a more robust basis for negotiating multi-GHG targets than alternative economic indexes. If the GWP remains a cornerstone of the future international climate policy, policy makers will have to rely on GWP-based instruments. GWP-based instruments leave the regulator with a one-dimensional command variable (e.g., the CO_2-equivalent target) to cope with an n-dimensional issue (n GHGs). The first-best outcome is thus not attainable with such instruments. The geometric interpretation of this distortion that we proposed in this chapter provides a synthetic measure of the GWP-related bias by capturing in one scalar the gap between first-best shadow prices and GWPs. In this comparison the differences among GHGs in terms of marginal abatement cost parameterization are taken into account. This measure can be used in a wide range of multi-gas modeling approaches to assess the GWP-related bias.

If the GWP is to be used in the design of quantity-based instruments, the level of the GWP-aggregated target has to be adapted to account for the bias caused by the use of an imperfect metric. Unless both the economic and ecological systems are satisfactorily approximated by linear systems, the second-best GWP-based targets differ from the CO_2-equivalence of the first-best abatements. Whether the second-best GWP-based target should be revised upward or downward relative to first-best GWP-aggregated abatement depends on the gap between shadow prices under the first-best and the second-best regimes.

Appendix: Linear Damage, Linear Atmospheric Behavior, and Linear Temperature Change

In addition to the assumptions made in section 17.2, suppose that decay rates in the atmosphere are constant and independent from one another and that temperature change is linear with respect to concentrations. We then can posit the following specifications:

$$\mathbf{f}(\mathbf{z}) = \mathbf{F}\mathbf{z}, \tag{17.24a}$$

$$\theta(\mathbf{z}) = \boldsymbol{\theta}'\mathbf{z}, \tag{17.24b}$$

where $\mathbf{F}$ is a diagonal $n \times n$ matrix with τ_j ($\tau_j > 0$ for all $j = 1, \ldots, n$) on the diagonal. $1/\tau_j$ is the average atmospheric lifetime of gas j. $\boldsymbol{\theta}$ is an n-vector, whose generic entry—θ_j—represents the marginal change in temperature caused by one additional unit of gas j in the atmosphere ($\theta_j > 0$ for all $j = 1, \ldots, n$). By assumptions (17.24a) and (17.24b), equation (17.1) gives

$$\text{GWP}_{j,k}(\hat{T}) = \frac{\theta_j \tau_k (1 - e^{-\tau_j \hat{T}})}{\theta_k \tau_j (1 - e^{-\tau_k \hat{T}})}. \tag{17.25}$$

In addition, suppose that damage is linear with respect to the temperature change so that $D(\theta) = \beta \theta$ (with $\theta > 0$). The first-best regime is governed by system (17.9) and (17.10), which under the assumptions above becomes in matrix form

$$\begin{pmatrix} \dot{\mathbf{z}}_t^* \\ \dot{\boldsymbol{\mu}}_t^* \end{pmatrix} = \begin{pmatrix} -\mathbf{F} & -\mathbf{C}^{-1} \\ 0 & \rho \mathbf{I}_n + \mathbf{F} \end{pmatrix} \begin{pmatrix} \mathbf{z}_t^* \\ \boldsymbol{\mu}_t^* \end{pmatrix} + \begin{pmatrix} \bar{\mathbf{e}} \\ -\beta \boldsymbol{\theta} \end{pmatrix}. \tag{17.26}$$

In solving the subsystem in $\boldsymbol{\mu}_t^*$ in this case, the shadow price of gas j is found to be constant over time, and equal to its steady-state value: $\mu_{jt}^* = \beta \theta_j/(\rho + \tau_j)$ for all $j = 1, \ldots, n$ and all t. As a consequence abatements in all gases are constant over time ($\mathbf{a}_t^* = \mathbf{C}^{-1} \boldsymbol{\mu}_t^*$). The shadow price ratio of gas i relative to gas k is thus also constant:

$$\frac{\mu_j^*}{\mu_k^*} = \frac{\theta_j(\rho + \tau_k)}{\theta_k(\rho + \tau_j)}. \tag{17.27}$$

From equations (17.25) and (17.27), it is evident that, in general, μ_j^*/μ_k^* and $\text{GWP}_{j,k}(\hat{T})$ are different. Under the assumptions discussed above, the difference between the shadow price ratio of any two gases and the respective GWP depends on the discount rate (ρ), the time horizon ($\hat{T}$) and the respective decay rates (τ_j and τ_k). Suppose, without loss of generality, that $\tau_j > \tau_k$ (gas j is shorter lived than gas k). Comparing equations (17.25) and (17.27), rearranging, and recognizing that the sign of $(\tau_j(1 - e^{-\tau_k \hat{T}}) - \tau_k(1 - e^{-\tau_j \hat{T}}))$ is the same as that of $(\tau_j - \tau_k)$ results in

$$\text{GWP}_{j,k}(\hat{T}) < \frac{\mu_j^*}{\mu_k^*} \Leftrightarrow \rho > \bar{\rho} = \frac{\tau_j \tau_k (e^{-\tau_k \hat{T}} - e^{-\tau_j \hat{T}})}{\tau_j(1 - e^{-\tau_k \hat{T}}) - \tau_k(1 - e^{-\tau_j \hat{T}})}. \tag{17.28}$$

In analyzing the expression of $\bar{\rho}$, it can be seen that $\bar{\rho}$ is decreasing with respect to $\hat{T}$.

As the equation of motion of $\boldsymbol{\mu}_t$ is independent of the level of atmospheric concentrations in the linear case, the subsystem in $\boldsymbol{\mu}_t$ is identical under the first-best and second-best regimes (see section 17.4). In the linear case we thus have $\boldsymbol{\mu}_t^* = \hat{\boldsymbol{\mu}}_t$ at all times.

Notes

This chapter was written while Stéphane De Cara was a Guest Research Scholar with the Forestry Program at IIASA. The hospitality of the Institute is gratefully acknowledged. The authors would like to thank Brian O'Neil, Michael Obersteiner, Eric Nævdal, and Richard Tol for helpful discussions. Thanks are also due to participants at EAERE Conference (Bremen 2005) and CESifo Venice Summer Institute (David Bradford Memorial Conference: "The Design of Climate Policy", 2005). The usual disclaimer applies.

1. The GWP of methane is as estimated in the 1995 IPCC Second Assessment Report (SAR). In its Third Assessment Report, the IPCC revised methane's GWP upward to 23 (Ramaswamy et al. 2001). However, the 1995 SAR GWPs are used for the verification of compliance with the Kyoto commitments (see also table 17.1).

2. GWPs can be (and have been) revised from time to time in the light of updated information on atmospheric composition and new scientific findings regarding atmospheric behavior and radiative efficiency of GHGs (e.g., see table 17.1). The GWP, however, remains a static concept based on current atmospheric and climate responses.

3. In the early 1990s it was widely accepted that CO_2 was to be the primary, if not the only, target in any action taken to combat climate change: "Whatever type of international agreement is reached during the next decade, it will probably only cover CO_2, not other climate gases.... Although agreements encompassing all climate gases could be

more efficient, practical considerations, will thus force governments, at least initially, to limit an agreement to CO_2." (Hoel 1991). This prediction was proved wrong by later developments in the climate negotiations.

4. In the case of very short-lived greenhouse gases however, this view does not hold (e.g., NO_x; see Shine et al. 2005a).

5. Vectors and matrices are denoted in bold lowercase and uppercase respectively. All vectors are column vectors. The transposed operator is denoted by a prime.

6. Assuming linear marginal abatement costs with respect to the whole vector of abatements greatly simplifies the calculation of the abatement supply (equation 17.8). More general assumptions on the abatement cost function are possible. The qualitative nature of the results would not be changed provided that the conditions for the implicit function theorem are met; that is, the Hessian matrix of $C(\mathbf{a})$ is positive definite.

7. The formulation does not account for technical progress in the abatement technology, neither by an exogenous cost-decreasing trend nor by a learning-by-doing process. Admittedly such consideration is important in deriving optimal trajectories, but it would increase the complexity of the subsequent developments without fundamentally changing the nature of the results.

8. Abatements costs and business-as-usual emissions are assumed to be such that abatements in each gas are nonnegative and strictly lower than business-as-usual emissions at any point in time. As we focus on interior solutions, we do not explicitly introduce a constraint on abatements

9. The norm defined by $\mathbf{C}^{-1}$ gives the length of any vector $\mathbf{u}$ in $\Re^n$ as $\|\mathbf{u}\| = \sqrt{\mathbf{u}'\mathbf{C}^{-1}\mathbf{u}}$. We can then define the cosine between any two vectors $\mathbf{u}$ and $\mathbf{v}$ in $\Re^n$ as $\cos(\mathbf{u}, \mathbf{v}) = \mathbf{u}'\mathbf{C}^{-1}\mathbf{v}/\|\mathbf{u}\|.\|\mathbf{v}\|$.

10. More general formulation of the abatement cost function involving nonlinear marginal abatement costs would require a first-order approximation of marginal abatement costs at each point in time. In that case the Hessian matrix of abatement costs would replace $\mathbf{C}$ without changing fundamentally the results.

References

Aaheim, H. A. 1999. Climate policy with multiple sources and sinks of greenhouse gases. *Environmental and Resource Economics* 14(3): 413–29.

Bradford, D. F. 2001. Global change: Time, money and tradeoffs. *Nature* 410(6829): 649–50.

Bradford, D. F., and K. Keller. 2000. Global warming potentials: A cost-effectiveness approach. Mimeo. Princeton University.

Börhinger, C., A. Löschel, and T. F. Rutherford. 2005. Efficiency gains from "what"-flexibility in climate policy: An integrated CGE assessment. Presented at Venice Summer Institute: David F. Bradford Memorial Conference on the Design of Climate Policy, Venice, Italy. CESifo, Munich.

De Cara, S., M. Houzé and P.-A. Jayet. 2005. Methane and nitrous oxide emissions from agriculture in the EU: A spatial assessment of sources and abatement costs. *Environmental and Resource Economics* 32(4): 551–83.

De Cara, S., and G. Rotillon. 2003. Multi-greenhouse gas international agreements. Working paper 50. INRA Economics, INRA ESR Toulouse-Grignon, France.

Demeritt, D. 2001. The construction of global warming and the politics of science. *Annals of the Association of American Geographers* 91(2): 307–37.

Eckaus, R. S. 1992. Comparing the effects of greenhouse gas emissions on global warming. *Energy Journal* 13(1): 25–35.

Fearnside, P. M. 2002. Time preference in global warming calculations: A proposal for a unified index. *Ecological Economics* 41(1): 21–31.

Førsund, F. R., and E. Nædval. 1998. Efficiency gains under exchange-rate emission trading. *Environmental and Resource Economics* 12(4): 403–23.

Fuglestvedt, J. S., T. K. Berntsen, O. Godal, R. Sausen, K. P. Shine, and T. Skodvin. 2003. Metrics of climate change: Assessing radiative forcing and emission indices. *Climatic Change* 58(3): 267–331.

Fuglestvedt, J. S., T. K. Berntsen, O. Godal, and T. Skodvin. 2000. Climate implications of GWP-based reductions in greenhouse gas emissions. *Geophysical Research Letters* 27(3): 409–12.

Hammit, J. K., A. Jain, J. L. Adams, and D. Wuebbles. 1996. A welfare-based index for assessing environmental effects of greenhouse-gas emissions. *Nature* 381: 301–303.

Hayhoe, K., A. Jain, H. Keshgi, and D. Wuebbles. 2000. Contribution of CH_4 to multi-gas reduction targets. In J. van Ham, ed., *Non-CO_2 Greenhouse Gases: Scientific Understanding, Control and Implementation*. Dordrecht: Kluwer Academic, pp. 425–32.

Hayhoe, K., A. Jain, H. Pitcher, C. MacCracken, M. Gibbs, D. Wuebbles, R. Harvey, and D. Kruger. 1999. Costs of multigreenhouse gas reduction targets for the USA. *Science* 286: 905–906.

Hoel, M. 1991. Efficient international agreements for reducing emissions of CO_2. *Energy Journal* 12: 93–107.

Houghton, J., Y. Ding, D. Griggs, M. Noguer, P. van der Linden, X. Dai, K. Maskell, and C. Johnson, eds. 2001. *Climate Change 2001: The Scientific Basis. Volume I of IPCC Third Assessment Report*. Cambridge: Cambridge University Press.

Houghton, J., G. J. Jenkins, and J. J. Ephraums, eds. 1990. *Scientific Assessment of Climate Change: Report of Working Group I. Volume I of IPCC First Assessment Report*. Cambridge: Cambridge University Press.

Houghton, J., L. Meira Filho, B. Callander, N. Harris, A. Kattenberg, and K. Maskell, eds. 1995. *Climate Change 1995: The Science of Climate Change. Volume I of IPCC Second Assessment Report*. Cambridge: Cambridge University Press.

Kandlikar, M. 1995. The relative role of trace gas emissions in greenhouse abatement policies. *Energy Policy* 23(10): 879–83.

Kandlikar, M. 1996. Indices for comparing greenhouse gas emissions: Integrating science and economics. *Energy Economics* 18: 265–81.

Kurosawa, A. 2004. Multigas reduction strategy under climate stabilization target. Contributed Paper 181. 7th International Conference on Greenhouse Gas Control Technologies, Vancouver, Canada.

Manne, A. S., and R. G. Richels. 2001. An alternative approach to establishing trade-offs among greenhouse gases. *Nature* 410: 675–77.

McCarl, B. A., and U. A. Schneider. 2001. Greenhouse gas mitigation in U.S. agriculture and forestry. *Science* 294: 2481–82.

Mendelsohn, R., W. Morrison, M. Schlesinger, and N. Andronova. 1998. Country-specific market impacts of climate change. *Climatic Change* 45(3–4): 553–69.

Michaelis, P. 1999. Sustainable greenhouse policies: The role of non-CO_2 gases. *Structural Change and Economic Dynamics* 10: 239–60.

Michaelowa, A. 2003. Limiting global cooling after global warming is over—Differentiating between short- and long-lived greenhouse gases. *OPEC Review* 27(4): 343–51.

Moslener, U., and T. Requate. 2007. Optimal abatement in dynamic multi-pollutant problems when pollutants can be complements or substitutes. *Journal of Economic Dynamics and Control* 31(7): 2293–2316.

O'Neil, B. C. 2000. The jury is still out on global warming potentials. An editorial comment. *Climatic Change* 44(4): 427–43.

O'Neil, B. C. 2003. Economics, natural science, and the costs of global warming potentials. An editorial comment. *Climatic Change* 58(3): 251–60.

Ramaswamy, V., O. Boucher, J. Haigh, D. Hauglustaine, J. Haywood, G. Myhre, T. Nakajima, G. Shi, and S. Solomon. 2001. Radiative forcing of climate change. In vol. 1 of Houghton et al. (2001), pp. 349–416.

Reilly, J. M., M. Babiker, and M. Mayer. 2001. Comparing greenhouse gases. Global Change Report 77. MIT, Cambridge.

Reilly, J. M., H. D. Jacoby, and R. G. Prinn. 2003. Multi-gas contributors to global climate change. Climate impacts and mitigation costs of non-CO_2 gases. Report, Pew Center on Global Climate Change, Arlington, VA.

Reilly, J. M., R. G. Prinn, J. Harnisch, J. Fitzmaurice, H. D. Jacoby, D. Kicklighter, J. Melillo, P. Stone, A. Sokolov, and C. Wang. 1999. Multi-gas assessment of the Kyoto Protocol. *Nature* 401: 549–55.

Reilly, J. M., and K. H. Richards. 1993. Climate change damage and the trace gas index issue. *Environmental and Resource Economics* 3: 41–61.

Sarofim, M. C., C. E. Forest, D. M. Reiner, and J. M. Reilly. 2005. Stabilization and global climate policy. *Global and Planetary Change* 47(2–4): 266–72.

Schmalensee, R. 1993. Comparing greenhouse gases for policy purposes. *Energy Journal* 14(1): 245–56.

Shine, K. P., T. K. Berntsen, J. S. Fuglestvedt, and R. Sausen. 2005a. Scientific issues in the design of metrics for inclusion of oxides of nitrogen in global climate agreements. *Proceedings of National Academy of Sciences USA* 102(44): 15768–73.

Shine, K. P., J. S. Fuglestvedt, K. Hailemariam, and N. Stuber. 2005b. Alternatives to the global warming potential for comparing climate impacts of emissions of greenhouse gases. *Climatic Change* 68(3): 281–302.

Smith, S. J., and T. M. L. Wigley. 2000. Global warming potentials. Part 1: Climatic implications of emissions reductions. *Climatic Change* 44(4): 445–57.

Tol, R. S. J. 2002a. Estimates of the damage costs of climate change: Part II: Dynamic estimates. *Environmental and Resource Economics* 21(2): 135–60.

Tol, R. S. J. 2002b. Fearnside's unified index for time preference: A comment. *Ecological Economics* 41(1): 33–34.

Tol, R. S. J. 2005. The marginal damage costs of carbon dioxide emissions: An assessment of the uncertainties. *Energy Policy* 33(16): 2064–74.

Tol, R. S. J., R. J. Heintz, and P. E. M. Lammers. 2003. Methane emission reduction: An application of FUND. *Climatic Change* 57(1–2): 71–98.

UNFCCC. 1997. Kyoto Protocol to the United Nation Framework Convention on Climate Change. COP3, UNFCCC (Climate Change Secretariat).

USEPA. 2006. Global mitigation of non-CO_2 greenhouse gases. Report EPA 430-R-06-005. US Environmental Protection Agency, Washington, DC.

Watson, R., M. Zinyowera, and R. Moss, eds. 1997. *The Regional Impacts of Climate Change: An Assessment of Vulnerability. A Special Report of IPCC Working Group II*. Cambridge: Cambridge University Press.

Weitzman, M. L. 1998. Why the far-distant future should be discounted at its lowest possible rate. *Journal of Environmental Economics and Management* 36(3): 201–208.

Index

Abatement targets
 and GPGP scheme, 43
 GWP-based, 371–74
 GWP-based (second-best), 374–78
Ad hoc Group on the Berlin Mandate, 80
Agriculture, and abatement costs (methane and nitrous oxide), 368–69, 377
Allocation of emission permits. *See* Emission permits
Allowance trading regime, and market equilibrium, 26. *See also* Global Public Good Purchase approach
Annex-B countries, 6, 38–39, 40, 43
 aid needed by, 92–93
 in alternate climate regime, 27–28
 and cap and trade programs, 90
 and carbon trading, 73–79
 in coalition-formation model, 145
 in comparison of GPGP and OFK, 52, 55, 56, 57–58
 effort level of, 51–52
 and property right regime, 18
 and viability condition, 39
Annex-B_{US} countries, in coalition formation model, 146, 147–48, 149, 150, 152, 153
Annex B-R countries, 335
 in cement trade model, 339, 341–42, 343, 344, 346–47, 348–49, 350, 351, 352, 353
Annex-I countries, 204
Armington specification, 333, 354, 356n.1
Asia Pacific Partnership for Clean Development and Climate, 87, 137
Atmospheric concentration ceiling, 307
Auctioning of emission permits, 254, 255
Australia, in Asia Pacific Partnership for Clean Development and Climate, 137

Banking, 255, 264–69. *See also* International Bank for Emissions Allowance Acquisition
Bargaining paradigm, shifting of, 87–88
Barrett, Scott, 179, 180–81
Baselines, 81
 developmental, 82
 as impractical to measure, 94
BAU (business as usual)
 and Global Public Good Purchase approach, 20, 21–22
 and intensity limits, 223, 230, 234
 and international cement trade, 339–42
 and "treaty BAU" quantities, 22–23
 updating of, 23
 also mentioned, 28–29, 108, 158, 343, 368
BAU allowances or quotas, 3, 31, 38, 39, 44
BAU trajectory, 19, 27
Berlin conference (1994), 76
Bilateral climate-friendly activities, 137–38
Bingaman-Specter resolution, 204
Bohr, Niels, quoted, 221
Bonn conference and agreement (2001), 13, 15, 145, 159n.6
Border-tax adjustments (BTA), for cement trade model, 335, 353
 BTA scenario, 347–52, 353–54
 no-BTA scenario, 343–47
Borrowing, 255, 264–69
Bottom-up approach, 211, 213
Bradford, David, 1
Brazil
 Compliance Fund proposed by, 91
 debt trap of, 81
 PLANTAR project in, 93
Budget proposal, definition of, 67

Burden distribution (sharing)
  and equity, 79–81
  and Kyoto vs. GPGP approach, 26
Business as usual. *See at* BAU
Byrd-Hagel resolution, 77, 204

California
  as emission-level leader, 138
  Regional Clean Air Incentives Market (RECLAIM) program of, 254
Canada, in cement-trade model, 344
Canadian Offset System, 138
Cap and trade approach, 76–77, 93–94
  and developing countries, 90
  and enforcement, 26
  equity issues in, 86–87
  vs. GPGP approach, 13, 20
  in Kyoto Protocol, 5, 13, 20, 26
  and transfers between nations, 5
Capture and sequestration of carbon (CCS), 273. *See also* Carbon sequestration
Carbon accumulation, and dynamic analysis, 5
Carbon dioxide. *See* CO_2
Carbon flow dynamics, 313–14
Carbon leakage, 300
  in cement-trade model, 6, 346–47, 352, 353, 354
  and emissions total, 25
  in Global Public Good Purchase system, 23–24
  as industry concern, 333
Carbon market, world, 89
Carbon Mitigation Initiative group, Princeton, 35
Carbon price(s)
  climate coalition effort as, 40
  control policy for, 50
  differentiation of, 92
  in equilibrium, 54
  in GPGP definitions, 39, 42, 45, 49, 53, 55
  heterogeneity of, 4, 6
  as inducement rather than driver, 89
  on international transport, 354
  and OFK, 46, 58
  and production costs, 92
  for second-best world, 77
  taxing of, 61 (*see also* Carbon price schemes)
  total, 48
  uncertainty about, 60–61, 80
  uniqueness of assumed, 6
  as world price, 6
Carbon sequestration, 273, 275–76, 292–98
  in model, 278–79, 286–87
  for large reservoir, 287–91
  for small reservoir, 291–92
Carbon sinks, 18, 273, 275–76, 278, 302n.2
  and Kyoto targets, 159n.6
Carbon tax schemes
  equity issues in, 86–87
  and fossil fuels pricing, 50, 51
  unpopularity of, 76
Carbon trading. *See also at* Emissions trading
  European Carbon Trading system, 95n.9
  in synergy, 83–86
Carbon value, and cement-trade model, 343
CDM. *See* Clean Development Mechanism
Ceiling. *See also* "Safety valve" ceiling
  on atmospheric concentration, 307
  on cumulative carbon emissions, 274–75, 280, 296
Cement trade, international, 333
  from "domestic excess capacities," 340, 355
  GEO-CEMSIM model of, 6, 334–39, 352–54
  in business-as usual scenario, 339–42
  with border-tax adjustments, 347
  with border-tax adjustments ("complete" vs. WTO), 348–52, 353–54
  without border-tax adjustment, 343–47
CEMSIM model, 334, 338–39. *See also* GEO-CEMSIM model of international cement trade
Chander-Tulkens (CT) solution, 174, 175, 179, 190–92
  and free-riding, 179
Chander-Tulkens transfer scheme, 122, 125, 127
Chicago Climate Exchange, 138
China, 151
  and allocations rules, 80
  in Asia–Pacific Partnership for Clean Development and Climate, 137
  in coalition-formation model, 144, 149–50, 152–53
  CO_2 emissions of (cement-trade model), 342

Index 387

cumulative energy investments in, 81
and emissions-GDP relationship, 244
and emissions trading, 77
and EU, 137
savings decline in, 81
and stringent targets, 194
in transfers model, 120, 128
and United States, 149, 151, 152, 153
Clean Air Act Amendments (1990), 204–205
Clean Development Mechanism (CDM), 25, 78, 203, 211, 213n.1
and cement sector, 343
GPGP approach contrasted with, 26
Climate blocs, 196
and future negotiations, 141
and Kyoto Protocol, 141
and membership rules, 193–96
regional and subglobal, 139–42
Climate change
and developing countries' priorities, 209
growing recognition of, 137
impact of vs. damage from (GWP), 364
infrastructure vulnerable to, 85–86
special difficulties with, 103–104
Climate change risks
approximating of, 307
and climate constraints, 316
and danger of delays, 311
global mean temperature as metric for, 307–308, 309, 326
and rate of change, 309
Climate coalition. *See* also Coalition
Climate control, as capital intensive, 86
Climate policies
and absolute vs. intensity limits on emissions, 221–23, 246
arguments on in literature, 223–27
under certainty, 226, 227–28
empirical test on, 235–45
and equivalence of absolute and intensity limits, 247–48
under uncertainty, 225, 226, 228–35
broader program than emission trading needed in, 212–213
carbon prices as inducement in, 89
cement-trade model of, 353–54
with border-tax adjustment, 347
with border tax adjustments ("complete" vs. WTO), 348–52, 353–54
without border-tax adjustment, 343–47

and danger
of delays, 311
social consensus on, 323
and developing countries, 75
and development baselines, 82
and economics, 201, 212, 220
and developing countries' needs, 209
flexibility in, 377
and global agreement vs. multiple agreements, 139
and individual countries' incentives and specificities, 137
and international agenda, 93–95
issues on (pace vs. means), 37
optimal (uncertainty on climate sensitivity with learning), 318–23
participation and target as dimensions on, 38
plan for resort to fossil fuels and renewables, 273–77, 300
and case of abundant renewable substitute, 286–92
and case of rare renewable substitute, 292–99
model of, 277–82
and optimal paths with no opportunity for abatement, 284–86
and pure Hotelling paths, 283–84
price-style vs. quantity-style, 7–8, 46–49, 50, 51, 58–59, 60–61
short-term abatement policy in, 327–28
tools outside carbon markets for, 89
and transfer schemes, 124–25 (*see also* Transfers between nations)
as unitary vs. divisive, 88
Climate quality
as luxury vs. inferior good, 191
as public good, 14
Climate regime, shifting the status of, 87–88
Climate sensitivity, 315–18, 328n.5
and EVPI, 324–26
need to know, 327
need to quantify, 323
uncertainty on, 309, 310–11, 315, 328n.5
with learning, 318–23
Climate stabilization, integrated assessment in context of, 308–11
CLIMNEG world simulation model, 105–107, 130–33, 189
Clubs, theory of, 193–94

Coalition formation, 128–30, 169–71
  axiomatic approach to, 171
  fundamental features of, 107
    noncooperation and full cooperation, 108–109, 128
    partial cooperation, 109–14, 128
  and game-theoretic analysis, 2, 139–40
  and multiple initiatives/agreements, 138
  and reactions of ex-partners, 191
  stability of, 114–17, 192
    and change of membership rules, 125–28, 129
    and transfer schemes, 104, 115, 118–25, 127–28, 128–29, 190–93
Coalition formation model, 144–46, 153–54, 170
  results from, 146–53
  theoretical framework of, 142–43, 159n.2
Coalition formation theory, 196
  and heterogeneity, 188
Coalition structures, 4, 110–14, 122, 147
  "grand coalition," 3–4, 4, 5, 165, 171 (see also "Grand coalition")
  multiple, 143
  rationale for game with, 175–178
  stability of, 143, 170 (see also Stability)
Coalition theory, and regional or subglobal climate blocs, 139–42
"Coalition unanimity" game, 141
Coase, R. M., and Coasean theory, 167, 168, 174, 188, 261
Cobb-Douglas technology, 154
Collective action, Foley's model of, 66–69, 70
Collective demand concept, 69
Collective rationality, 180
Command and control regimes
  improvement on, 16
  and incidence issues, 31
Commitment, quantity vs. procedural, 46
Common architecture, and need for capital, 86
Communication, among players, 171
Competition module (GEO model), 355
Competitive collective consumption with contributions (CCCC) equilibrium, 68–69
  with fixed contributions, 69–70
  and GPGP approach, 71–73
Compliance Fund, proposed by Brazil, 91

Conjoint market, for local and greenhouse gas emissions, 82–83
Conscription, and public goods, 14
"Consensus treaty," 180
Consumption, demand for, 66–68
Consumption module (CEMSIM model), 354
Contestable GPGP equilibrium(a), 71, 73
  nonstrategic, 73
Cooperation, 171, 217–218
  and coalition formation, 108–14, 128 (see also at Coalition)
  communication as determinant of, 171
  as concern of economics, 201
  and current action, 205–206
  economic rationale for, 166–67, 181, 188
  importance of questioned, 203, 205, 206
  in repeated negotiations, 212
  and technological change, 219
  unilateral action as step toward, 204
Cooperative games or game theory, 167–68, 188
  core of, 168
  y-core of, 182n.6
  vs. stability theory, 172
COP, 15, 19, 27, 35, 69, 77
Core concept, 168, 175
Cost-benefit approach, and GWP issue, 364, 366
Cost-effectiveness approach, and GWP issue, 364–65
Cost-efficiency analysis, 310, 322
Cost-efficiency optimal control integrated assessment model, 311–18
Cost sharing, in GPGP approach, 20, 26, 70
CO_2, 20, 365, 366, 381n.3
  and intensity limits, 223, 235, 246 (see also Intensity limits)
  and Kyoto targets, 362
  as reference gas, 362, 364, 365, 371, 372, 374, 376 (see also Global Warming Potential index)
Cournot problem, 52
CT. See at Chander-Tulkens
CWS model, 105–107, 129

Debt trap, of Brazil, 81
Decay process, 368, 378
Decentralized production, 68
Delhi Declaration on Sustainable Development and Climate Change, 88

Index

Demand, for private and collective consumption, 66–68
Developing countries (LDCs)
 aid needed by, 93
 and cap and trade programs, 90
 carbon pricing in, 77–78
 climate change as low priority for, 209
 and Clinton indexing proposal, 223
 creating incentives in, 206
 criteria for contributions from, 191
 crucial importance of, 76
 demand for infrastructures in, 88
 emission compensation difficult in, 87
 emission-GDP correlation for, 238
 as failing to endorse emission limits, 201
 and grandfathering principle, 79
 "impatience" of, 107
 incentives linked to assistance for, 90–91
 industrialized countries' trade with, 209
 intensity limits for (Bush administration), 224
 and national policies, 202
 as necessary to climate-policy effectiveness, 111
 participation of, 3
 silence on targets for, 76
 structural transitions of, 87–88
 technology export to, 206, 209, 211, 213
 and timing of climate change issue, 75
Development and climate synergies, 82
Discount rates, long-run, 37
Distributional effects
 within countries, 27
 internationally, 27–32
Distribution of burden, and tax harmonization, 49
Domestic climate activities, 137–38, 213
 in absence of international agreement, 211–12
 focus on, 202, 205
 wide variety of, 210
Dynamic analysis or setting, 8
 and carbon accumulation, 5–6
 minimal architecture in, 89–92
 and multi-period allocation of emission permits, 254–55, 257–64
 tax increase in, 48–49

East European countries, and Kyoto Protocol, 225
"Ecological subsidy," 120

"Ecological surplus," 174, 175
Economics
 and climate change policy, 201, 212, 220
 and developing countries' needs, 209
 and global warming potential, 359
 and second-best vs. first-best, 77
Efficiency, 190
 and cooperation, 181, 188
 for group of counties, 166
 internal and external, 170
 and market for pollution permits, 194
 stability for, 166, 181
Eighth Conference of Parties (2003), 88
Electricity market reforms, in synergy, 83–86
Emission abatements, and GPGP, 3, 19–20. See also Global Public Good Purchase approach
Emission permits
 initial allocation of, 253–54
 multi-period (dynamic) allocation of, 254–56, 269–70
 alternative allocations, 257–61
 through banking or borrowing, 255, 264–69
 market model of, 256–57
 numerical illustration of, 261–64, 268
Emission reduction purchase agreements (ERPA), 93
Emissions trading, 201. See also Carbon trading question of present pursuit of, 206–10
Emissions trading markets, 137–38, 139
Emissions trading program, and question of who sells, 207
Emissions trading scheme (ETS)
 of Europe, 203–204, 205, 218, 254, 356n.6
 of UK, 222, 249n.2
Energy-intensive industries, special treatment of, 91–92
Enforcement
 and Kyoto vs. GPGP approach, 26–27
 self-enforcement, 179–81
Environmental externalities, 168
Environmental problems, and property rights, 15–16
"Equilibrium binding agreement" rule, 141
Equity principles, 79–81
ERPA (emission reduction purchase agreement), 93

European Carbon Trading system, 95n.9
European Union (EU)
 and China, 137
 climate policy goals of, 308
 in coalition-formation model, 144, 146, 148–49, 150, 151, 152–53
 and ETS, 202, 203–204, 205, 218, 254, 356n.6
 and Kyoto Protocol, 146, 189, 203–204, 218
 membership rules of, 126
 and national allocation of permits, 269
 and price of allowances, 207, 213n.4
Exclusive membership rule, 105, 126, 127, 129, 193, 194
Exhaustible resource, tax incidence for, 59–60
Expected value of perfect information (EVPI), 323–326
Externality(ies), 166–67. *See also* Positive externalities
 carbon emissions as, 109
 and cooperation, 188
 and "ecological surplus," 174
 games with, 168–69
 and IEA, 166, 168
Externality model, multilateral, 173, 183n.21

Fairness, 180, 190
Favela regime, 86, 88
FCCC (Framework Convention on Climate Control), 13, 19, 69, 76, 87–88, 218, 269
FEEM-RICE model, 138, 144–46, 154–55, 170, 189–90
 emissions in, 145
 emission trading in, 158–59
 and knowledge spillovers, 157–59
 standard without induced technical change, 155–57
 with technical change induced, 157
Flexible Kyoto arrangement (FK), 41–42
 open (OFK), 42–43, 46, 49, 50
 and GPGP, 50, 52–58
Flexible Kyoto equilibrium, 54
Foley's model of collective action, applied to climate change, 66–69, 70
Fossil fuels
 plan for use of with renewables, 273–77, 300
 and case of abundant renewable substitute, 286–92

and case of rare renewable substitute, 292–99
 model of, 277–82
 and optimal paths with no opportunity for abatement, 284–86
 and pure Hotelling paths, 283–84
 taxation on, 48–49
Framework Convention on Climate Control (FCCC), 13, 19, 69, 76, 87, 218, 269
Free-riding, 39–43, 196
 by coalitions on each other, 195, 196
 and coalition structure, 114
 and differentiation under Kyoto Protocol, 196–97
 and Global Public Good Purchase approach, 21
 and national governments, 70
 vs. non-compliance, 183n.27
 preference revelation vs. nonparticipatory, 178
 research needed on, 130
 and stability, 178–79, 183n.25
 and transfers, 122, 129
FSU (former Soviet Union)
 in coalition-formation model, 144, 150, 151, 152–53
 and emissions-GDP relationship, 244, 245
Futures trading, of IBEAA, 24

Game(s)
 cooperative, 167–68
 with externalities and the core, 168–69
 with particular coalition structure, 175–78
 in partition function form, 169–70, 182n.12
Game-theoretic analysis, 2, 104, 128
 and bargaining power, 118
 and coalition formation, 139–40
 and GEO model, 337
GDP
 in CWS model, 105, 106, 119, 120
 and intensity limits, 222, 226, 234, 235, 246 (*see also* Intensity limits)
 and price versus quantity debate, 8
GEO-CEMSIM model of international cement trade, 6, 334–39, 352–54
 in BAU scenario, 339–42
 with border-tax adjustments, 347
 with border-tax adjustments ("complete" vs. WTO), 348–52, 353–54

Index

without border-tax adjustments, 343–47
modules of, 354–56
GHG equivalence rule, 359, 361, 373, 377
GHG index(es), 360, 361, 364–65, 367, 371, 372, 375, 377
and non-CO_2 gases, 365–66
uncertainties in, 366, 367
Global Fast Track Assessment, 309
Global mean temperature rise. *See* Temperature rise, global mean
Global Public Good Purchase approach (GPGP), 3, 8, 13, 13–22, 38–39
assessment of, 37–38
with CCCC approach, 71
and financing of public good, 44–45
institutional structure underlying, 65
and Kyoto-style system, 25–27, 45, 50–51
and measurements, 61n.2
noncooperative equilibrium of, 69
and OFK, 50, 52–58
and participation problem, 39–43
and price versus quantity policy debate, 58–59
and ratchet effect, 43–44
and "treaty BAU 'imports,'" 22–23
updating of in long run, 23–25
Global public good purchase plus taxation (GPGPPT), 45–46
and tax harmonization, 59
Global Warming Potential (GWP) index, 7, 359–61, 362, 377–79
abatement targets based on, 371–74
abatement targets based on (second-best), 374–78
and alternative economic GHG indexes, 364–65
economic shortcomings of, 362–64
vs. optimal abatement, 371
paradox of, 360
success of explained, 365–67, 378
revision of, 381n.2
GPGP. *See* Global Public Good Purchase approach
"Grand (world) coalition," 3–4, 4, 165, 171
and coalition formation theory, 178
core-stable, 5
efficiency of, 170
World Trade Organization compared with, 195
Grandfathering, 79, 254, 255, 257, 269, 270n.3
fixed, 257–258
rolling (updated), 258, 260–61, 262, 266

Greenhouse gas emission control
and conjoint market, 82–83
global public good purchase approach to, 13, 18–22, 38–39 (*see also* Global Public Good Purchase approach)
Greenhouse gases (GHG), and global warming potential, 359. *See also* Multi-greenhouse gas issue; *at* GHG

Harmonized tax. *See* Taxation, harmonized
Heterogeneity
and absolute vs. intensity emission limits, 240
and extensive form games, 192
sources of, 189–90
and theory of coalition formation, 188
Hotelling model of exhaustion, 274–75
Hotelling paths, 283–84, 288–89, 293, 298, 300

IBEAA (International Bank for Emissions Allowance Acquisition), 19–20, 24–25, 27, 29, 38, 46, 69
IEA. *See* International environmental agreement
Incentives
and coalition formation, 140
in model, 144, 153
economic, 139
and self-enforcement, 179
Incentive structure, and full-cooperation scenario, 109
Incidence issues, 27–35. *See also* Tax incidence
Indexed limits, 229–33
India, 83–84
in Asia Pacific Partnership for Clean Development and Climate, 137
coal emissions in, 82–83
Konkan railway in, 85–86
Individual rationality, 180
Information value, in ranking of uncertainties, 323–27
Infrastructure
climate vulnerability of, 85–86
developing countries' demand for, 88
Initial allocation of emission permits, 253–54
Integrated assessment
in context of climate stabilization, 308–11
probabilistic, 309–10

Integrated assessment model, 104, 105–107, 128, 154, 273
  cost-efficiency optimal control, 311–18
Intensity limits, 221–23, 246
  arguments on in literature, 223–27
  under certainty, 226, 227–28
  empirical tests on, 235
    historical forecasts used in, 240–45
    historical time series data in, 235–40
    two sets of experiments compared, 245
  and equivalence to absolute limits, 247–48
  examples of, 249n.1
  as indexed, 249n.4
  under uncertainty, 225, 226, 228–29
    and indexed limits, 229–33
    measurement issues in, 234–35
    and optimal indexation, 233–34, 240
Intergovernmental Panel on Climate Change (IPCC)
  and carbon sequestration, 273
  Global Warming Potential Index of, 7, 359, 359–60, 362
International agreement, question of necessity of, 202–206
International Bank for Emissions Allowance Acquisition (IBEAA), 19–20, 24–25, 25, 27, 29, 38, 46, 69
International Climate Change Taskforce, 308
"International Climate Policy after COP6" (conference, Hamburg 2001), 35
International cooperation. *See* Cooperation
International emissions trading, question of present pursuit of, 206–10
International environmental agreement (IEA), 167
  core property in, 172
  and externalities, 166, 168
  free riders of, 121 (*see also* Free-riding)
  and game-theoretic analysis, 2
  and high participation, 111
  voluntary, 121
  y-characteristic function of, 175
  y-core solutions for, 173–75
International permit trade, 218–19. *See also at* Emissions trading
International treaties. *See* Treaties
Intertemporal arbitrage, 363
Intertemporal competitive pricing, 9
Intertemporal transfer of permits. *See* Multi-period allocation of emission permits

IPCC. *See* Intergovernmental Panel on Climate Change

Japan
  in Asia Pacific Partnership for Clean Development and Climate, 137
  in cement-trade model, 344
  in coalition-formation model, 144, 148–49, 150, 151, 152–53
  as committed to Kyoto targets, 146, 189
  voluntary domestic emissions trading scheme in, 138
Johannesburg Declaration, 88
Jurisdiction formation, 193–94

Kyoto arrangements
  flexible (FK), 41–42 (*see also* Flexible Kyoto arrangements)
  and free-riding, 40
  and quotas, 62n.8
Kyoto coalition, members of free to leave, 146
Kyoto equilibrium, flexible, 54
"Kyoto forever" hypothesis, 145, 160n.7
Kyoto-like variants, 38
  and carbon sequestration, 275–76
  and harmonized taxation, 49–50
  and participation problem, 41–43
  and Global Public Good Purchase (GPGP) approach, 45, 50–51
Kyoto Protocol, 103
  aggregate quotas in, 181
  and capital flows, 208
  cap and trade approach of, 5, 13, 20, 26
  and carbon market, 89
  and control of price vs. quantity, 8
  and differentiation among countries, 6, 196–97
    country-specific targets, 187, 196
    emission targets in, 224, 225
  and FEEM-RICE model, 158
  multi-gas, 366
  enriched interpretations of, 5–6
  and European Union, 146, 189, 203–204, 218
  fixed caps in, 221
  and Global Public Good Purchase (GPGP) approach, 25–27, 38, 45, 50
  and global warming potentials, 359, 362
  and greenhouse gases variety, 159n.6
  and incidence issues, 32
  and industry concerns, 333

Index

and measurements, 61n.2
and minimal architecture, 9
and property rights, 15, 17
and quantity policy, 47
and ratchet effect, 62n.13
reinterpretation and amendment of, 86–87, 94
and trade, 6
and transversal debates, 3–9
and United States, 103, 111, 201, 202–203, 212
 and European ratification, 203–204, 218
 on income transfers, 62n.16
unlikelihood of signing of by all relevant countries, 141

LDCs. See Developing countries
Leakage, carbon. See Carbon leakage
Learning, and uncertainty on climate sensitivity, 318–23
"Legally binding," 217
 and GWP concept, 359
 misperception of, 91
Limits on emissions, absolute vs. intensity, 221–23, 246
 arguments on in literature, 223–27
 under certainty, 226, 227–28
 empirical tests on, 235
  historical forecasts used in, 240–45
  historical time series data used in, 235–40
  two sets of experiments compared, 245
 and equivalence of absolute and intensity limits, 247–48
 under uncertainty, 225, 226, 228–29
  and indexed limits, 229–33
  measurement issues in, 234–35
  and optimal indexation, 233–34, 240
Linear damage, atmospheric behavior, and temperature change, 379–81
Local emissions, conjoint market for, 82–83
Long-term goals, and immediate pressures, 92–93
Lump-sum transfers, 5

Market(s)
 conjoint, 82–83
 for emission permits, 253 (see also Emission permits)
 for pollution permits, 194
  externalities from, 195

Marrakech conference (2001), 159n.6
Measurements, in GPGP scheme and Kyoto Protocol, 61
Membership rules
 changing of, 125–28, 129
 and climate-blocs, 193–96
 and stable coalition structures, 141
Methane, 362, 365, 367, 368, 368–69, 377–79
Millennium Declaration, UN Millennium Summit (2000), 88
Minimal architecture, in dynamic world, 89–92
Montreal Protocol on Substances That Deplete the Ozone Layer, 187
Moral motivation, in transfer schemes, 118–22, 129. See also Normative criteria
Motivation
 in change of membership rules, 125–27
 in transfer schemes, 118–22, 129
Multigreenhouse gas issues
 and GHG index, 377
 and GWP concept, 360, 361
 optimal abatement path for, 367–71
Multilateral externality model, 173, 183n.21
Multi-period (dynamic) allocation of emission permits, 254–56, 269–70
 alternative allocations, 257–61
 through banking or borrowing, 255, 264–69
 market model of, 256–57
 numerical illustration of, 261–64, 268
Multiple climate coalitions (initiatives or agreements), 141–42
 and coalition formation, 138
 vs. global agreements, 139
Multiple coalition structures, 143

Negotiation paradigm, broadening of, 88
New investment module (GEO model), 355–56
New Mexico, and voluntary emissions trading, 138
No BTA scenario, in cement-trade model, 343–47
"No Cap but Trade Approach to Greenhouse Gas Control, A" (Bradford), 35
Noncooperative approach to IEAs, 169
Noncooperative game-theoretic model, 104

Noncooperative game theory, 188
Noncooperative GPGP equilibrium, associated with fixed contributions, 65, 69–70
Noncooperative theory of coalition formation, 139–40
Normative criteria, 79–81
and "norm cascade," 195
vs. stability, 191–93
North-South misunderstandings (1988 to 2005), intellectual sources of, 75–77

Oceanic storage, 302n.4
ODA, assistance from, 93
OECD nations, 238, 240, 244, 250n.9
in Europe, 245
Open flexible Kyoto variant, 42–43, 46, 49, 50
and GPGP approach, 50, 52–58
"Open membership" rule of game, 125–26, 129, 141, 146, 193
Optimal climate policy under uncertainty on climate sensitivity, 318–23
Optimal indexation, 233–34, 240
Optimal multi-gas abatement path, 367–71
Ozone depletion potential (ODP) index, and GWP, 365

PANE-equilibrium concept, 144, 169
Participation assistance programs, 90
Participation problem, 39–43, 188–89
and climate blocs, 194
and LDCs, 3
Pataki, George, 205
Performance criteria, targets based on, 90
Permit allocation. *See* Emission permits
PLANTAR project, Brazil, 93
POLES model, 338, 343
Policies on climate. *See* Climate policies
Positive externalities, 196. *See also* Externality(ies),
in coalition formation, 110–11, 114
and extensive form games, 192
nonexcludability from, 103
Post-Kyoto schemes, 3
Price cap, 91
Price equalization, global, 202
Price mechanisms, as emission caps, 206–207
Price versus quantity issue, 7–8, 46–49, 50, 51

and Global Public Good Purchase approach, 58–59
Private property. *See* Property rights
Production, decentralized, 68
Property rights, 15–18
emission permits as, 253
Protectionism, of border-tax adjustment, 350, 353
Public goods, 14–15. *See also* Global Public Good Purchase approach
climate quality as, 5, 14
and environmental externalities, 168
financing of (GPGP), 44–45
and free-riding, 178
voluntary provision of, 103

Quantity policy
and GPGPPT vs. Kyoto, 51
as incentive system (Berlin 1994), 76
Kyoto Protocol as, 47
Quantity vs. price issue, 7–8, 46–49, 50, 51
and Global Public Good Purchase approach, 58–59
Quota(s), 47
as allowances vs. rights, 80
and Kyoto Protocol or scheme, 47, 62n.8
Quota differentiation, 5

Radiative forcing, and GWP, 360, 362, 363, 364, 377
Ratchet effect, 43–44, 46
and harmonized taxation, 49–50
and Kyoto Protocol, 62n.13
Rationality
collective, 180
individual, 180
Rawlsian maximin rule, 120
R&D, 87
China's investment in, 151
in coalition-formation model, 152
in FEEM-RICE model, 155, 156
subsidies on, 219
RECLAIM program, 211
Regional cooperation, 84–85, 89
Regional Greenhouse Gas Initiative (RGGI), US, 138, 205
Regional or subglobal climate blocs, 139–42
vs. global agreements, 139
Renewable energy resources, plan for use of with fossil fuels, 273–77, 300

Index 395

and case of abundant renewable
  substitute, 286–92
and case of rare renewable substitute,
  292–99
model of, 277–82
and optimal paths with no opportunity
  for abatement, 284–86
and pure Hotelling paths, 283–84
Research. *See also* R&D
  and coalition formation, 129–30
  on emission permit allocation, 270
Reservoirs for carbon sequestration, 273
  large, 287–91
  small, 291–92, 301
RESPONSE Θ and RESPONSE model
  family, 311–12, 315, 318, 319, 323
Retirement and retrofitting module
  (CEMSIM model), 355
Reviews of national actions, 202, 211,
  213
RICE model, 144, 147, 154, 156, 189, 313
Rio Declaration on the Environment,
  principle 9 of, 84
Russia
  in cement-trade model, 339, 341–42, 343,
    344, 346
  emissions trading with, 203
  and Kyoto Protocol, 62n.8, 103, 146, 189,
    225

Safe landing analysis, 310
"Safety valve" ceiling
  on abatement costs, 221, 222, 246
  on price of allowances, 25, 207, 208–209,
    212, 214n.10, 218
Schlessinger, James, 82
Secular growth baselines, projecting of,
  81
Self-enforcement, 179–81
Self-regeneration, 278–79
Selten trick, 52, 53
Sensitivity analysis or study, 310, 317–18
Sequential decision-making with learning
  framework, 311
Sequestration, carbon. *See* Carbon
  sequestration
Sink credits, 203
Sinks, carbon, 18, 273, 275–76, 278, 302n.2
  and Kyoto targets, 159n.6
Social planner problem, 281–82, 287, 295,
  301

and multi-gas abatement, 369–70
and second-best GWP-based abatement
  targets, 375
South Korea, in Asia Pacific Partnership
  for Clean Development and Climate,
  137
Soviet economy, and ratchet effect, 44
Speed limits, 89
Spillovers
  and border-tax adjustment, 352, 353
  in FEEM-RICE model, 157–59
Stability, 4–5, 171–72, 181
  alternative concepts of, 172–73
  of coalition formation, 114–117
  and change of membership rules, 125–
    28, 129
  and transfers, 104, 115, 118–25, 127–28,
    128–29, 190–93
  of coalition structure, 143, 170
  and free-riding, 178–79, 183n.25
  of GPGP and OFK, 43
  internal and external, 114–15, 125, 143,
    169, 172
  internal-external vs. core, 5
  intracoalition, 143
  and subglobal blocs, 194
  of treaty, 105
Stability analysis, and coalition-formation
  model, 147
Stability and Growth Pact, EU, 217
Static competitive model, and taxation,
  48
Stern Review, 37
Stockholm, UN conference on Human
  Environment at (1972), 75, 81
Subglobal climate blocs, 139–42
  vs. global agreements, 139
Sulfur Allowance Trading Program, US
  EPA 254
Superadditivity, in coalition formation,
  110, 114
Synergies
  development and climate, 82
  electricity market reforms and revenues
    from carbon trading, 83–86

Tabula rasa myth, 77–86
Taxation, 45–46
  on carbon prices, 61 (*see also* Carbon tax
    schemes)
  on fossil fuel, 48–49

Taxation, harmonized, 33–35, 46, 49, 51
 and Global Public Good Purchase Plus Taxation, 59
 vs. Kyoto-compatible schemes, 49–50
 partial, 47
Tax incidence, 9. *See also* Incidence issues
 for exhaustible resource, 59–60
 uncertainty on, 60–61
Technical change, in FEEM-RICE model, 157
Technological change, acceleration of, 219–20
Technological progress, and borrowing of emission permits, 266
Technology(ies)
 for cement industry, 338
 and "complementarities," 196
 emission-permit funds for, 190
 excessive prices for, 201
 export of to developing countries, 206, 209, 211, 213
 long-term development of, 212
 retirement and retrofitting module for (EMSIM model), 355
Technology standards, in emission allocations, 258, 259–61, 262, 266, 268, 269
Temperature rise, global mean
 and climate risks, 307–308, 309, 326
 and climate sensitivity, 316–17
 model of, 314
Theory of clubs, 193–94
Timing, 7
 and developing countries, 75
 and Global Public Good Purchase approach, 24
 and price vs. quantity issue, 7–8
Tolerable windows analysis, 310
Trade, international, 333–34
 and carbon leakage, 333
 model of (GEO), 334 (*see also* GEO-CEMSIM model of international cement trade)
Trade issue, 6
Transfers between nations, 5, 104
 and stability of coalition formation, 104, 115, 118–25, 127–28, 128–29, 190–93
Transparency, in GPGP approach, 13, 43–44
Transversal debates, 3–9
Treaty(ies), 165
 breadth vs. depth of, 180
 compliance with, 188

 different legal statuses of, 217
 lowest common denominator in, 217
 stability of, 105
 voluntary participation required for, 103
 "Treaty BAU 'imports'" 22–23
Two-coalition structure, 4

Uncertainty(ies)
 on carbon prices, 60–61, 80
 and choice between absolute and intensity limits, 225, 226, 228–35
 on climate dynamics and risks, 308, 309
 on climate sensitivity, 309, 310–11, 315, 328n.5
 with learning, 318–23
 in cost-efficiency analysis, 310
 in GHG index, 366, 367
 and quantitative objectives, 47
 ranking of, 323–27
 about rate constraint, 328
UN conference on Human Environment, Stockholm (1972), 75, 81
UN Security Council, membership rules of, 126
Unilateral approach, 204
United Kingdom, Emissions Trading Scheme of, 222, 249n.2
United States
 in Asia–Pacific Partnership for Clean Development and Climate, 137
 Bush administration target for emissions intensity, 222, 223–24, 225, 249n.5
 and China, 149, 151, 152, 153
 in coalition-formation model, 144, 147–48, 150, 152–53, 153
 emission ratio of, 120
 and global welfare, 153
 and Kyoto Protocol, 103, 111, 201, 202–203, 212
 and European ratification, 203–204, 218
 as lacking mandatory emission policy, 202
 and price of allowances, 207, 213n.5
 Senate resolutions in, 77, 204
 state and local initiatives in, 138, 205
 and stringent targets, 194
 and transversal debates, 3
 treaty law of, 205

Variable cost module (CEMSIM model), 355
Venice Summer Institute of CESifo, 1

Index

Walras's law, 68
World Energy Conference (WEC) (1979), 82
World Summit on Sustainable Development (2002), 88
World Trade Organization (WTO)
  and border-tax adjustment, 347, 348–49, 350–52
  and climate-bloc formation, 195
  and differentiation of targets and carbon prices, 92
  and emission permits, 219
  membership rules of, 126
Worldwide emissions, and Kyoto vs. GPGP approach, 25